Lifestyle and Nature

Lifestyle and Nature

Integrating Nature Technology to Sustainable Lifestyles

edited by

Ryuzo Furukawa

PAN STANFORD PUBLISHING

Published by

Pan Stanford Publishing Pte. Ltd.
Penthouse Level, Suntec Tower 3
8 Temasek Boulevard
Singapore 038988

Email: editorial@panstanford.com
Web: www.panstanford.com

British Library Cataloguing-in-Publication Data
A catalogue record for this book is available from the British Library.

Lifestyle and Nature: Integrating Nature Technology to Sustainable Lifestyles

ISBN 978-981-4774-62-8 (Hardcover)
ISBN 978-1-315-14687-4 (eBook)

Contents

Preface

We have come up with two types of limits. The first is the degradation of our planet's natural environment. One major source of global environmental problems is unsustainable levels of human activity, which are draining the planet's energy and resources, decreasing biological diversity, and raising the risks of food and water shortages, as well as the risks associated with climate change, notably global warming. These phenomena have, in turn, begun to have a large and direct impact on human activity. The second "limit" is our decreasing appetite for material consumption. You could call it the limit of capitalism. For over 10,000 years, the community was firmly connected to the overwhelmingly powerful Mother Nature, and individuals existed within this system. However, due to industrialization, people broke away from the community and from nature, and the expansion of human activity began to erode communities and nature. It is necessary that we find solutions to both these limits simultaneously.

Technology is meant to enrich our lives. What we need to do now is design the kind of lifestyles that provide spiritual affluence despite the increasingly severe environmental constraints that are predicted for the future. We then need to identify the technological components necessary to support such lifestyles and generate ways to fully recycle resources with minimum use of energy. The answers we seek can be found in nature, and in the only truly sustainable social activities on earth. We need to apply the filter of sustainability in redesigning our lifestyles. We use the term *nature technology* to refer to the overall process of creating such technologies.

The Nature Technology Creation System is already giving rise to numerous new technologies. In 2009, we were given the opportunity to set up a Nature Technology Research Consortium as part of the Japan Manufacturing Conference (Monozukuri Nippon Conference), which, with Nikkan Kogyo Shinbun as secretary, enjoys the participation of more than 2000 corporations. In this forum, we have been able to discuss the relationship between new technologies and ways of living using the Nature Technology Creation System and to promote concrete, new business initiatives.

Meanwhile, the SEMSaT (Strategic Environmental Management and Sustainable Technology Solution), which Dr. Emile H. Ishida initiated in 2005 at the Graduate School of Environmental Studies, Tohoku University—a course which mainly targets people already working—allowed us to embark on numerous trails of the actual creation and implementation of business systems based on the notion of nature technology in the context of corporations, government organizations, and nongovernmental organizations. Furthermore, from 2012 to 2016, I had the opportunity to participate in the Innovative Materials Engineering Based on Biological Diversity Project (a new academic field under the auspices of the Ministry of Education, Culture, Sports, Science and Technology) initiated by Dr. Masatsugu Shimomura. The output from this project became one of the main contents in this book.

In order to achieve sustainability under the harsh environmental constraints projected for the future, each region in Japan must give serious thought to the important values that need to be maintained and develop new businesses and policies that enable the creation of an independently sustainable and spiritually affluent lifestyle that makes the most of regional characteristics. A number of local municipalities, including Toyooka City (Hyogo Prefecture), Kitakami City (Iwate Prefecture), Akita City (Akita Prefecture), Okinoerabu Island (Kagoshima Prefecture), and Shima City (Mie Prefecture), have started to implement proof-of-concept projects to change lifestyles, developed by interviews with 90-year-olds (the method to find spiritually affluent lifestyle concepts from old lifestyles living with nature. and lifestyle design method using backcasting. From 2015 to 2018, the JST-RISTEX decided to adopt "Creating a Fountain of Future Lifestyle Ideas" as one of its projects, and I became a project leader of this project. This project also became one of the main contents in this book. So we could introduce many cases using lifestyle design method using backcasting, interview with 90-year-olds, and the Nature Technology Creation System in this book.

In order to connect the new lifestyles to new technologies and nature, we applied ontology engineering and Bio-TRIZ (the method to find technology learning from nature). By applying ontology engineering, we developed action decomposition tree method to depict lifestyles, to match technology needed including nature technology, and to evaluate them, supported by Dr. Riichiro Mizoguchi, professor at Japan Advanced Institute of Science and Technology. We demonstrate how the use of the Bio-TRIZ Database can support the application of serendipitously discovered biological functions to engineering technology, solve various technological contradictions, and help design a sustainable lifestyle.

We hope this book will find use as a text in university curricula as well as in other programs dealing with environmental technology and lifestyle innovation. We hope that businesspersons and NPO or other activists who are eager to solve problems assailing our society will also explore new ideas from this book for the future sustainable society.

Ryuzo Furukawa
Tokyo, Japan
November 2018

Acknowledgments

I am grateful to many people for their help during the preparation of this book. I can't thank individually all the people who helped this research and shared ideas with me. But I want to single out a few people.

I thank Dr. Emile H. Ishida, since 2005 at the Graduate School of Environmental Studies, Tohoku University. He gave me the research theme of nature technology and lifestyle innovation. He has been a continuous source of stimulation and encouragement.

I thank Dr. Riichiro Mizoguchi, professor of Japan Advanced Institute of Science and Technology, for advice about ontology engineering. I thank Dr. Masatsugu Shimomura, who gave me the chance to research in the project "Innovative Materials Engineering Based on Biological Diversity." I thank all the project members (the "Innovative Materials Engineering Based on Biological Diversity Project" and "Creating a Fountain of Future Lifestyle Ideas" by JST-RISTEX) and the authors Dr. Takeshi Yamauchi, Dr. Hidetoshi Kobayashi, Dr. Toru Kobayashi, Dr. Yuko Suto, Dr. Ryo Kohsaka, Dr. Yoshinori Fujiwara, Dr. Yuta Uchiyama, Ms. Masae Mitsuhashi to discuss and complete this book.

The following colleagues have provided knowledge in many helpful discussions: the members of Nature Technology Research Consortium as part of the Japan Manufacturing Conference (Monozukuri Nippon conference), Dr. Yuko Kishikami, who developed action decomposition tree method using ontology engineering, Dr. Hiroyuki Fukuyama, Dr. Mekumi Tomimura, Dr. Mizue Kama, Ms. Yukimi Matsuda, Mr. Yusuke Wakabayashi, Mr. Noriyasu Kawase, Mr. Kevin Muhamad Lukman, Ms. Rini Purwanti, Mr. Yuta Saito, Ms. Rei Sugawara, Mr. Tatsuya Fujii, and my students at SEMSaT, Tohoku University.

I thank more than 500 people (around 90 years old) whom I interviewed about old lifestyles. They gave me tons of precious phrases about what is affluent lifestyle. I thank more than 210 people in the project "Creating a Fountain of Future Lifestyle Ideas" by JST-RISTEX especially in Toyooka City (Hyogo Prefecture), Kitakami City (Iwate Prefecture), Okinoerabu Island (Kagoshima Prefecture), and Shima City (Mie Prefecture), who provided enthusiastic support and assistance to achieve this lifestyle innovation in local areas over many years. Many of the facts, ideas, and lessons in this book have come from these hundreds of collaborators.

Finally, I wish to express my gratitude to my family, Hiroko, Ryosei, Ruriko, who always take great care to keep the best possible environment for me to work.

Ryuzo Furukawa

Chapter 1

Environmental Degradation and the Need to Change Our Lifestyles

Emile H. Ishida
Earth Village Research Lab. LLC, Okinoerabu Island, Kagoshima Z/C 891-9222, Japan
emile.h.ishida@gmail.com

Right now, the human race is coming up against two types of limits. The first is the degradation of our planet's natural environment. One major source of global environmental problems is unsustainable levels of human activity, which are draining the planet's energy and resources, decreasing biological diversity, and raising the risks of food and water shortages, as well as the risks associated with climate change, notably global warming. These phenomena have, in turn, begun to have a large and direct impact on human activity. The second "limit" is our decreasing appetite for material consumption. You could call it the limit of capitalism. For over 10,000 years, Japanese society was based on animism. The community was firmly connected to an overwhelmingly powerful Mother Nature, and individuals (people) existed within this system. However, due to industrialization, people broke away from the community and from nature (i.e., joined a capitalist society), and the expansion of human activity began to erode communities and nature. It is necessary that we find solutions to both these limits simultaneously. One solution

Lifestyle and Nature: Integrating Nature Technology to Sustainable Lifestyles
Edited by Ryuzo Furukawa

ISBN 978-981-4774-62-8 (Hardcover), 978-1-315-14687-4 (eBook)
www.panstanford.com

is creating fulfilling lifestyles and ways of life that work even in a world with harsh environmental constraints, and that reconnect individuals to nature and the community.

1.1 What Are the Current Issues?

In the year 2045, supercomputers equipped with artificial intelligence (AI) will control our planet. The intelligence of these computers will surpass that of humans, and the world will experience an event called "the singularity." Biological limitations and weaknesses such as disease and aging will be eliminated; we will even be able to cheat death. This is the vision laid out by Ray Kurzweil in his book *The Singularity Is Near: When Humans Transcend Biology*. Kurzweil says, "The computational capacity needed to emulate human intelligence will be available in less than two decades." The mid-15th century saw the emergence of what are often called the three great inventions of the Renaissance: the nautical compass, firearms, and, finally, the printing press, which was developed by the German printer Johannes Gutenberg. It subsequently took four centuries for printing technology to spread fully to the masses. When, in the second half of the 19th century, the telephone was invented, one quarter of all households in the West had a phone within 50 years. Mobile phone technology became widely adopted in just seven years—blogs and social media, in three years (Fig. 1.1).

As the speed of this evolution increases, development costs decrease in proportional measure. One invention leads to another, which leads to another, over ever shorter time spans. This is what Kurzweil calls the Law of Accelerating Returns. Kurzweil was hired by Google, which also purchased the British AI company DeepMind for 500 million US dollars. The company announced it was making a strategic shift from "mobile first" to "AI first." Apple acquired the California-based firm Emotient, and is eagerly hiring experts in fields related to AI. In January 2016, Toyota established the Toyota Research Institute, a new entity based in Silicon Valley. Between 2016 and 2020, it intends to invest roughly a billion

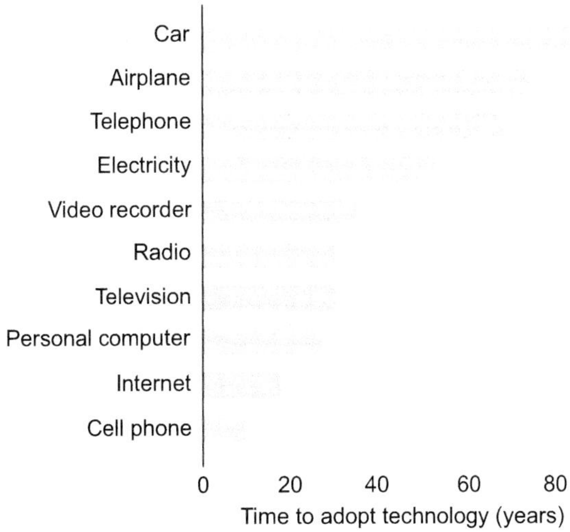

Note: Adoption refers to time for penetration of 50 percent of the population.
Source: Donay 2014. Human development report 2015 (UNDP)
http://hdr.undp.org/sites/default/files/2015_human_development_report.pdf

Figure 1.1 Speed of adoption of new technologies in the United States.

US dollars in cutting-edge research and development related to artificial intelligence.

And there's much more to the story. Right now, average human life expectancy is increasing by five hours each day. Genome analysis and regenerative medicine, regenerating organs using 3D printing technology, nanomachines that move through the human body dispensing cancer-fighting drugs, robots that can perform any surgery no matter how difficult. In 2045, average life expectancy will surpass 100 (NHK Special: Next World, Part 2).

It's also said that information technology will play a leading role in the next generation of money and finance. Rather than money managed and printed by nations or central banks, virtual currencies such as Bitcoin, which are stored as data and managed in the form of blockchains, will redefine the meaning of trust in finance, and even the meaning of money itself.

The Institute of Electrical and Electronics Engineers (IEEE) has estimated that by 2040, 75% of the vehicles on ordinary roads will be autonomous vehicles, and by the mid-21st century, vehicles that drive themselves will be the norm. As it becomes more and more common for advanced autonomous vehicles to take the road, traffic signs and signals will decrease, and it's possible that many people will stop getting driving licenses altogether. Other reports, including one by Barclays, are predicting that, in 2040, the car market will be 40% smaller than it is today, and perhaps car makers will have been acquired by companies like Google and Apple. This is exactly what we witnessed during the spread of the automobile in the early 20th century. Horse-drawn carriages soon disappeared from the streets, as did horses.

So, a new society will be born. AI will take the place of people in key areas, humans will live over 100 years on average, virtual currencies on the internet will take center stage, and thanks to autonomous vehicles, there will be no traffic accidents. But is this society actually coming?

At the very least, a wide range of technologies will continue to advance in that direction, and the national strategies of many countries will be aimed at leading us to that future. Putting aside the question of whether this society will arrive around 2040 or 2045 (frankly, I think it will come earlier), it seems certain that if we can solve the environmental problem of dwindling energy and natural resources, we will get there.

But there's an important question we should ask ourselves: Is this the kind of society we truly want? Now a few commentators are sounding warning bells about what this kind of society will mean for human dignity. Despite that, we could find almost no formal studies that ask if this is kind of the society we want. And if you interview the scientists or businesses who are advancing the technologies of today, and ask them if this is the kind of society they want, their responses indicate they think it is a stupid question. But they've only thought about what kind of influence these technologies will have on society from one angle. The mass media, too, without exception, automatically praises things that are safer or more convenient. It would seem that they do not consider it important to discuss whether these things are really necessary for society in the first place.

The essential purpose of technology is to make people's lives better, and by doing so make society better. So the market has continued to invest in technology that makes things more comfortable, more convenient. AI, long lifespans, virtual currency, autonomous vehicles, all of these have been pursued from the angle of making life more comfortable and convenient. This seems like a perfectly reasonable, logical approach, but it has been changing the very definition of "affluence" in our era. We must reconsider that definition and look at things from a different viewpoint. It is a fact that the capitalist society of the world's industrialized nations is approaching its limit, as is the affluent people can draw from the material possessions created by this society. The meaning of affluence has been changing greatly, and clear signs of that have begun to appear. The affluence that society needs going forward is not material affluence, but spiritual affluence. Unfortunately, a new value system based around this idea cannot come into being if we follow the path we are currently on.

Now, there is something we must consider. In a world where environmental constraints are growing ever harsher, what exactly is the spiritual affluence we need, and what are the technologies and services necessary to bring it into being? We need to find clear answers.

1.2 The Two Limits

It is not an exaggeration to say that the structure of Japanese society, beginning 15,000 years ago, continuing through the over-10,000-year span of the Jomon Period, and going right up until Japan's full-fledged industrialization in the early 1970s, was based on animism. The divine was everywhere, in everything. Shinto has its "eight million gods," and one of the foundational ideas of Japanese Buddhism is that everything—mountains, rivers, grass, trees, land—has the nature of the Buddha. Nature was abundant; the country has never been less than 50% covered with forest. These factors evidently gave birth to the kind of societal structure in which the community is firmly connected to an overwhelmingly powerful Mother Nature, and individuals belong to the community.

However, after Japan's rapid economic expansion began, the societal structure that took shape starting in the late 1970s caused great changes. People began to break away from the natural world and the community. In this era of industrialization and rapid economic growth, mass production and mass consumption became virtues. It was an era of boundless expansion. Efficient use of time became highly valued, and affluence depended on one's material possessions. Accordingly, money now underpinned all demand. If you have money, you don't have to be part of some annoying community. When you need something, you can use money to purchase that service. The same thing happens to your relationship with nature. Now, you can go to a supermarket and obtain all the food you want—all you need is money. You don't have to labor out in nature, drawing on your (and your neighbors' and friends') knowledge and skills, to get that food. As long as you have money. If you want to go somewhere, you can. You can travel anywhere, and enjoy the beautiful nature there to the fullest—all you need is money.

This is how people broke away from nature and the bonds of community, and it went hand in hand with how people began to head towards unsustainable growth (Fig. 1.2). The people who broke away forgot their bonds with nature and their community. They lost all understanding of where the food they ate and goods they consumed came from, and how these things were produced. People began to feel boundless desire for material things, and so they began to produce and consume boundlessly, and the degradation of nature

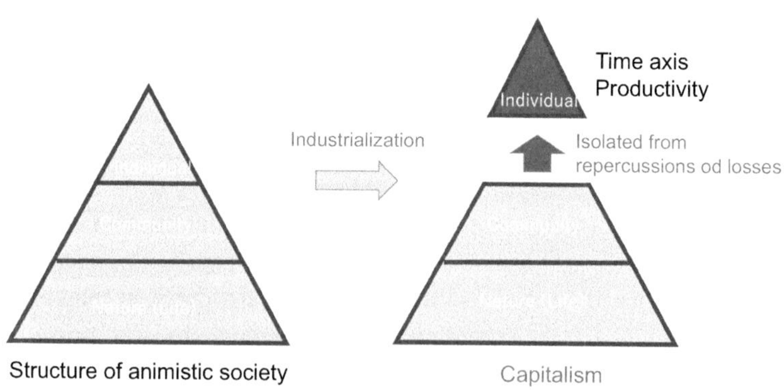

Figure 1.2 How people in industrialized societies are alienated.

and community began. Excessive exploitation of nature damaged the environment and gave rise to global environmental problems. People became excessively pampered and viewed community as a negative thing that oppressed individuals and suppressed individuality. The idea of helping one another came to be seen as old-fashioned, a relic from another age, and was cast aside. This is our current situation. To sum up, the capitalist system, which is based on the premise of boundless expansion and development, would seem to be reaching its bounds. The environmental problems this system has helped to generate are already beginning to present challenges to our civilization. Moreover, people who are already materially affluent will lose their appetite for consumption, the whole basis of capitalism, leading to a nightmare scenario where products cannot be sold. In short, capitalist society is coming to a standstill, and this has given rise to the two limits we have mentioned—the degradation of our planet's natural environment, and our decreasing appetite for material consumption—occurring simultaneously. We now need ideas that will deal with these two limits simultaneously.

1.3 Environmental Degradation (Expansion of Human Activity)

Leaving aside social science types of issues, we currently face seven types of danger. These dangers are related to the depletion of energy, the depletion of natural resources, the loss of biodiversity, the supply of water, the supply of food, rapidly increasing populations, and climate change, notably global warming. It is important to understand that the reason these have become dangers is the unsustainable expansion of human activity. Put another way, expansion of human activity is the essential problem facing the environment, whether we talk about global warming or issues related to resources. And the big, overarching problem is that if human activity continues expanding at its current pace, we then risk having all of these issues becoming extremely dangerous for our future. The question now is how we can use technology as a tool to halt or slow this expansion. In terms of manufacturing and

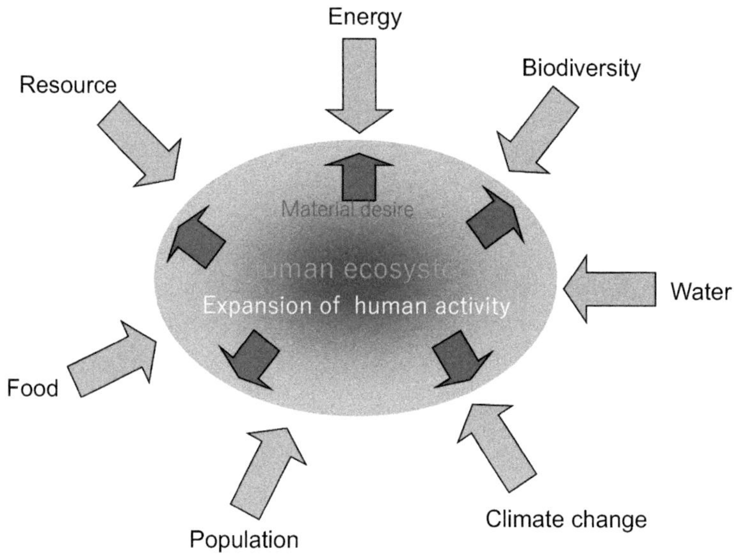

Figure 1.3 The problem facing our environment is expansion of human activity.

industry, the greatest danger is that boundless expansion of "input" (exploitation of the planet) leads to a similar increase in "output" (emissions out into the planet). This would be a brutal one-two punch that would hinder the planet's ability to renew and sustain itself in various ways, and that would lead to the acceleration of the depletion of nonrenewable resources (Fig. 1.3).

For example, let's think about risks related to energy. Around the year 1960, a Japanese person's energy consumption per day was approximately 28,000 kcal. Then, the country's economy grew rapidly, and Japan entered the club of industrialized nations. In 1975, energy consumption per day was approximately 92,000 kcal; in 2009, approximately 114,000 kcal; and in 2015, approximately 127,000 kcal. (In the US, by the way, people consume roughly 240,000 kcal/day.) The amount of energy the human body needs to live is something like 2400 kcal/day (basal metabolic rate $\times$ 1.5), so the average Japanese person consumes over 50 times that much energy. What is that much energy needed for? Well, think about moving 10 kilometers. Walking that distance burns 308 kcal of energy, and cycling it burns 118 kcal, but a car running on gasoline consumes 8,670 kcal of energy over that distance, more than 30 times as much as walking. And the energy from fossil fuels is inefficient, much of it being converted into heat, which gets released into the atmosphere. This inefficient (underground-resource) energy plays an essential role in our lifestyles. Our continuing small gains in comfort and convenience have caused geometric growth in the impact on the environment—this was not the original problem, but it has become a significant danger.

Moreover, global warming, which has begun to have a significant impact on our lives, is being caused by these dangers. Perhaps it will be easier to understand how if we go into detail about the indicators of environmental degradation.

After the onset of the Industrial Revolution, which began in Britain in the 18th century, industrialization, fueled by energy taken from fossil fuels, led to a geometric expansion of human activity. In the 50-year period between 1950 and 2000, the world's population increased by 2.4 times to 6.1 billion, and GDP increased 8.1 times to 30.9 trillion US dollars, meaning that per capita GDP increased 3.4 times. The number of vehicles on the road increased by 10.3 times, to 723 million, while annual usage of oil increased by 7.3 times, wheat by 4.1 times, steel by 4.3 times, and electricity by 21 times. If human activity continues its expansion, then all these factors—resources, energy, water, food, population increases, climate change—become extremely likely to act as triggers that will "pop the balloon" of our civilization, leading to its rapid downfall. We cannot say which of those dangers poses the greatest risk (although at the moment it seems likely to be global warming, i.e., climate change). Small changes in the overall balance could lead to any of these potentially posing the greatest danger.

There are climate change specialists, food specialists, energy specialists, resource specialists. There are lots of specialists in each of the domains where humanity faces grave risks. However, if specialists formulate solutions that use only their narrow field of expertise, we will never decide on anything. Solutions that come out of a limited point of view will always be lacking in key ways. When we think about solutions to environmental problems, considering the exploitation of our planet's environment is unavoidable. People in manufacturing and industry, as well as average citizens, are rapidly becoming more concerned about environmental problems, but this hardly ever translates into concrete action. We suspect the reason is that people do not think enough about exactly how the environment is being exploited, and that there are areas where they have not clearly teased out what the issues actually are. At the very least, if we want to solve our environmental problems, it will rest in finding an approach that will somehow convince us, from both a material and a psychological perspective, to stop the geometric expansion of human activity. We must think of the technologies required to do that.

In terms of technology, there is the issue that the expansion of human activity is causing two problems that go hand in hand: the rapid increase of environmental input and the expansion of environmental output.

The increase in input is leading to an expansion of output that surpasses the extent to which the earth can renew itself—i.e., that is unsustainable. The earth's ability to restore itself is deteriorating at the same time the burdens being placed on it are increasing. Decreasing biodiversity is one marked example of this. This decrease in biodiversity is directly linked to a degradation of the beneficial "environmental services" (a.k.a. ecosystem services) that we derive from our planet. If these environmental services degrade enough, we may not even be able to get the clean water we need to drink to survive. Can we make water? We can cause a reaction in a test tube that splits a drop of water into oxygen and hydrogen. But what about making huge quantities of water for industry, agriculture, and for us to drink? Unfortunately, with our current technology, it is impossible, even if we mobilized all the resources we had, to produce artificial water efficiently in a way that does not use energy. Meanwhile, the sun heats up seawater, which evaporates and becomes clouds, which run up against mountains and release rain, which falls down upon us—water we can use. This is a "service" that is offered to us completely free. It does not cost a cent.

This water is also stored in the trees and the soil, or it ends up deep underground, and then, after dozens or hundreds or even tens of thousands of years, this water, full of minerals and delicious to drink, emerges above ground (this is an "adjustment service" provided by the environment). The rich greenery on mountainsides, made possible by abundant water, makes our hearts happy (a "cultural service" provided by the environment). Water alone offers us all these natural services, free of charge. Food, fuel, wood ... I think you understand my point. Without these kinds of services, it will be impossible for us to maintain our current lifestyle. In 1997, the economist Robert Costanza calculated what the approximate value of the world's ecosystem services would be if you put it in monetary terms (internalizing externalities). He put the number at 33 trillion US dollars. At the time, total global GDP amounted to 18 trillion US dollars. We do not pay a single dollar for these ecosystem services; in fact, business activities could not create them. 33 trillion dollars for these natural materials, while the world only generates 18 trillion dollars of goods and services a year. If today's business and industrial systems had to pay compensation for these services, they would not be able to.

We must understand that our current society was created using resources that exist because of biodiversity. We must also understand that this biodiversity, an indispensable part of maintaining our way of life, is rapidly decreasing. We are in the midst what is being called the Sixth Extinction. Current rates of extinction are 1000 to 10,000 times faster than they were in the recent past, and it has been estimated that between the years 1990 and 2020, 5–15% of all species on our planet will have gone extinct. Furthermore, the Living Planet Index (LPI) estimates that between 1970 and 2010, the total population of animals decreased by 52%. Land animal populations decreased by 39%, freshwater animal populations by 76%, and seawater populations by 39%. Declines this rapid are unprecedented in human history.

In addition, decreasing biodiversity means that the earth is less capable of absorbing greenhouse gases, which speeds up global warming. However, it is sad to say that despite the fact that the discussion around biodiversity has concluded that we must conserve the natural environment, people are still not truly aware of what a vital issue this is for the human race's very existence. In 2010, the tenth meeting of the Conference of the Parties to the UN's Convention on Biological Diversity (COP 10) was held in Japan. At the conference, the convention recommended that 2011–2020 be designated the United Nations Decade on Biodiversity, a proposal that was eventually adopted, but with biodiversity continuing to deteriorate, particularly in the tropical zones of Central and South America, it is necessary to remind ourselves yet again of the gravity of the situation.

Second, the absolute quantity of nonrenewable inputs we need is rapidly attenuating, and the probability is high that our current industrial structure is not sustainable. The decrease in the amount of fossil fuels and precious metals we have is a big part of this. For example, "peak oil" is the point at which the extraction rate of oil will reach its maximum. We are currently debating when this occurred or will occur—some people say around the year 2030—but in the International Energy Agency's 2010 World Energy Outlook, it declared that peak oil had already been reached in 2006. Going forward, the rate of oil production will decrease by a few percent each year, and although there will be short-term price fluctuations, the long-term trend will be a gradual increase in the price of oil. There is a good chance that the end of our civilization's dependence on oil will become necessary much sooner than expected.

In the November 24, 2010, edition of the *New York Times*, an article called "Is 'Peak Oil' Behind Us?" appeared. In *Le Monde diplomatique* on November 12 of the same year, there was an article titled "The End of Cheap Oil," which concluded, "There is no time for denial. Governments and communities need to start adapting now." Then, though, the "shale revolution"—the full-scale production of shale oil and shale gas—got underway in the US. The price of natural gas in particular plummeted, and the economic effects were wide-ranging. However, once these resources are used up, what will we be able to pass on to the next generation? Although research into alternative, renewable energy sources, in particular natural renewable energy, is advancing, in 2012 in Japan, just 4.2% of energy came from renewable sources (ISEP 2014 Renewables Global Status Report). In Germany, renewables accounted for 23.4% of energy use, although a standard household in Germany pays close to $300 extra a year to support the introduction of renewable energy and the challenges the country has faced increasing renewable energy adoption are numerous.

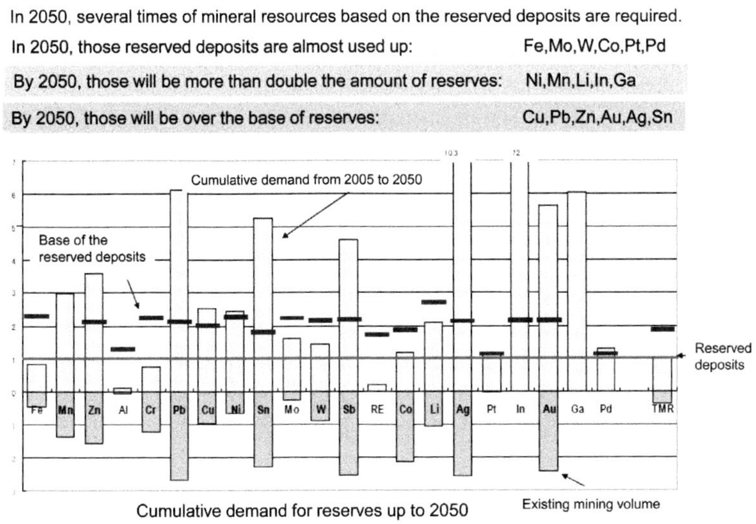

Source: "Global Resource Constraints Wall by 2050", Press presentation by National Institute for Materials Science (2007.02.15).

Figure 1.4 Cumulative amount of metal resources between 2000 and 2050.

Moreover, the depletion of organic metals is progressing rapidly, beginning with rare metals, notably precious metals. Annual consumption of metals between 1960 and 1970 amounted to about 380 billion tons. Now, annual consumption is 560 billion tons. In February 2007, the Natural Institute for Materials Science, an independent administrative institution in Japan, published a forecast of demand for and reserves of existing resources (Fig. 1.4). They found that by the year 2050, the cumulative use of many metals may exceed existing reserves.

According to this report, the number of metal reserves we have left (Reserve Life Index: reserves [confirmed reserves]/annual production) would last 120 years at 1970 usage rates. At today's rates, they will last 80 years, but if our use of metals keeps increasing at its current rate, in 2040 we will only have ten years' worth of reserves left.

The study investigated the amount of resources that we can expect to be consumed going forward, as per capita GDP increases in the BRICs, and they become industrialized resource-consuming societies in the same way Japan did. The researchers found that by 2050, the current reserves of even metals that are considered fairly abundant today, such as iron and platinum, are expected to be completely used up. Lead and silver around 2015; gold, antimony and tin around 2020; zinc around 2025; copper, manganese, nickel and molybdenum around 2030; lithium and tungsten around 2035; cobalt and palladium around 2045. Consumption of indium and gallium has already greatly surpassed available reserves.

Much of the consumption of non-ferrous metals is in cars and goods related to the IT industry. These are indispensable resources. In particular, rare metals, which are essential to high-tech goods such as hybrid cars and solar batteries, are at the heart of the global competitiveness of Japan's manufacturing industry. These are indispensable raw materials. Without iron, our current societal infrastructure would not exist. Without platinum, fuel cell vehicles (FCVs) could not run. Cemented carbide tools require tungsten, and high-performance batteries all use lithium. The circuits and parts that make the IT industry possible could not be made without copper, gold, or gallium.

In November 2014, the UN's Intergovernmental Panel on Climate Change (IPCC) published its Fifth Assessment Report. Though its findings were expected, they were still shocking compared to the Forth Assessment Report (2007). The level of carbon dioxide in the atmosphere before global industrialization was 280 ppm. In 2011, that figure was 391 ppm. For the past 65 million years, the number had naturally fluctuated between 180 and 300 ppm. We were now clearly far above that. And it is obvious that this increase is due to the use of fossil fuels, which is to say it is obvious that this increase is manmade. In May 2013, the Mauna Loa Observatory in Hawaii recorded levels above 400 ppm. At the same time, carbon dioxide, but also methane and dinitronic oxides (i.e., all greenhouse gases), had increased to levels beyond anything in the past 800,000 years. In Japan, too, atmospheric carbon dioxide levels in excess of 400 ppm were recorded in April 2014 at observation stations on Minamitorishima and other places.

The amount of greenhouse gases being emitted increased from 38 Gt in 1990 to 49 Gt in 2015—almost double the 1970 figure of 27 Gt. There is no doubt that global warming is occurring. Between

1980 and 2012, the earth's surface temperature rose by 0.85 degrees Celsius. If you look at temperatures for each of the past three decades, and compare them to the temperatures of each decade starting in 1850, you can see that global warming is accelerating. Japan is currently warming at a pace of about 1.15 degrees Celsius a century.

In addition, the oceans are absorbing roughly 30% of manmade carbon dioxide emissions, which means they are acidifying. On the basis of observations done since 1961, we can tell that the average temperature of the oceans (at least down to 3000 meters) has been rising, and that the oceans absorb over 90% of heat from climate systems, which is further raising temperatures at the ocean's surface. As a result, the temperature of the oceans across the world is rising at an average rate of 0.51 degrees Celsius a century. The seas around Japan are warming at a rate of 1.08 degrees Celsius a century. The costs of this are becoming apparent. In the near future, in Japan's coastal seas, acidified water will flow down from the north, and warmer water will flow from the south. Some people are saying that this will kill all of the coral in Japan's coastal waters. Also due to global warming, it has been observed that most land regions are experiencing long-term changes in patterns of precipitation. This will make life harder, particularly in tropical and subtropical zones. The number of regions experiencing longer and drier droughts is on the rise. In Japan, short, torrential rainfalls of 50–80 mm or more are clearly increasing, and it should obvious to everyone in Japan that the number of extremely hot (35 degrees Celsius or more) days has increased dramatically between the early 1930s and today.

The IPCC's Fifth Assessment Report (AR5) ran four simulations. In one scenario, where things continue as is, it found that in 2100 the global temperature will be 4.8 degrees Celsius higher than in 2000 (1981–2000). Even in a scenario where disparities between regions shrink, the global population decreases, economies rapidly change into service economies, and people's material desires decrease, the temperature in 2100 will still be 1.7 degrees Celsius higher than in 1990. What impact should that information have on us? Well, if the temperature rises just 1 degree Celsius higher than what it was between 1986 and 2005, the incidence of "extreme weather events" such as heat waves and floods increases. With an increase of 2 degrees, the natural systems of the Arctic Ocean and the world's coral reefs will be in danger. With an increase of 3 degrees or more, ice sheets will vanish, and the risk is high that this will lead to major, irreversible changes to our planet. An increase of 2 degrees is therefore seen as the tipping point that we must avoid reaching. By 2050, we need to reduce our greenhouse gas emissions to 40–70% lower than 2010 levels, and by 2100, reach net zero or net negative emissions. If we cannot do this, then in Japan, the temperature will rise 3.5–6.4 degrees Celsius, precipitation will increase 9–16%, sea levels will rise 60–63 cm, the cost of flood damages will triple, 83–85% of current coastline sand will vanish, and tidal flats will also decrease by 12%.

Humans are inertial creatures. We all want to keep doing the same things tomorrow that we did today. However, if we bury our heads in the proverbial sand, if we continue to pretend that nothing is wrong, then things are going to head in a direction that, eventually, we will be unable to control. In October 2006, the Stern Review on the Economics of Climate Change, prepared by Her Majesty's Treasury of the United Kingdom, was released. This report outlined the dangers and costs to the global economy if measures were not taken to combat climate change. It concluded that these costs would equal at least 5% of global GDP, and perhaps even 20% if a wider range of impacts were included. On the other hand, the report states, measures could be taken, costing about 1% of global GDP, to keep concentrations of carbon dioxide at 500–550 ppm, and that making this investment over the next 10 or 20 years would have an influence on our climate from the mid-21st century all the way to the 22nd century. If we believe that what is important for the human race is our responsibility to future generations of our fellow humans, then his words make perfect sense.

The AR5 powerfully drives home the point to us yet again that our planet is a limited resource. How can we deal with these problems that we ourselves have created? Simulations show that, if things continue as is, around the year 2028, the concentration of greenhouse gases in the atmosphere will be double what it was before the Industrial Revolution, and the landmark 2 degrees Celsius increase in average global temperature will occur. Meanwhile, the International Energy Agency (IEA) reported that we reached peak oil in 2006. Which is to say, we believe that the collapse of current climate systems that is expected to be triggered by a 2 degree increase will lead to an era where drilling for oil will no longer make economic sense. That will be a good thing, right? Finally saying goodbye to fossil fuels, we will use new sources of clean energy to create a new world, won't we?

Whether we can create a paradigm shift away from "double restriction" (the one-two punch of environmental burdens that surpass the environment's ability to renew itself, and the depletion of nonrenewable energy sources and other resources) will depend on whether we can create new systems of business and industry. Admittedly, debating the essential issues and coming up with solutions is not going to be easy. Some industries and business will have a harder time than others. However, it is fair to say that we do not think the solution lies in the continued use of current business models. We must understand the specific factors that make up double restriction, and then think about how to (1) adapt our business models accordingly, (2) create the business models of tomorrow that will cause the paradigm shifts we need, and (3) make sure (1) and (2) are connected.

1.4 Our Decreasing Appetite for Material Consumption

1.4.1 The Limits of the Consumer Society

A special section in the March 21, 2013, edition of the *Yomiuri Shimbun* called "Yomigaere Nippon" (Rise Again, Japan) featured an article titled "Mono hoshigaranu wakamono" ("The Young People Who Don't Want Things"). There have been lots of articles and magazine features and books about decreasing consumption, starting with, famous for the line that "buying a car is stupid."

As previously mentioned, when Japan industrialized, moving away from an animistic society to a capitalist one, individuals began to break away from nature and the community. In other words, society became a consumer society. And once this happens, a consumer society remains a consumer society, even in bad economic times, and the technology that has sprung up to support it speeds up the consumer cycle, which demands constant evolution. However, the wealth of the generation that was singing the praises of capitalism during Japan's bubble era is fading away, and that former affluence has not been handed down to the younger generations. Many people have begun to doubt the merits of a consumer society, and are growing tired of this constant consumption.

The promise of "A chicken in every pot and a car in every garage" often credited to US President Herbert Hoover exemplifies 1920s America. Today, we can see the drawbacks of that way of thinking. In the 1970s in Japan, fast food shops and convenience stores began to spread, and the movement of people from the countryside into the city accelerated. The result was the breakdown of local community bonds, and people living as self-contained individuals. During the era of rapid economic growth that began in the 1970s, it was seen as uncool and old-fashioned to worry about being wasteful, and the consumer society was idealized. However, even back then, some people could feel that this attitude was inviting the destruction of the environment. More ordinary people began to have doubts about consumerism, and their appetite for material consumption decreased. We could begin to see that capitalism, which is based on the principle of limitless growth and expansion, does in fact have limits. Production and consumption are essentially a single thing. In between production and consumption, there are lots of large "gaps" of space and time—tertiary industry would be a good example—that make us think they are not one thing, but the limits of consumerism are in fact the exact same thing as the limits of production.

1.4.2 Two Reasons We Feel a Sense of Stagnation

Why does it seem as if Japan has reached its limit as a capitalist society so suddenly? No one would dispute that Japan today has great abundance, even compared with other industrialized nations. Towns are clean and well maintained, there is lots of greenery, the food is good. People are kind, you feel safe walking around late at night, and virtually no one owns a gun. And yet, a feeling of stagnation hangs over the country. A survey of 24,000 people across 23 countries conducted by Ipsos in 2010 asked people how confident they felt about the future. Of all countries, Japanese were the least confident (14%), and therefore the most worried (86%), about the future (IPSOS 2010 http://www.ipsos.com). A survey of the attitudes of young Japanese adults by Macromill (http://www.macromill.com/r_data/20110107shinseijin/index.html) found that 90% were worried about the future. A separate survey of junior high school students by Sony Life (http://www.sonylife.co.jp/info/pdf/120106_syouraichousa.pdf) found that 69.4% of them were worried about the future 10 years from now, and 77.6% said they "had thought about the future." It is clear that young people are deeply concerned about the future. The fact that people have gone too far away from nature and community is certainly part of what is stirring up these worries, but two major reasons why people so worried are (1) the murkiness of the path forward—i.e., uncertainty about what will happen if we continue living the way we have—and (2) a large shift in the value attached to material things, a value that is at the heart of why certain individuals have continued to consume more and more.

1.4.3 The Murkiness of the Path Forward

A survey of 10,000 people by the Nomura Research Institute (NRI) yielded some fascinating results (Questionnaire Survey of 10,000 Consumers, NRI, 2006, 2009, 2012). The graph shows people who said they were "very worried" about their finances, and as you can see the percentage of people who felt that way increased significantly between 2006 and 2009 for every age group from 10–19 to 60–69: 24% of people between the ages of 10 and 19, and about 35% of people in their 20s said they were very worried. Even though these are people who haven't started working, or have just started their careers, they feel great unease about the future. This is probably because, being in their teens or twenties, they don't yet have a solid plan for their lives, a clear image of what they will be. Put another way, they don't have a specific reason to feel worried, but they sense that something is standing in their way. People who were between the ages of 10 and 29 in 2006 didn't really have firsthand knowledge of Japan's bubble era. The country entered a period of stable economic growth in December 1973. Then, from 1991 to the second half of 1993, Japan went into recession, and ever since the country has been stuck with low growth. People used to talk about Japan's "lost decade." Now they talk about a lost 20 years. During the period of growth, there was an asset price bubble that lasted from December 1986 to February 1991—this is what is known as the "bubble era." Abundant capital funded the vigorous redevelopment of big cities, and land prices were driven ever higher. Also during the bubble era, companies focused on using their financial assets to generate profits. Financial engineering took off, and both businesses and individuals invested in land, stocks, bonds, and other assets with the aim of making big profits. They also used their money to go on buying sprees of luxury foreign vehicles, paintings and antiques at art auctions, and so on. Consumerism overheated.

These were good times, and companies were posting good results. They expanded their businesses, which led to a severe shortage of qualified people, which led to companies competing furiously over freshly minted graduates. Once they gave these students job offers, in order that they did not get lured away by other companies, they would send them on "training" excursions either somewhere in Japan or abroad where other companies wouldn't be able to reach them—these became known as "isolation trips." In 1991, the ratio of job offers to jobseekers was 1.4 to 1, and a recruiting survey found that in the same year, that same ratio was 2.86 to 1 among college graduates. The 10–29 age group today, however, knows only the long period of sluggish economic growth following the bursting of the bubble. It is noteworthy that although they have been born and raised in a society that is "stagnant," not moving, at the same time their sense of unease is growing sharply with each passing year. As for people in their 40s and 50s, over 40% reported feeling great unease about their finances in 2012. They are old enough to remember the bubble era, and many of them are still in the prime of life, with large responsibilities on their shoulders, including the need to support their families. When they compare today to the bubble era, it makes them all the more worried. At the same time, and we will discuss this again later, they look at Japan's stagnant economic growth, low birthrate, and aging population, and they cannot see a solution to these problems. Meanwhile, people in their 60s showed a spike in unease about their finances. The

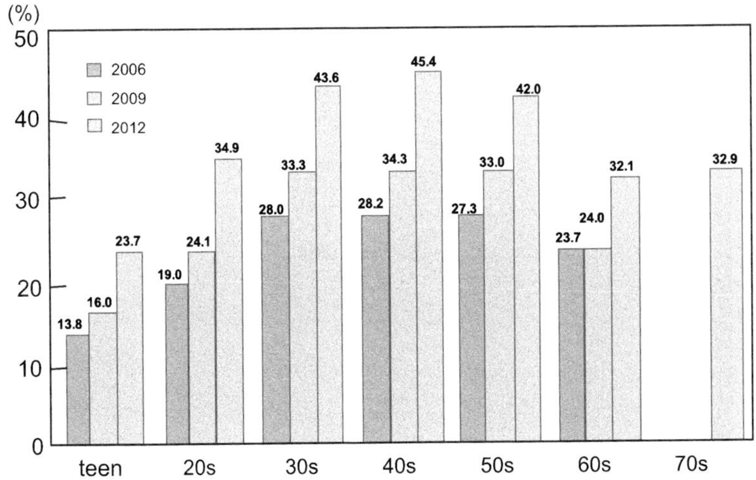

Note: Respondents aged 70 and above were not polled previous to the 2012 survey NRI Questionnaire Survey of 10,000 Consumers (2006, 2009, 2012).

Source: Nomura Research Institute "Why won't the Japanese go out and buy things?" Toyo Keizai Shimbun 2013.

Figure 1.5 How do you think your finances will be when you become elderly?–People who answered "I'm very worried about that" (by age cohort).

previous two times the survey was held (2006, 2009), less than 25% had reported strong worry about their finances—in 2012, over 30% did. This is despite the fact that most of them are receiving pensions, and so we would therefore imagine them to be living well in their golden years. A similar percentage of people in their 70s are also worried. We presume their worries stem from the fact that, as people are living longer than ever, it is uncertain whether, in an era of weak economic growth, the government will continue being able to afford paying their pensions (Fig. 1.5).

This murkiness about the way forward is not limited to Japan. If you look at the growth rates of different countries, you see that they have been declining gradually in the EU, Japan, South Korea, and the US. These are all becoming mature societies. That said, at critical junctures, Japan's growth rate has declined significantly, and it now has the lowest growth rate of any industrialized nation (2015 growth rates: Japan 0.47%, France 1.14%, Germany 1.45%,

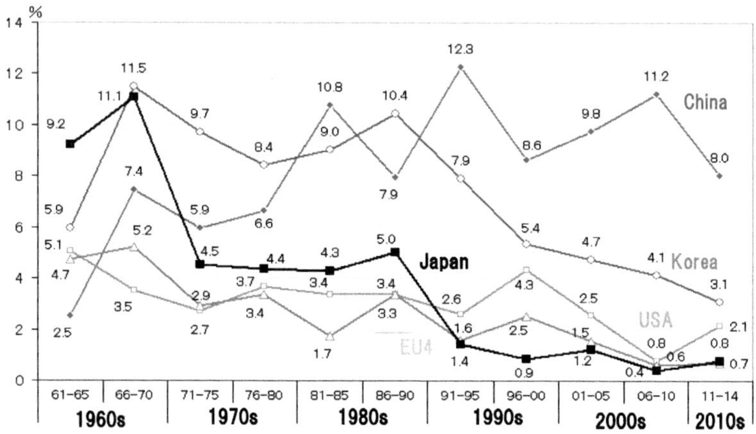

Note: Average annual economic growth by decade is the mean of growth rates (year-on-year real GDP % increase) for the years in that decade.
EU = 4 UK, Germany (West Germany before 1990), France, Italy

Source: World Bank Development Indicators: Growth of output (2015.01.30) (For West Germany before 1990, OECD data is used); Cabinet Office of the Prime Minister of Japan; etc. (http://www2.ttcn.ne.jp/honkawa/4500.html).

Figure 1.6 Economic growth rate by country.

UK 2.25%, US 2.43%). During its period of high economic growth that began in the 1960s, with the exception of two falls precipitated by the oil crises of 1973 and 1979, Japan maintained a growth rate of over 4%. All of that changed when the bubble burst. The growth rate collapsed, and it had not recovered when the collapse of Lehman Brothers triggered a global financial crisis, causing further damage. The weakness of the economy continues to this day (Fig. 1.6).

A capitalist society is a society where things flow from strong places into weak places. The competition is who can find these weak places first. At first, there are always weak places close by to strong places, and this flow happens easily. Think about the Britain of the Industrial Revolution in the second half of the 18th century.

As the world globalized, these weak places became harder to find. In order to search for them, beginning in the 1970s, having an "information society" based on information technology became essential. When a weak place was found, the world's businesses instantly crowded in. A good example would be how car makers invaded China. Now, as the world's markets become more uniform, or come close to saturation, these weak places became harder than ever to see. This is essentially what a mature society is.

So a new axis was added: finance. "Financial capitalism" has its strengths and weaknesses, but the reckless application of it led to the global financial crisis. A society using technology to become more information based is not unique to our modern, financial capitalist society. Take the early, hunter-gatherer form of society. Through the use of tools, it changed from a system where people hunted and gathered as individuals to one where they hunted and gathered in groups. This was obviously more efficient, and in order to preserve these groups, it was necessary to come up with stories and rules—in other words, information—that would convey intentions and keep the groups together. Cave paintings are an example of how these societies matured, and began to place value on culture. If we define a civilization as an accumulation of technology, and a culture as an accumulation of wisdom, then we can see humanity moving from civilization to culture. The emergence of agricultural societies continued this trend.

With the sun as their main energy source, people cultivated grain, and if they managed to produce a stable supply, then new towns—new points of consumption—emerged to consume the excess, or systems to distribute the grain grew. It was an era where the world became more information based. Ultimately, these became cultural societies, with arts, crafts, performances, and so on. Our modern society, which emerged in the second half of the 18th century, has developed in the same way. Mass production began, powered by fossil fuels, and, in a similar process to the one we just described, railways and other transportation systems were developed to distribute goods to points of consumption. Then, information and communication technology meant to find more points of consumption developed rapidly. Now, the market is becoming mature and saturated. The time is ripe to create the next concept of cultural value.

For a sign of this maturation, we can look at recent trends in GDP. With the exception of China and the US, per capita GDP fell from 2014 to 2015. We can also see that Japan in particular has failed to recover from the aftereffects of the global financial crisis, with per

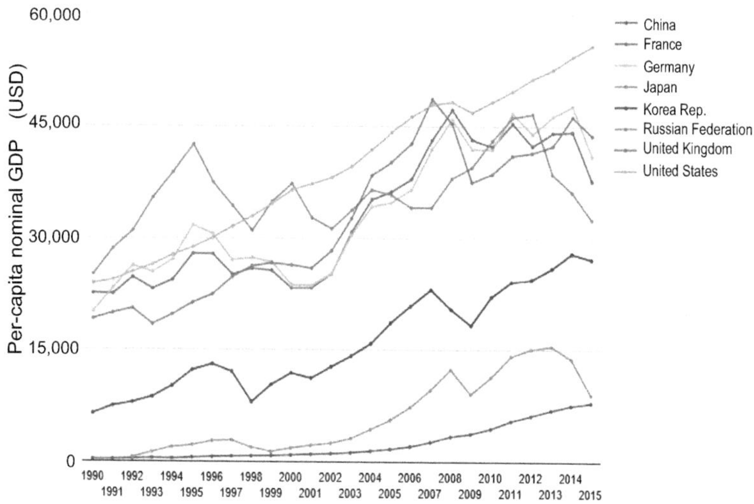

Source: http://www.globalnote.jp/p-data-g/?dno=8870&post_no=1339)

Figure 1.7 Per-capita nominal GDP (IMF statistics).

capita GDP falling significantly compared to other countries since 2012 (Fig. 1.7).

One important part of the murkiness of Japan's future is that it became an aging, low-birthrate society before any other country on earth, and that this trend is only accelerating. Japan's population reached a peak in 2004, and then the decline began. It is estimated that in 2060 the country will have just 90 million people. In 2013, the working population made up 62.1% of the total population. It is estimated that in 2060 figure will be 50.9%. Over the same time span, the percentage of the population 65 and older will increase from 25.1% to 39.9%, and the total fertility rate will fall from 1.41 to 1.35 (Fig. 1.8). In this strikingly graying society, how will industry be maintained? How can ballooning healthcare costs be reined in? The minimum age for receiving a full pension was raised from 60 to 65, and now people are beginning to discuss raising it to 70. Amid all this, how can we preserve our way of life? There seems to be no clear path, no clear answers—in other words, the way forward is murky, and this is an important part of the sense of stagnation hanging over Japan.

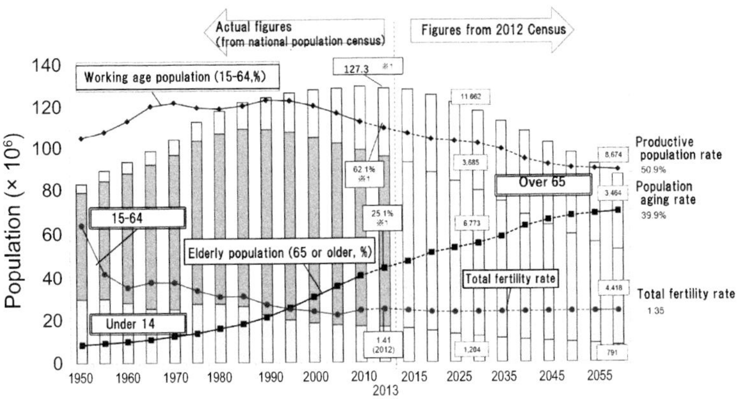

Sources: Ministry of Internal Affairs and Communications census figures and population estimates; National Institute of Population and Social Security Research, "Population Projection for Japan: 2011–2060" (Jan. 2012) estimated fertility rates and mortality rates (populations at Oct. 1 of each year); Ministry of Health, Labour and Welfare, "Population Trends of Japan".

Figure 1.8 Population change in Japan. In recent years, Japan's population went flat—now it has begun to decline. In 2060, it's estimated that Japan will have 90 million people, almost 40% of whom are senior (65 or older) citizens.

1.4.4 A Change in How Things Are Valued

A large change in how people valued material things began in the mid-1980s. People started valuing spiritual affluence over material affluence. This trend continued in cities across Japan, and currently spiritual affluence sits 30 percentage points higher than material affluence (Fig. 1.9). So, what are the businesses that will generate spiritual affluence? Well, despite this increasing recognition that societies need spiritual affluence, not much has been created. There seems to be a misunderstanding that technology is something that creates material affluence, whereas the creation of spiritual affluence falls to tertiary industry. There is a lack of awareness that technology, too, can generate spiritual affluence. If we look at what kinds of items have made up what proportions of final consumption expenditure, we see that durable goods (cars, appliances, etc.) and semidurable goods (clothing, shoes, textiles, etc.) have been declining since 2000, nondurable goods (food and similar essentials) have stayed steady, and services are on the rise. There is a clear trend away from material possessions. Actually,

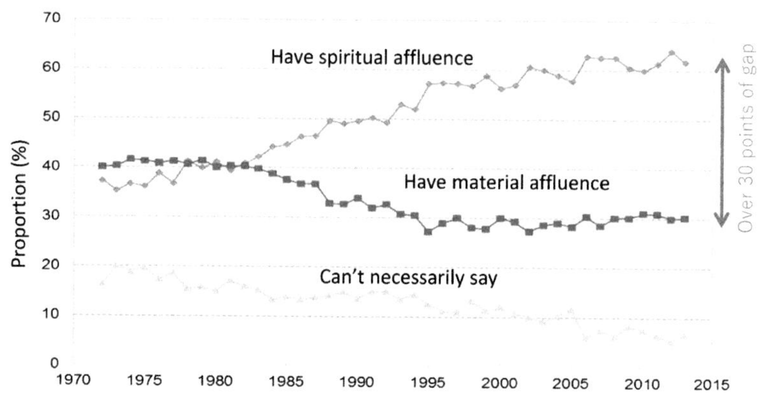

Source: Cabinet Office "Opinion survey for People's Life" 2013.

Figure 1.9 Emotional/material fulfillment awareness survey.

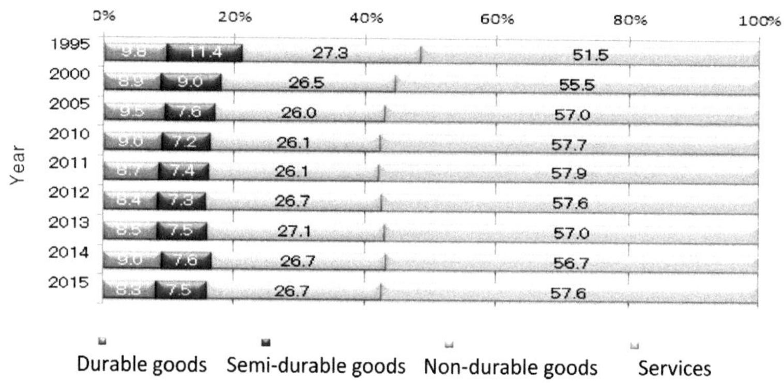

Source: Cabinet Office, Government of Japan, Quarterly Estimates of GDP 2016
http://jp.gdfreak.com/public/detail/jp010010001010100002/3

Figure 1.10 Changes in the types of nominal domestic final consumption expenditure of households.

calling it a "trend away" may be fairly misleading. What we are in fact seeing is a decreased appetite for material consumption, for material affluence itself (Fig. 1.10).

In a sense, this is only natural. After the Second World War, Japan emerged from the ashes to achieve astonishing economic growth. The television, the washing machine, and the refrigerator became society's "three sacred treasures," and people worked furiously to

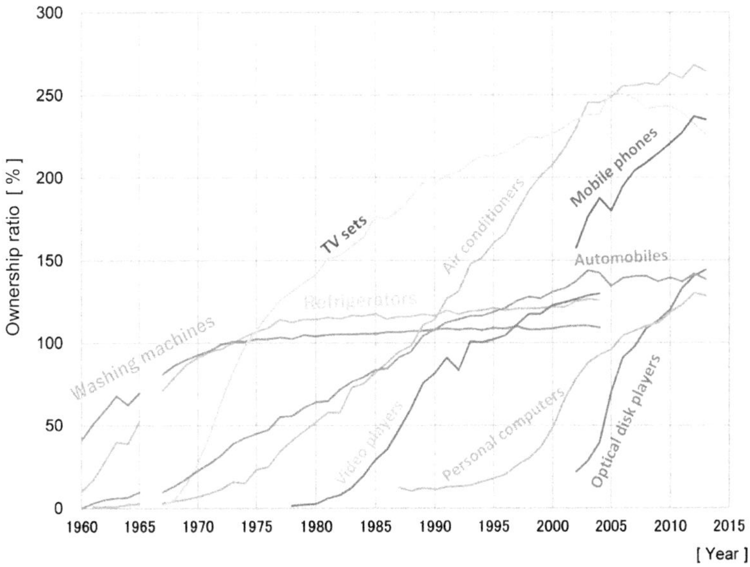

Note: 1966 data is not available for washing machines, refrigerators, air conditioners and automobiles.

Source: Japan Center for Climate Change Actions
http://www.jccca.org/home_section/homesection01.html

Figure 1.11 Ownership of technology by household.

be able to afford them someday. By the 1970s, 100% of Japanese households had acquired them (Fig. 1.11). The next items people longed for were cars and air conditioners. Nowadays these things are everywhere. Children today simply take them from granted. The performance of these items has been raised to the point where there is almost no room for improvement. The only place to go from here is away from material things. Take home appliances. Households have more than one of each, their performance is about as good as it's going to get, and they are perfectly eco-friendly. In a world like this, if a new product comes out and doesn't offer some sort of novel feature or design that truly appeals to people, then it's only natural that people will not feel a strong desire to buy it.

Appliance makers are still clinging to the production of refrigerators and washing machines and televisions, but in recent years this system has been shaken to its foundations. These companies do not have the next "three sacred treasures" to offer us. They are stuck

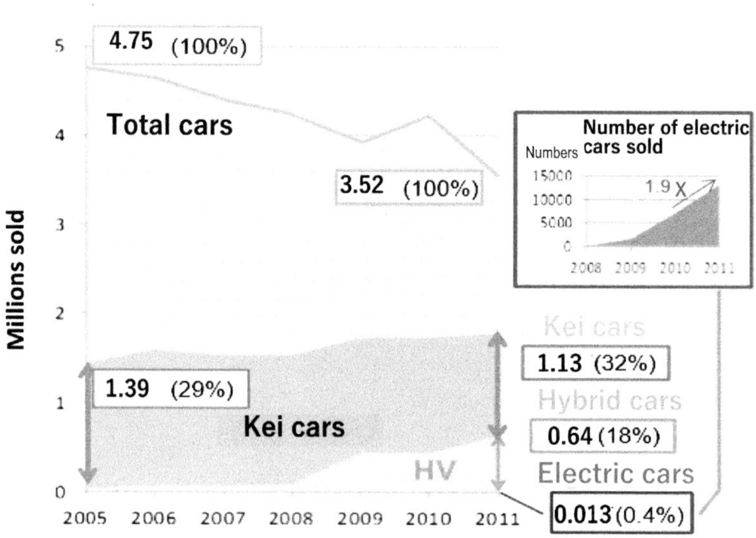

*Totals for all cars and kei cars do not include hybrid or electric vehicles.
Source: Japan Automobile Manufactures Association, INC. Next generation vehicle promotion center (http://www.cev-pc.or.jp/NGVPC/data/index.html).

Figure 1.12 Trend in number of vehicles sold by type in Japan.

making the same old products. They do not want to believe what they are seeing. The next product to follow this path will probably be cars. Automobiles can only become so eco-friendly. Research into things like electric cars and fuel cell vehicles is advancing, but think about the other side. If you're just going around your neighborhood, riding a bicycle is cooler and more convenient. Getting a parking space in the city takes time, and it costs money, and you can't drive if you drink, and you're expected to take out a loan to buy this thing? These are the feelings of the times. At the very least, we can be sure that the number of passenger vehicles sold in Japan will continue to decline (Fig. 1.12).

We can summarize this change in the way material things are valued as follows. In Japan, disposable incomes continued to increase until the early 1990s on the back of strong economic growth. On the other hand, the average propensity to consume (APC), a measure of what proportion of disposable income is being

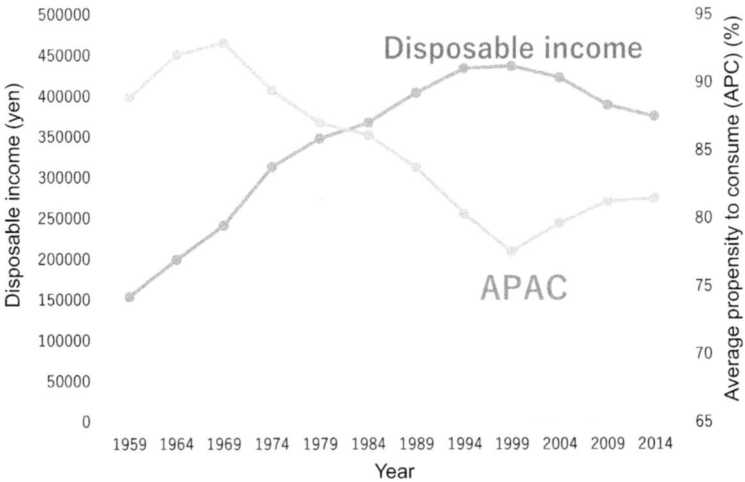

Source: Graph created using data from household survey by Ministry of Internal Affairs and Communications.

Figure 1.13 Disposal income and APC.

spent, passed its peak in the early 1970s. Disposable incomes were rising, but people were spending less of it on consumption. In 1975, the ownership rate of the three sacred treasures surpassed 100%, and in 1990, cars and air conditioners surpassed 100%. This was a continuous trend towards people having their desire for material possessions satisfied.

We also see that as disposable income peaks and begins to decline, the average propensity to consume rises. Yet there are problems with concluding this happened because of an economic recovery (Fig. 1.13). If we want to get a sense of how the attitudes of consumers are changing—what they think about life circumstances, increasing their incomes, employment situations, when durable goods are purchased, and so on—we can look at changes in the consumer confidence index score. After the end of the bubble era, the score dipped below 50 (a score of 50 or above means that consumers are generally confident about the economy) and stayed there until 2000. Around that time, after Japan's so-called lost decade, the economy suddenly picked up, and consumer confidence rose somewhat until about 2005, when

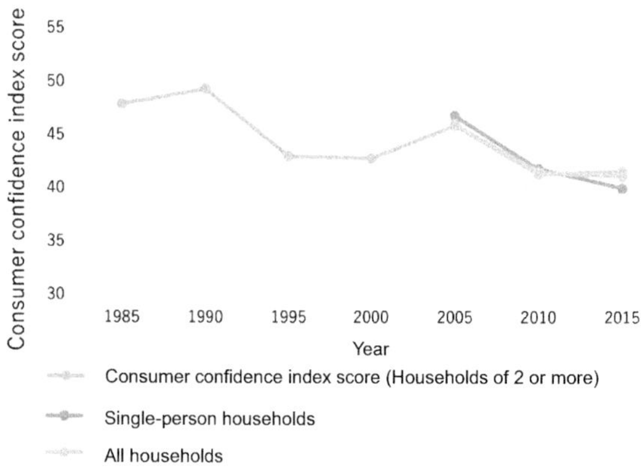

Source: Graph created using data from household survey by Ministry of Internal Affairs and Communications.

Figure 1.14 Household consumer confidence index score.

it began falling again. The decline has been even greater among single-person households. Consumer confidence decoupled from average propensity to consume, which has been rising since 2000 (Fig. 1.14). The reason for this is probably that, although people do not have great expectations for the future, as their disposable income falls they have no choice but to spend a greater share of it on consumption. This is certainly one factor creating the sense of stagnation in Japan.

A dilemma created by consumption is another factor in this sense of stagnation. We are talking about the "world of consumption" of children. One way to think about this is that kids do not produce anything; in a sense, they are not full-fledged members of society. But that makes them great targets for today's consumer culture. Young people have the potential to be the leading consumers of the future. Consumer culture has expanded to the point where children are exposed to it at school, ostensibly the opposite of a consumer environment. There is only so much we can do at this point to block information flooding in from our consumer society, or set rules or regulations on our children. They look up to celebrities on television and the heroes of cartoons more than their parents and teachers. The structure of our society is having a large impact on how the changes in value placed on material possessions in an information-based society are playing out. We are in a consumer society, and consumption requires money. As the number of unemployed and part-time workers rises, and the graying of Japan continues, the people who do have full-time jobs will have to work tirelessly in order to drive consumption. This is part of why we are seeing the phenomenon we discussed previously, of disposable incomes falling as the propensity to consume rises. The workers act as the cogs turning the wheels of the consumer market economy.

1.4.5 The Concept of Consumption

The people who were separated from nature and the community because of industrialization followed a path of excessive material consumption and unsustainable expansion. This led to the erosion of interpersonal bonds and degradation of the environment. Then, these people who had consumed boundlessly began to feel a decrease in their appetite for material consumption. They also began to feel that the future was murky, that they did not value material possessions as much, and that they felt a sense of stagnation hanging over them. We have already laid this out. This society is based on production and consumption, and centered on material possessions. Convenience is the same thing as fulfillment, and in a society structured like that, the key is the principle of boundless expansion.

In his 1970 work, *The Consumer Society: Myths and Structure*, Jean Baudrillard says that consumption is language. Consumption is about how something can be used, or who owns it. In the natural world, things are simply what they are made of. In the economic world, we start talking about something's economic value. And in a consumer society, we talk about how useful things are from a social perspective. We add more and more signs.

Furthermore, consumption does not simply serve as a display of individual or group power (conspicuous consumption). Baudrillard thought the things you consume are signs of your individuality, your originality—what differentiates you from others—and that consumption exists at the juncture between production systems and the structural logic of individualization/differentiation. So, when you buy an item, which is a sign of what differentiates you from other people, this is language. It is communication. The crucial thing is that the meanings associated with the item are part of a code, and in order to effectively express your individuality, you must master this code and its shifting meanings. People think this way about certain semidurable goods (e.g., fashion) and services, but most durable and semidurable goods have not advanced to this "encoding" level in people's minds. Most people are still focused on the economic value of things. They cannot escape this way of thinking. Their inability is, in essence, an inability to see beyond the idea of material affluence—they may want spiritual affluence, but lack the signs and language to attain it. Manufacturers take the same things for granted. Mass production based on mass consumption, done as efficiently as possible. With this approach, it will be extremely difficult to "encode" their businesses.

To clear away the current feeling of stagnation plaguing Japan, we need a new form of consumption (and production—and if it involves signs or language, then so much the better). What is it? To figure that out, we need to think not simply about how to revolutionize production and consumption, but also about what our civilization as a whole requires.

1.5 The Lifestyle Changes We Must Make

1.5.1 Breaking Away, Expanding Unsustainably

The people who broke away from nature and community as a result of industrialization are the people who started our unsustainable expansion. Regardless of whether these people were aware of what they were doing, they sought to acquire everything for themselves, and their pursuit of limitless convenience led to geometric expansion of the impact on the environment. At the same time, breaking the bonds of community meant that each individual had to provide entirely for himself or herself, giving rise to geometric expansion in the consumption of materials. It was the ascendancy of the individual, and the erosion of nature and the community (Fig. 1.15).

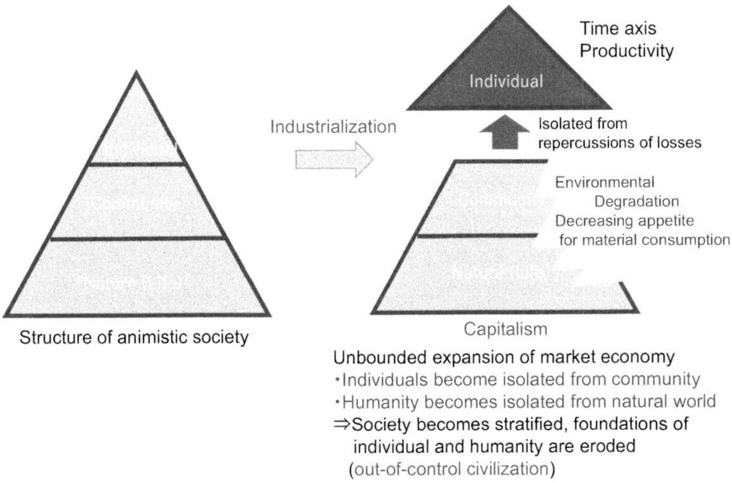

Figure 1.15 The expansion (breaking away) of the individual eroded nature and community, and gave rise to the two limits.

By all rights, the negative feedback that our actions engendered should have put a stop to our recklessness, but we were not able to get away from the idea that nature was limitless—or perhaps in our competitive society, where efficiency is everything, it is simply impossible to stop our expansion.

Now though, many measures and policies are being enacted regarding the environment. Which ones will prove effective? Which ones will be able to stop the expansion of the individual? These are crucial questions we need to be asking if we want to create sustainable societies.

In Japan today, practically everything is becoming more eco-friendly. It has come to the point where if your product does not consider the environment, the market will not invest in it. Naturally, some of this is mere greenwashing (deceptive marketing to make a product seem more environmentally friendly than it actually is), but at the very least, we can see that environmentally unfriendly products do not succeed in the market, and people are watching the activities of companies very closely. For example, look at household energy consumption. In 1965, electricity usage made up 22.8% of total household energy consumption, whereas in 2008 it made up

50.1%. (In that time, Japan's population increased by 1.29 times, and total household energy consumption increased by 2.2 times.)

Two appliances, air conditioners (25.2%) and refrigerators (16.1%), make up over 40% of total household electricity consumption. In the past 15 years, air conditioners have become 40% more efficient, and refrigerators have become 80% more efficient. In other words, a refrigerator today can run on just 20% the energy of a refrigerator from 15 years ago. This is the kind of world-class, cutting-edge technology that Japan prides itself on. Home appliances have been greatly improved from earlier generations. The standby power consumption of home appliances has fallen 71% between 1999 and 2008—it is estimated that standby power consumption makes up about 6% of all household energy consumption. Similar things are happening with televisions, LEDs, and cars. We are entering the golden age of eco-friendly automobiles, instigated by hybrid cars, but now including everything such as Japan's "kei" (light) four-wheeled vehicles. Everything is getting green, and in our country, where people are highly conscious of these issues, phony "greenwashed" products are almost unheard of. We are creating a series of world-leading eco technologies, and they are getting out into the market. Meanwhile, ordinary citizens have an extremely high awareness about issues concerning the environment, even compared to other advanced nations, and in a survey we found that 90% of people are concerned about the environment, and 70% of people are either taking action now or feel that they need to take action.

Eco-friendly products are coming to market, and there is a citizenry that is putting their high awareness of environmental issues into action. It would seem that Japan should be able to dramatically improve its natural environment, but unfortunately, the impact on the environment by the residential sector continues to increase. It grew by 47% between 1990 and 2014. The situation is worse than ever. Even though the population in Japan has begun to decline, the number of households is actually increasing. If we consider that, it seems certain that environmental degradation will continue. It is the same for total greenhouse gases. If we leave off the effect of global recession caused by the subprime mortgage crisis in

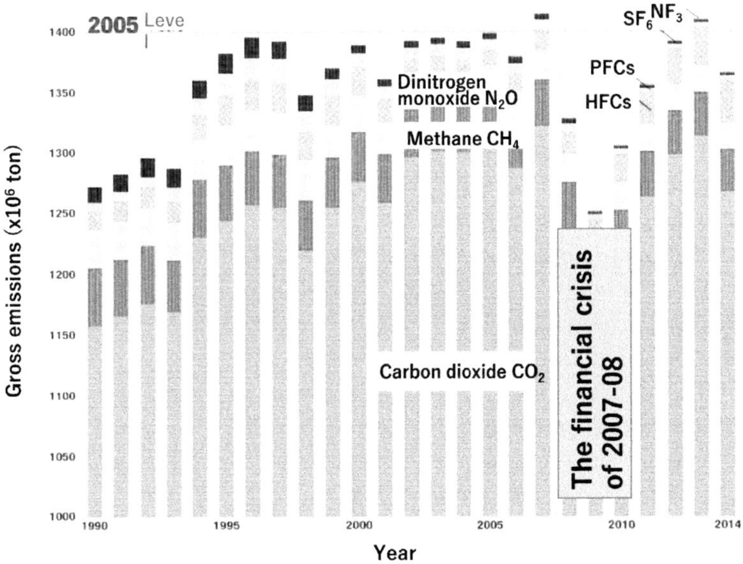

Source: Japan Center for Climate Change Actions http://www.jccca.org/chart/chart04_01.html

Figure 1.16 Transition of greenhouse emission burden (Japan).

2007, emissions in Japan have been steady at around 1.4 billion tons (Fig. 1.16). There does not seem to be a way to lower that number significantly.

Despite fantastic eco-friendly products on the market and highly conscious citizens, the destruction of our environment continues. We call this the "Eco Dilemma."

We must now gain an understanding of what creates this Eco Dilemma, and steer the ship in a very different direction. Businesses are investing in eco-friendly products, and ordinary people are strongly aware of the importance of the environment, and still the environment is deteriorating. Why does this happen? If we don't understand that, then even well-intentioned technologies will be useless—on the contrary, they will only spur on further environmental degradation. Our investigations have found the structure that gives rise to the Eco Dilemma is made up of two factors. First, eco-friendly products seem to give people an excuse to indulge. We humans are not strong creatures. The air conditioner is efficient, so you get one more for the bedroom. As TVs become more and more energy efficient, you buy bigger and bigger models. You have an eco-friendly car, so you feel OK driving it longer distances. Many things have fanned the flames of this. There was the 1000 yen all-you-can-drive pass for the country's expressways (although that was abolished on June 6, 2011), as well as the home appliance "eco point" system that ran from May 2009 to March 2011. The government had extended these benefits, so people really felt like they had a good excuse to take advantage of them. But the result was greater environmental impact. The government put its stamp of approval on this indulgence.

The second factor is that the proper way to use eco technologies is often not properly explained. On the contrary, you come across lots of people saying that because something is eco-friendly, you can use it however, and as often, as you want.

In 2010, 25.19 million flat screen televisions were sold in Japan, an 85% increase on the previous year, as well as 8.24 million air-conditioning units. These were record-setting numbers, and if you consider the fact that Japan had about 50 million households in 2010, it meant that about half of all households bought a flat screen television that year. Moreover, the number of single-person households continues to increase, as does the total number of households. (In 2010, the average number of people in a household was 2.47.) People in Japan already own over 2.4 televisions per household, 2.5 air-conditioning units per household, and even 1.3 refrigerators per household. We can already see that it is a delusion to think that if all households adopt eco-friendly appliances, environmental impacts will decrease. Consumption is expanding, without any signs of stopping.

Even if the power consumption of all appliances is reduced, if the overall number of appliances in each household increases, and the overall number of households also increases, then the overall environmental impact will not decrease but increase. More households means increased numbers of items used in everyday life, starting with televisions, refrigerators, and so on. Despite this, or perhaps because of this, roughly 40% of Japanese people report some sort of stress related to material things, whether it's the simple fact of owning them or the feeling of being surrounded by material things in everyday life. And yet, televisions and refrigerators are not just getting more energy efficient, but they are also getting bigger. A/C units are getting more energy efficient, so people are putting one in each room. Families are getting smaller, which means more households. This is where we are headed.

Of course, making products and services more eco-friendly is extremely important, but at the same time, we must not forget to ask what kind of lifestyle we are creating for ourselves as we use these products. If we look at global consumption, we see that while the fuel efficiency of cars is rising each year, the total number of cars owned is rising faster. Between 1950 and 2000, the world's population increased 2.5 times over, from 2.52 billion to 6.06 billion. Over the same span, the number of registered vehicles grew more than tenfold (70 to 720 million), fuel consumption more than sevenfold, and electricity consumption more than twenty-one-fold.

With this kind of approach, even if we make great technology, overall it will not help to lessen our environmental impact. If this continues to happen, it is possible that technology itself will become seen as something bad. This is not the first time this sort of dilemma has been discussed. In 1865, William Stanley Jevons published a book called *The Coal Question*. As steam engines became more common, people began to feel worried about the depletion of coal. Many of them anticipated increases in the steam engine's thermal efficiency, which would prolong the lifespan of coal reserves, but Jevons refuted this line of argument. While it is true that if a single steam engine's efficiency was raised, its coal consumption would decrease, if steam engines in general become more efficient, they would become more appealing, and therefore more widely adopted. Jevons showed that the increased rate of adoption of these higher-efficiency engines would outweigh the gains in efficiency themselves. There are cases like this, where something is correct on the micro scale, but if you try to apply it on a macro scale, it doesn't work. If we try to increase eco-efficiency on an individual basis, we will trigger a large reaction in the reverse direction, and increase environmental impacts. That plan will backfire.

A debate in the 1990s related to increasingly efficient information and communication technology (ICT) followed the same pattern. The "constant time budget hypothesis" states that people spend, on average, about 75 minutes in transit each day, regardless of their culture or income level. People with low incomes generally get around by walking (not having any other options), while people with high incomes get around in cars and planes. However, whatever method people use to get around, their travel time averages out to about 75 minutes per day. It's not about distance, it's about time. If a person works from home to cut down on travel time, he will spend more time in transit for other purposes, and then, when a traffic jam lets up, and he's able to drive faster, he'll drive a longer distance because he feels he's being "efficient," releasing more CO_2 into the atmosphere. Another example of a backfire. The evolution of IT may have made managing work more efficient, but it has not decreased environmental impacts, and it has made regular workers even busier. No matter how efficient we become, we do not seem to get more free time in return.

Why do we have to ponder this Eco Dilemma? It's true that eco-friendly technologies have been widely adopted, and that the importance of the environment has been instilled in people—in Japan, close to 90% of the population. This is a huge help. However, if we simply continue to come up with new technologies that replace the old ones, people will feel justified in indulging in them, and we will not get environmental impacts under control. In the October 12, 2012, edition of the *Nikkei*, it was reported that the Japanese government's eco point system had reduced CO_2 by only 1/13 the amount it had hoped to. This number symbolizes the Eco Dilemma. Today's technology and services are made to meet the needs of people who have broken away from nature and community, who require unlimited convenience. These technologies and services will never reconnect these people to nature, and they will never forge community bonds. In other words, technology and services made for individuals will not contribute to the creation of a sustainable society.

1.5.2 Thinking about a Sustainable Society

In 1987, the World Commission on Environment and Development, chaired by Norwegian Prime Minister Gro Harlem Brundtland, released Our Common Future (also known as the Brundtland Report). A central theme was sustainable development, which the report described as "development that meets the needs of the present without compromising the ability of future generations to meet their own needs." It also discussed how environment and development are not in opposition—they coexist. Also, it described a sustainable society as one where sustainability is achieved through continual sustainable development.

If we take that way of thinking as our basis, it is obvious that a society of individuals who are no longer connected to nature or the community will never be able to establish a sustainable society. The basis of a sustainable society will be reconnecting these people to nature and community. That said, we cannot return to an animistic society. There are certainly many things we should learn from animistic societies, but rather than going back, we will re-fix these bonds in a more timely way. As we said before, technologies and services that simply trade one thing for a newer thing—such as replacing air conditioners with green air conditioners—will only incite individuals to consume more. That approach does not make sense. Now, we need to change our very way of life. And investing in the new technologies and services these new lifestyles require will generate exactly the innovation we need (Fig. 1.17). These lifestyles must be exciting, appealing, and fulfilling. They must be lifestyles where, even in harsh environmental constraints, we can feel nature, and create a new form of community.

We said that we cannot go back to an animistic society, and that is because spiritual affluence is largely influenced by the human desire for their reversible value of life.

Human desire is limitless. No matter how much we are satisfied, we always desire more. Consequently, the market is always investing in new products, resulting in unsustainable expansion. Mobile phones use huge quantities of metals that we fear may be running out. One ton's worth of mobile phones (about 12,500 units) contains 400 g of gold, and 2.3 kg of silver. "Before these metals are depleted,

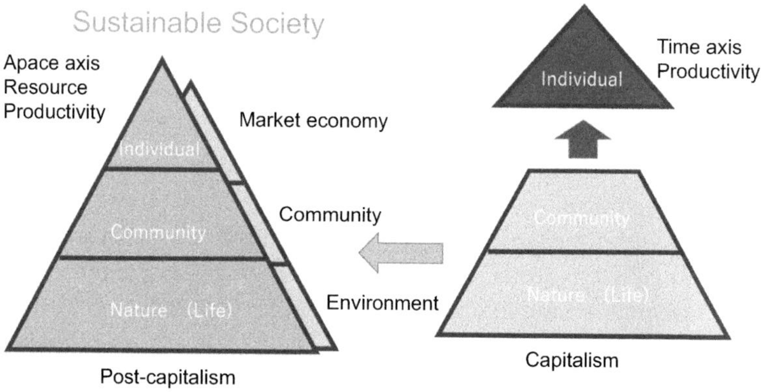

Figure 1.17 Sustainable society will be reconnecting people to nature and community but not back to the animistic society.

we ought to give up our cell phones! Twenty years ago, almost no one carried a mobile phone, and everything was fine, so we should able to stop using them whenever we want." But the fact is, once people have a mobile phone, they can't give it up.

Now think about household energy consumption. As incomes increase, so too do gasoline consumption and energy consumption. After Japan's economic bubble burst, from around 1996, we see a decrease in disposable incomes. You would think that household energy consumption and gasoline consumption would have decreased, but unfortunately even as incomes fell, energy consumption increased (Fig. 1.18). Why does this sort of phenomenon occur? We believe it is because we place great value on the irreversible value of life. This means that once we have some form of comfort or convenience, we cannot easily abandon it. Only humans exhibit this kind of desire.

Once we get a mobile phone, we start wanting a more convenient mobile phone. Once we get a car, we start wanting a more comfortable car. And if we try to give up a convenience we have, it hurts us, and makes us sad.

The fact that this desire for the irreversible value of life exists means that, even though animistic societies were very close to sustainable societies, we cannot go back to them. We can learn things from animistic societies, but we simply cannot go back. The basis of

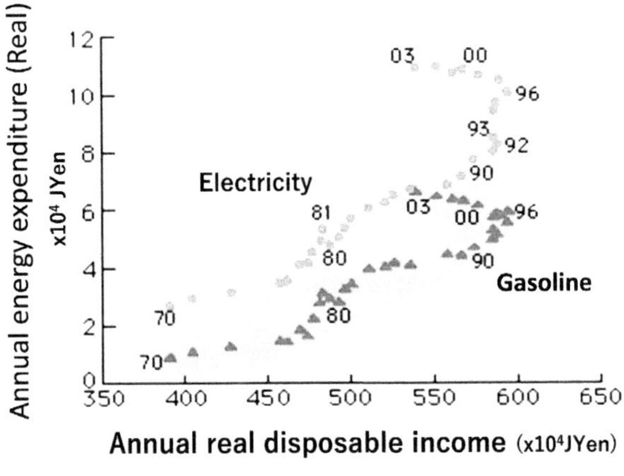

Figure 1.18 The connection between disposable income and energy.

reconnecting individuals to nature and community is affirming our desire for the irreversible value of life.

1.5.3 The Signs Are Already Here

To reconnect individuals with nature and the community, should we wipe the slate clean and start over from scratch? No. This reconnection is already happening. In the mid-1980s in Japan, people began to value spiritual affluence over material affluence, and the gap between the two continues to widen—it is currently at 30 points (Fig. 1.19). Many young people are leading the way. They prefer riding their bicycles to owning a car. They have no resistance to the idea of going to a flea market and exchanging items with others. They are into the idea of spending the weekend doing outdoor activities. They are starting to think it's cool to take care of old items, even if they're not "vintage." Home gardens are a big new thing. We are seeing the signs that forming bonds with nature and the community is becoming cool.

The 2011 Great East Japan Earthquake and the resulting tsunami also created new bonds between individuals and the community. Because of the disaster, transportation was cut off, and countless facilities were damaged by the tsunami. Access to food, electricity, water, and gas were lost. Many people who had been affected by

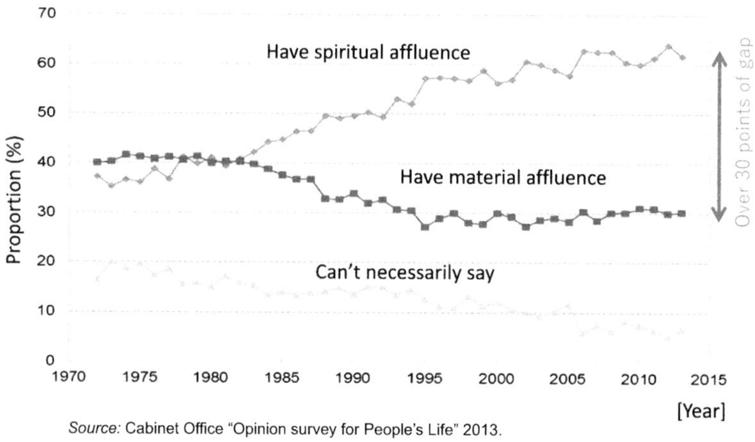

Source: Cabinet Office "Opinion survey for People's Life" 2013.

Figure 1.19 Spiritual/material affluence awareness survey.

the disaster took quick action: some offered up all the food in their homes to someone else, while others gave a little bit to everyone around them. It created a group of people who were truly grateful even for a single rice ball to eat each day. People lined up for hours in the snow for small numbers of supplies, but no one said a word of complaint. On the contrary, there were smiles all around. I was a victim of the disaster myself, and I've never felt such pride to be part of a group of people. This group already has some knowledge of what needs to be done to connect with nature and the community, and we are already beginning to move in that direction. We are moving beyond the civilizational saturation of modern society, and beginning to move towards valuing culture.

These signs show that this new way of life, which affirms the principle of continuous improvement in quality of life while forging connections with nature and community, is already emerging to a significant degree. However, innovation does not just come from the bottom up. We also need to be pulled in the right direction. In *The Theory of Economic Development*, Joseph Schumpeter says that if business does not educate the citizenry, then innovation will not happen. What we need now are both bottom-up approaches and approaches that rely on pulls from business and government that will help to revolutionize lifestyles. Both are vital.

Acknowledgment

This work was supported by JSPS KAKENHI Grant number JP24120008 in Scientific Research on Innovative Areas: "Innovative Materials Engineering Based on Biological Diversity."

References

Baudrillard, J. *The Consumer Society: Myths and Structures*, Sage, London, England (1998).

Chambers, N., Simmons, C., and Wackernagel, M. *Ekorojikaru futtopurinto no katsuyou: chikyuu ikko bun no kurashi he* (*Sharing Nature's Interest: Ecological Footprints as an Indicator of Sustainability*), trans. Michi Goto. Intershift (2005).

Constanza, R., et al. The value of the world's ecosystem services and natural capital, *Nature*, **387**, 253–260 (1997).

Digital Europe 2003. Making the network: steps towards a sustainable network world (2003).

Energy Whitepaper 2010. Ministry of Economy, Trade and Industry. Agency for Natural Resources and Energy (2010).

Ferguson, N. *Manee no shinkashi* (*The Ascent of Money*), trans. Senna, Osamu, Hayakawa Shobo, Tokyo (2009).

Foliensatzzur Energie-Info, Erneuerbare Energien und das EEG: Zahlen, Fakten, Grafiken (2013); https://www.bdew.de/internet.nsf/id/17DF3FA36BF264EBC1257B0A003EE8B8/$file/Foliensatz_Energie-Info-EE-und-das-EEG2013_31.01.2013.pdf.

Gesell, S. *Jiyuchi to jiyuukahei ni yoru shizenntekikeizaichitsujo* (*Die natürliche Wirtschaftstordnung durch Freiland und Freigeld*), trans. Aida, Shinichi, Pal Shuppan, Tokyo (2007).

Harada, K. Demand and supply of rare materials, *Japan Society of Material Cycles and Waste Management*, **20**(2), 49–58 (2009).

Harada, K., Shimada, M., and Ijima, K. 2050nen no kinzoku shiyouryou (Forecast of metal usage in 2050), *The Journal of the Japan Institute of Metals*, **71**(10), 831–839 (2007).

Hiroi, Y. *Posuto Shihonshugi* (Post-capitalism), Iwanamishinnsho, Japan (2015).

International Energy Agency. World Energy Outlook 2007, IEA Publications (2008).

Ishida, E. H., and Furukawa, R. *Nature Technology*, Springer, Japan (2013).

Ishida, H. *Channeling the Forces of Nature*, Tohoku University Press, Miyagi (2009).

Ishida, H. *Hikarikagayaku mirai ga Okinoerabushima ni atta!* (The brilliant future was in Okinoerabu Island), Springer, Japan (2013).

Ishida, H., Furukawa, R., and Dentsu Grand Design Laboratory. *Kimi ga otona ni naru koroni. Kankyou mo hito mo yutaka ni suru kurashi no katachi* (By the time you grow up: a way of living that makes both the environment and people rich), Nikkan Kougyou Shimbun, Tokyo (2010).

Kurzweil, R. *The Singularity Is Near*, Duckworth Overlook, UK (2006).

Living Planet Report 2012. Worldwide Fund for Nature, Japan (2016).

Ministry of the Environment and British Embassy in Japan. *Sutan rebyu: kikou hendoy no keizaigaku* (Stern review: the economics of climate change) (2007). <http://www-iam.nies.go.jp/aim/stern/Stern ReviewES(JP). pdf>.

Peart, S. *Jebonzu no keizaigaku* (*The Economics of W. S. Jevons*), trans. Ishida, Haruo, Sekiya, Kizaburou, and Kurita, Zenkichi, Taga Shuppan, Tokyo (2006).

Shimada, H. *Kenshohi sedai no kenkyu* (A study of an anti-consumption generation), Toyo Keizai Shinposha, Tokyo (2009).

Stern, N. Review on the economics of climate change, Office of Climate Change, Japan (2006).

Sugiyama, M. Jio-enjiniaringu gaisetsu (Overview of geo-engineering), Central Research Institute of Electric Power Industry, Tokyo (2010).

Sugiyama, M. *Kikou kougaku nyuumon* (Introduction to climate engineering), Nikkan Kougyou Shimbunsha, Tokyo (2011).

Takahashi, M. *Bunka hendo no naka no kodomo* (Children in a changing culture), Tohshindou, Tokyo (2002).

Taniguchi, M. *Metaru wozu* (Metal wars), Toyo Keizai, Tokyo (2008).

Toichi, T. *21 seiki no enerugi chiseigaku* (Energy geopolitics for 21st century), Sankei Shinbun Syuppan (2007).

Uzawa, H., and Uchibashi, K. *Hajimatte iru mirai* (The future that has already begun), Iwanami Shoten Publishers, Tokyo (2009).

World Energy Outlook 2010. International Energy Agency, Paris (2010).

Yamamoto, R. *Ondanka jigoku: dasshutsu no shinario* (Hell of the global warming: scenario for escape), Diamond (2007).

Chapter 2

Nature Technology Creation System

Emile H. Ishida
Earth Village Research Lab. LLC, Okinoerabu Island, Kagoshima Z/C 891-9222, Japan
emile.h.ishida@gmail.com

Technology is meant to enrich our lives. What we need to do now is design the kind of lifestyles that provide spiritual affluence despite the increasingly severe environmental constraints that are predicted for the future. We then need to identify the technological components necessary to support such lifestyles and generate ways to fully recycle resources with minimum use of energy. The answers we seek can be found in nature, and in the only truly sustainable social activities on earth. We need to apply the filter of sustainability in redesigning our lifestyles. We use the term *nature technology* to refer to the overall process of creating such technologies. The Nature Technology Creation System is already giving rise to numerous new technologies.

2.1 The Two Doors to Technology

There are two paths to a new technology approach. One is the academic advantage, which has shaped how we view technology

Lifestyle and Nature: Integrating Nature Technology to Sustainable Lifestyles
Edited by Ryuzo Furukawa

ISBN 978-981-4774-62-8 (Hardcover), 978-1-315-14687-4 (eBook)
www.panstanford.com

since the Industrial Revolution and is the product of societies that are dependent on fossil fuel. Academic advantage is the foundation of today's society and leads to the development of technologies and materials that are not found in nature. Certainly, there are some things we enjoy that have resulted from academic advantage, evident by the fact that this technology is supporting our current civilization. Because we create new technologies and concepts that have never existed on earth, it is imperative that our methods and technologies are sufficiently tested before they become common practice. The need for a new approach to technology is evident when we consider that the current view of technology is what has shaped our current lifestyle with technologies such as computers, cars, and fibers.

Chlorofluorocarbon (CFC) gasses were one of the most significant and most acclaimed inventions of the 20th century, but they are destroying the ozone layer. It is only now that we realize that these CFC gases are hundreds, if not thousands, of times worse for the environment than carbon dioxide, and is the main greenhouse gas contributing to global warming. Plastic also has a huge environmental impact because it does not biodegrade.

The current focus must be on nanotechnology. Considering that even bacteria are approximately 20 nm in size, materials created at a level of several nanometers can easily be absorbed into cells. To do this, we must ask how nanoparticles, which previously didn't exist on this earth, affect the human body. What influence do they have on biodiversity? The research is underway, and we can only hope for more and faster results. That said, the creation of such nanoparticles must be carefully considered to avoid negative impacts similar to what happened with CFCs. The effects such a technology would have on humans, the ecosystem, and biodiversity must be sufficiently assessed, as well as the amount of energy and resources that would be required.

The opposite of academic advantage and is known as natural advantage.

2.2 Considering Natural Advantage

Natural advantage takes a completely different approach from that of academic advantage, making it possible to create technology that learns from the perfect cycle of nature, and has benefitted from repeated and unlimited testing throughout the 4.6 billion years of earth's history.

Pre-Industrial Revolution technology, such as the Japanese tatara method of ironcasting, which has appeared in the animated film *Princess Mononoke*, was all based on lessons from nature. This steel-manufacturing method involves making steel with only natural iron sand from weathered rocks. It was capable of producing iron that is still unsurpassed in terms of strength and flexibility. The tatara method is also used to make the world-renowned Japanese sword.

Kenko Yoshida wrote in the 55th verse of Tsurezuregusa, "When building a house, one should focus on the summer." This concept is evident in the houses of monsoonal Japan which are designed with raised floors and made from wood, soil, and paper to be comfortable in the summer months. The houses are spacious and have clay walls that are literally works of art due to the craftsmanship of the plasterers. These clay walls act as a natural air-conditioner with insulating and cooling properties as well as controlling humidity. The plastered finish of the wall also has antibacterial properties that prevent mold from growing, allowing the wall to remain pure white indefinitely. Lacquer can be described as natural plastic as it possesses a lifespan and functionality that far exceed those properties of present-day plastic. Silk fabrics also boast antibacterial properties and are flexible and capable of blocking ultraviolet rays. Carpenters of temples and shrines from the Edo period were capable of simulating conditions 100 years into the future and used clever tricks of their trade to create technology of such precision that it is almost impossible to reproduce with today's electronic devices. Unfortunately, these methods were hidden behind the veil of tradition and these state-of-the-art technologies and engineering methods were classified as trade secrets to be passed down to the next generation of apprentices. And the spotlight fell on technology that drew on the mineral resources and ate away at the assets of the 4.6 billion year old earth.

In order to bring about the next phase of selection, we hope to propose a brand-new manufacturing philosophy where nature is viewed with a scientific perspective and technologies are redesigned with the needs of the human ecosystem in mind.

By understanding how the gecko darts across the ceiling, it may be possible to completely change our fundamental concept of adhesion. Using the mechanism that enables autumn leaves to fall from the trees, it may be possible to create electrical appliances that have pieces which break free once the lifespan of these parts ends. If we can understand the mechanism that makes the shell of an abalone so durable, we may be able to make a new material that draws on the properties of iron, ceramics, and plastics and is rust-proof, strong, and flexible.

2.3 Nature's Unique Closed-Loop Society

While we continue to talk about and strive toward a closed-loop society, we are no closer to achieving it; in fact, we are further than ever. Material inputs continue to increase exponentially. If we cannot create a closed-loop society ourselves, then perhaps imitating something else may be a shortcut. The only perfect closed-loop system we know of that functions on a minimum of energy is nature. We must once again make natural systems our point of reference. The system we call nature has been in place for the last 3.8 billion years and will continue into the future. It is so closely linked with the earth's cycles that we could even consider that this perfect cycle dates, in fact, to the origins of the earth over 4.6 billion years ago.

Homo sapiens have been a part of the ecosystem for 200,000 years, but our existence is now threatened. I believe that nature is basically intelligent, and it would not be an overstatement to say that that intelligence has certain moral values. Society continues to search for synthetic substances that do not exist naturally on the earth, all in the name of science and technology. This has resulted in the invention of around 30 million different chemicals the likes of fluorocarbons and plastics. As a result, we are trapped in a negative-feedback loop that further exacerbates environmental problems. Now, more than ever, we must reevaluate our actions from a scientific perspective. By looking at the form present in nature, physical and chemical performance, ecological features, and ethical significance, we are given a treasure chest of wisdom.

2.4 Nature Achieves a Perfect Cycle with Minimal Energy Input

What form does a perfect natural cycle exhibit, and why can't we create a closed-loop society?

The natural cycles of nature can teach us two very significant lessons. One of these is the mechanism that allows nature to function with minimal energy.

How does a gecko dart across the ceiling? Why don't fish crash into each other when they swim in a school? Why does the shell of an abalone not break when hit with a hammer? Why doesn't a snail's shell get dirty? Why are the temperature and humidity in termites' nests always constant? Why are spiders' webs stronger than steel? Why are maggots living in feces immune to harmful bacteria? How do silkworm cocoons protect the larvae from ultraviolet rays and bacteria? There many such amazing adaptations found in nature. Yet, nature does this all without rare and expensive metals. Nature does not need ultra-high temperatures or complex chemical treatments; rather, ordinary materials are used with their natural properties at normal temperatures. As a result, nature functions as a perfect cycle with the minimal energy consumption. By observing the wonders of nature from a scientific perspective, we can discover new approaches to technology. By leveraging the gifts of nature and the sun, we may be able to transition from a society reliant on fossilfuels to a natural society that uses high-performance technologies with extremely low environmental impacts.

2.5 Nature Boasts the Perfect Cycle Yet Is Self-serving

There are many other lessons that can also be learned from the natural cycle—the way the system works as a whole, for one. This is not to say that each individual part of the system is altruistic and consciously doing its part for the benefit of the whole. The praying mantis, for instance, named for its habit of lifting its front legs as though in prayer, does not share its food with its fellow mantis. In actual fact, it is not unusual for the praying mantis to eat its neighbor. Darwin said that survival of the fittest was the key to evolution. In other words, the result of fighting for one's benefit, for one's survival, produces the perfect cycle. This is a very important concept. Adam Smith assumed that people's egotistic nature optimized the whole (as the invisible hand of God). The cycle of nature continues to create a perfect cycle guided by the desires of living things.

Why is it, then, that natural ecosystems can be selfish and, at the same time, sustainable, yet the selfish nature of human beings threatens to destroy our civilization? What can account for this difference? One likely reason is that, unlike humans, the rest of nature does not have as high of a capacity to manipulate resources to satisfy desires. Dependence on the immediate natural system results in population control based on limited resources—a perfect cycle. The egotistical part is essentially weeded out naturally, and the resulting seeds have the accumulated result of continuing to sustain the ecosystem. So, while individual organisms in nature are selfish, nature as a whole is an unselfish construct.

What about humans? We are also selfish. An indication that our natural instinct tends toward selfishness is that in Buddhism we will be forgiven for doing something selfish if we follow it by an unselfish act. Our reliance on natural resources to create technology to satisfy our desires has led to the destruction of nature. Development such as this, that ignores and evades natural selection, can only be possible with the assumption of infinite natural resources. Faced now with the reality of finite resources, the situation has become desperate and we are on the verge of natural selection (see Diagram 2.3). It is this selection—possibly the first and the last—that will lead to the collapse of civilization.

Now we need to learn from the mistakes we made in thinking that we could control nature. We must come up with new ways to leverage both nature and technology. We must find a new perspective on technology that takes nature into account and avoids destroying civilization. We must find a way to bring about a process of selection that will create a new civilization of life.

2.6 Nature Technology: A New System of Manufacturing

To be able to pass on the blessings of Earth's finite resources to the next generation, we urgently and thoroughly need to reconsider the relationship between the global environment and human beings. The time has come to create a new mold for manufacturing and living which acknowledges the reality that we, humans, possess "genes of desire"—that is, we do not willingly let go of modern conveniences and comforts once gained—and which, with the smallest possible impact on the global environment, responds to this reality.

Nature technology is one such response, a new system of manufacturing which intelligently utilizes the amazing powers of nature. Nature technology does not, as is the case in a conventional approach to technology, create things that are not found on the surface of the Earth; it is, rather, an entirely new way of approaching manufacturing and living which looks at nature—functioning in perfect cycles and with a minimal input of energy based on 4.6 billion years of natural selection—from a scientific viewpoint and redesigns it as new forms of technology.

A virtually infinite number of living beings exist in nature. Nature does not use oil or other fossil fuels to support these different forms of life. Mainly, all living things on Earth are connected through the clever utilization of abundant energy reaching the Earth from the Sun. Or, the great forces of the Earth, such as tectonic shifts

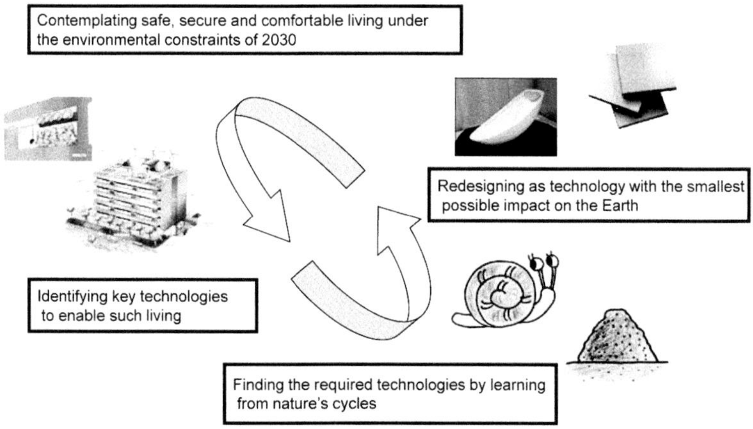

Figure 2.1 Nature technology, a new system for manufacturing.

and diagenesis turn organic life forms into resources and energy. In nature, we find an astonishing (to us humans, that is) number of fascinating seeds of technology. Nature's cycles are the fruit of wisdom developed through life's 4.6 billion year history. Nature technology is a new attempt at using the skills we human beings have acquired as tool makers and users to learn from, and bring to life, these amazing powers of nature.

Today we must, through the intelligent use of technology, make new proposals for ways and forms of living. Technology is needed that can make statements about future forms of living. These are also, we may say, the kinds of technology that—as is required in our age—assume responsibility for lifestyles.

As its point of departure, nature technology (Fig. 2.1) learns from nature's perfect cycles driven with a minimal input of energy. However, this does not mean merely substituting a technology learning from nature for a conventional technology. It must, also, involve a shift to a way of living in which we do no longer think of, for example, an air conditioner as a "thing/device," but rather focus on the "action/function" of controlling indoor air quality. In this way, we may satisfy people's desires for convenience and comfort—once gained so difficult to let go of–not in a materialistic way, but rather by fulfilling immaterial, or spiritual, urges.

In more concrete terms, we must envision wholesome, fulfilling lifestyles under the severe environmental constraints of 2030 using the backcasting methodology (details will be discussed in Chapter 4) and then distill the technologies required to make this vision come true. The important point is to take lifestyles as the point of departure for this process. This is fundamentally different from a conventional approach to technology development. In conventional development methodologies, a forecasting approach is most often taken, which first creates an image of the technology and then considers how to introduce it into a lifestyle. This approach cannot break free from wasteful thinking based on the notion that "to live humane lives, people must be materially prosperous," and thus ends up triggering eco-dilemmas. It is crucial that we first use backcasting to create a clear picture of a lifestyle, and then contemplate the technologies needed. This is the only way to develop technology which assumes responsibility for lifestyles while also ensuring that the prosperity needed to live spiritually affluent lives is secured.

When the elements of technology to aim for have been clarified, the next step is to search for these in nature. Obviously, this is, as mentioned above, important because nature has created perfect cycles with minimal energy inputs, but furthermore, there is a need to incorporate a view of nature into technology itself.

2.7 Can We Learn from Nature?

We can learn two major lessons from nature. One is the fact that nature, through a continuous process of natural selection, has achieved the optimum form of evolution adapted to the various local and global environments through time and has done this in a system of perfect cycles driven with a minimal input of energy. This is indeed the "sustainable society" which humanity, despite lofty ideals, is still far from achieving—created by nature with the use of solar energy and the abundance of things found on Earth's surface. We humans need to re-learn the mechanisms and systems—as well as the social process of creative destruction—found in nature. A second important lesson we can learn from nature is that modern technology, created as a result of the Industrial Revolution starting in Great Britain, succeeded only by removing itself and thus humans from nature, and as a result is intricately linked to the emergence of global environmental problems.

2.8 Nature Drives Perfect Cycles with a Minimal Input of Energy

In nature technology, the idea of mimicking nature is of great importance. The notion of biomimetics—the science of mimicking living things—was first proposed in the 1950s by neurophysiologist and inventor Dr. Otto Schmitt. On the basis of his research on the nervous system of squids, Schmitt in 1934 invented the so-called Schmitt trigger, an input circuit system for digital circuits which arguably is the foundation for present electronic devices. Of course, centuries before this, Leonardo da Vinci left us many sketches which, based on the mimicking of a bat's flying skills, show how to fly through the air, and finally in 1903, the Wright brothers famously completed their first flight. (For quite a while after that, people would argue that "machines cannot possible fly through the air" and refused to acknowledge their great achievement).

There are three large categories when we talk of biomimetics. The first is the mimicking of shapes and forms. The skin of a shark is covered by tiny protuberances with a toothlike structure called dermal denticles (Fig. 2.2), and these help lower water resistance, making it possible for the shark to swim fast. By mimicking

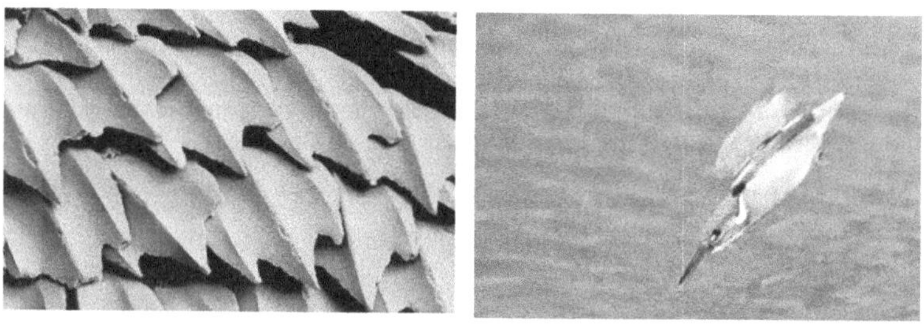

Figure 2.2 A shark's dermal denticles (left) (photo: Sue Lindsay © Australian Museum) and the kingfisher (right), a bird able to dive into water at top speed.

Figure 2.3 The water-repellent lotus leaf is covered by small protuberances (photos: © Dr. Osamu Takai, Nagoya University).

this mechanism, swimwear with reduced water resistance was developed. Mimicking the water-repelling mechanism of a lotus leaf helped give birth to external wall material that doesn't stain easily (Fig. 2.3), and the shape of the kingfisher's beak was inspiration for the shape of the front car of the 500-series Shinkansen Bullet Train (Fig. 2.2).

The second category is the mimicking of function. Insects do not have an immune system such as that of humans, but when facing an attack by bacteria, they immediately defend themselves by making a protein in their bodies called defense-related protein. From this protein, an anticancer drug and other medicinal products are about to see the light of day. Research is also underway on a material that, using the mechanism of leaves dropping from deciduous trees, disintegrates and falls apart once its lifetime has been reached.

Finally, there is the mimicking of systems. How do schools of fish form, and why do the fish not collide with each other in such schools (Fig. 2.4)? Why are birds able to fly in beautiful formations? Understanding such systems may help us change traffic systems or information networks.

An even broader concept than biomimetics is nature mimicking. For example, mimicking the mechanism behind the way in which hot magma cools and turns into volcanic rock is the idea behind ceramics. Nature—both below and above the Earth's surface—has created large and perfect cycles. Looking at these through the lens of global environmental issues and ways of living is the essence of nature technology.

Let us take a look at some concrete examples of this kind of technology. If we look, first, at nature technology that contributes

Figure 2.4 A school of bigeye jackfish (photo: © Sea Dream Okinoerabu).

directly to solving global environmental issues, the honeycomb structure of a bee nest is a good example: it is light but of great strength and is used as structural material in airplanes, boats, and residential houses (Fig. 2.5). Pine cones will close when moisture is high and open up when it is low in order to allow for the seeds kept inside to disperse. On the basis of this principle, smart clothing which automatically regulates temperature was developed. When moisture content in the air is high, the gaps between fibers will

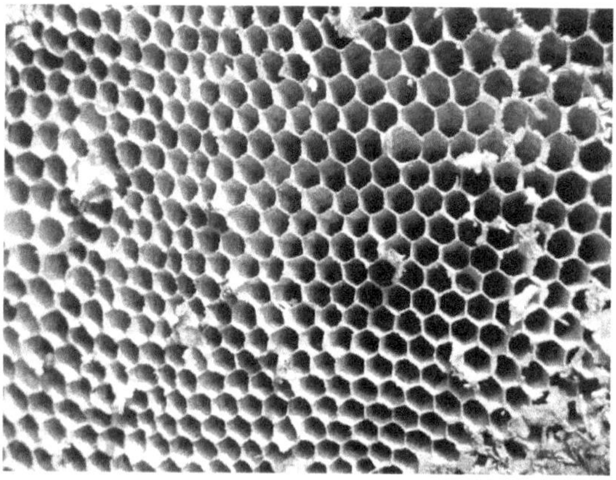

Figure 2.5 The honeycomb structure of a honeybee nest.

Figure 2.6 The hump of the humpback whale.

open up, and when it is low, they will contract, thus regulating temperature.

The humpback whale, as the name suggests, has humps on its fins (Fig. 2.6), and research has revealed that these humps reduce water resistance significantly. A prototype wind generator applying this structure of a humpback whale fin has been developed. When it comes to electricity generation, owls also make a contribution. The flight feathers of an owl have ragged, saw-like edges called serrations which eliminate the sound of flight. This mechanism was applied in pantographs (current collectors) for the 500-series Shinkansen Bullet Train, resulting in a noise reduction of more than 30%. Since sound is energy, less noise means that energy is being used more efficiently, and for this reason, the same mechanism is being studied for potential use in wind generators.

In the sea, seaweed rolls gently with the movement of the sea, and a generator utilizing such gentle flow energy to generate electricity has also been manufactured as a trial. Furthermore, amazing findings have been made recently in the research of dragonflies. The wings of a dragonfly are not, as with eagles or pigeons, smooth, but have a rugged structure. Dragonflies use them to fly skillfully even when there is no wind (Fig. 2.7). Actually, the mechanism behind

Figure 2.7 Dragonflies have rugged wings.

this is not yet fully understood, but researchers have discovered that the rugged shape of the wing creates air turbulence (much like whirlpools), which creates an effect similar to that of a ball bearing. It appears that this allows for air to be transported over these "bearings" as if on a conveyor belt, thus giving the dragonfly its needed uplift. Using this mechanism, it is possible to make a wind generator that generates energy even in a slight breeze (for details, see Chapter 12). Seeds of the tropical plant Alsomitra will fly long distances in tropical rainforests with no wind, and their shape has been adopted in hang gliders and, in the past, in airplanes (Fig. 2.8).

Velcro tape, known all over the world today, is a registered trademark of Japanese chemicals company Kuraray. The Swiss national George de Mestral gained hints from the cockleburs sticking to his clothes as he went hunting and first produced Velcro commercially in 1955. Cockleburs have numerous small hook-like structures on their surface which hitch on the fiber of clothes or a dog's fur (Fig. 2.9). De Mestral reproduced this mechanism using a special type of nylon. Kuraray bought the patent of the invention in 1964, and after being used to cover the headrests on the Shinkansen Bullet Train starting operations in the same year, the utilization of Velcro expanded rapidly.

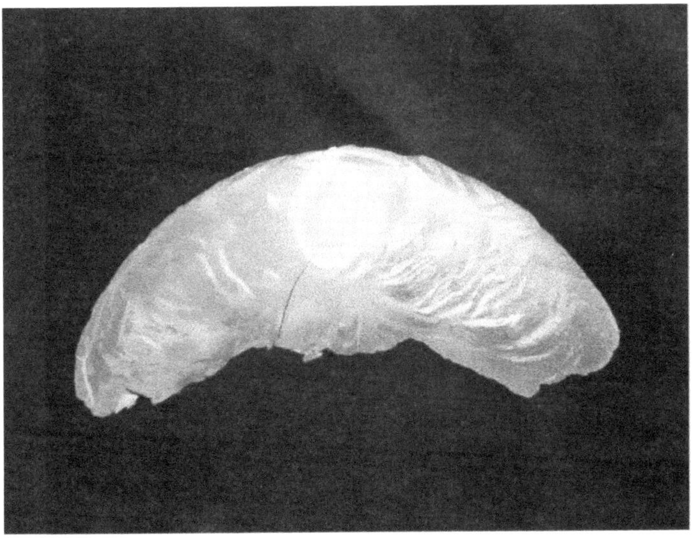

Figure 2.8 A seed from the Alsomitra plant.

Figure 2.9 The cocklebur, which led to the invention of Velcro.

Figure 2.10 The fog-collecting ground beetle, which collects water from the air with its body.

Due to the effect of population increase and global warming, a scramble for water has already begun. Here, the unique ways of collecting water mastered by the fog-collecting ground beetle living in deserts may come to be of help (Fig. 2.10). In the Namib Desert in Africa, fog covers the landscape every morning. The fog-collecting ground beetle does not waste this opportunity; facing the wind at just the right angle, it turns upside down with its back facing in the direction of the wind. On its back it has numerous hydrophilic protuberances with which it attracts the moisture in the fog. This moisture gathers and gradually turns into droplets of water. As the weight of the droplets increases, they run down hydrophobic channels between the protuberances and enter the beetle's mouth. Experiments are underway that, using this mechanism, set up hydrophilic and hydrophobic areas on the surface of buildings with the aim of making water from air.

Cellulose, a type of carbohydrate found in wood and fiber, is a tricky substance, indigestible not only to humans but to all animals except termites. It happens, though, to be about as strong as iron. It is actually cellulose which creates the unique jelly-like texture

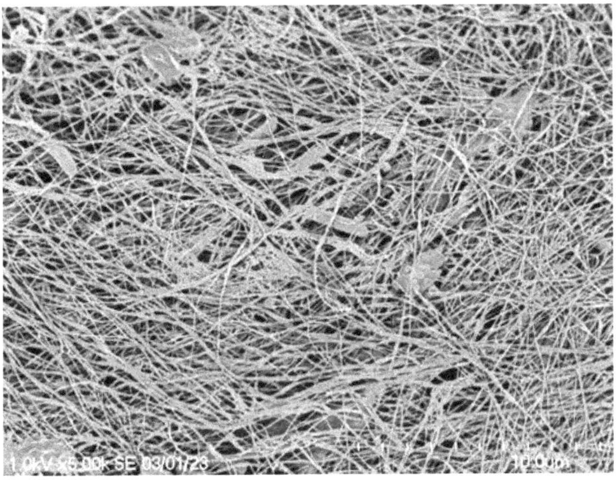

Figure 2.11 A microscopic image of the tiny cellulose fibers found in nata de coco (photo: © Dr. Hiroyuki Yano, Research Institute for Sustainable Humanosphere, Kyoto University).

of nata de coco (a popular Filipino dessert). In nata de coco, the cellulose is distributed in a meshlike pattern, and special filters have been created by removing 99% of the water of which nata de coco is composed and utilizing the 1% that is cellulose. Since the diameter of nata de coco cellulose is smaller than the wavelength of visible rays of light, it looks transparent to the human eye (Fig. 2.11). Flexible organic electroluminescence displays that can be made by inserting synthetic resin into this meshlike cellulose are also catching the world's attention.

Another example comes from the bottom of the sea. In the deep sea, numerous spots called black smokers have been discovered from which water heated by magma gushes out together with hydrogen sulfide and heavy metals. Near these spots, shells called calyptogena have been found in large numbers. With the help of sulfur bacteria, the shells transform hydrogen sulfide—a deadly poison to humans—into energy. It is conceivable that if we could learn to utilize this mechanism, a new source of energy could be created.

Yet other examples include new approaches to adhesion learned from geckos or from a bacteria called *Caulobacter crescentus*;

Figure 2.12 Barnacles, capable of underwater adhesion.

underwater adhesion learned from blue mussels or barnacles (Fig. 2.12), a relative of crabs and lobsters; spider thread as soft as silk yet as strong as steel; the shell of an abalone so strong that even a hammer cannot break it; the ability of the morpho butterfly to generate colors without pigments; the ability of the sandfish to move through desert sand as if it were swimming; the skills of a cupedidae beetle to detect a forest fire from several tens of kilometers away; the ultrasensitive sensing ability of a cricket which allows it to sense predators from movements of the air; and so on. Numerous attempts at developing technology that harnesses the ingenious mechanisms found in nature are taking place today.

For more details and examples, see, for example, our website Sugoi shizen no shouruumu (a showroom of nature's wonders) http://nature-sr.com/.

Here we have touched on but a few examples, but nature, in this way, does not need to use scarce resources such as rare metals or rare earth; it produces multitudes of life forms at normal temperature and pressure with materials found in abundance aboveground and creates perfect cycles. It is no exaggeration to say that nature is a treasure house of technologies. However, we have unfortunately only rarely, so far, studied the mechanisms and systems maintaining nature's cycles from the viewpoint of engineering. In conventional engineering, the main approach has been to work with underground resources or fossil fuels at high temperature and high pressure to make things not found in nature, and only very rarely have attempts been made to look at nature's cycles or natural functions from an engineering perspective. Therefore, in order to promote such an approach, the creation of a database with applicability for engineering is much needed, but the conversion technologies needed to make the huge existing volumes of data on biology applicable to the world of engineering are only now entering a study phase, and we can, thus, hardly say that actual ways of applying such technologies have been established.

Let us at take a look at trends in different countries around the world. In October 2010, the 10th meeting of the Conference of the Parties to the Convention on Biological Diversity (COP10) was held in Nagoya, Japan, where the main topics discussed included post-2010 targets for the protection of biodiversity, an international agreement on the access to and fair and equitable sharing of the benefits arising from use of genetic resources, and operational rules for the principles of liability and redress adopted in the previous Cartagena Protocol. When it came to specific detail, however, the participating countries prioritized national interest—much in the same way as with climate change negotiations—and no major progress was achieved. Preceding this COP10 conference, the German government in 2009, as a follow-up to COP9 held in Bonn in 2008, initiated a Business and Biodiversity Initiative dubbed "Biodiversity in Good Company." The website of this initiative includes examples of biomimetic technologies that have already been developed and marketed. Also, Germany through financial support helped establish the Biomimetics Association and is vigorously promoting cross-disciplinary academic research in the field. Apparently, as part of national strategy, biomimetics is being as supported as a new field with future potential. In October 2012, with Germany as a driving force, the first meeting of a process to establish ISO standards for biomimetics was held (ISO/TC266 Biomimetics).

In the United States, the science writer Janine Benyus in the latter half of the 1990s put forward the idea of "biomimicry," which, with her publication of *Biomimicry: Innovation Inspired by Nature*, rapidly caught on. Benyus and others later established a nonprofit organization, The Biomimicry Institute, and a consulting service, The Biomimicry Guild, and from these efforts a new industry is beginning to take shape. Toward the end of 2010, an economic evaluation of such efforts was published which estimated that biomimicry has the potential to become a USD300 billion industry creating 1.6 million new jobs within 15 years.

In England, the BIONIS Network, with corporations and universities as main participants, was established in 2002. From around 2000, the number of academic papers on biomimetics rapidly started increasing and by 2012 had reached more than 1600, thus indicating that a new trend is being created.

In Japan, the Ministry of Education, Culture, Sports, Science and Technology (MEXT), as part of its 21st Century Centers of Excellence initiative, has adopted programs concerning "Biomimetic Manufacturing" and "New Uses of Biological Resources," and in the "Declaration on Biodiversity," launched in 2009 by the Keidanren (Japanese Business Federation), article 5 mentions, "We will aim to create new industries, ways of living, and culture inspired by biodiversity," and goes on to explain this in detail in the following way: "Learning from the profound and boundless blessings of nature, as well as from tradition and the wisdom of our forefathers, we will, through the advancement of corporate activity with low impact on biodiversity and the development of environmental technology, aim to create sustainable industry, ways of living, and culture." Meanwhile, the corporate world has also launched some related products. Comparing with other nations, however, in which biomimetics is regarded as a whole new academic field and where scenarios are being pursued to use it as a trigger for the creation of new industries, Japan is still lagging behind, and it seems to us that actions taken have been rather haphazard and disjointed. Against this background, there are recent positive recent developments. In 2012, a program called "Innovative Materials Engineering Based on Biological Diversity" (chaired by Masatsugu Shimomura, Tohoku University) was started as one of the new academic fields under the auspices of MEXT, and in 2009, The Nature Technology Research Consortium (coordinated by Emile H. Ishida,

Tohoku University) began activities as a research association under the "Japan Manufacturing Conference" (MONODZUKURI Nippon Conference). The former works mainly with scholars, the latter with people from the corporate world, but the purpose of both initiatives is the same, and through the cooperation of these two groups, it is our hope that biomimetics research in Japan, somewhat belatedly, will now pick up speed. Taking environmental constraints into account and pursuing the issue of lifestyles are new values we hope to add to establish an innovative new system for technological development. In any case, we strongly hope that a country like Japan, which has manufacturing as one of its foundations, will urgently, and as a national strategy, pursue new kinds of manufacturing which intelligently harness the amazing powers of nature. It is in this context that the Ministry of the Environment decided in April 2013 to initiate a policy planning and institutional design program which—with the concept of "a civilization of environment and life" as its point of departure—works towards the creation of a sustainable society in which greater emphasis is placed on spiritual affluence than on materialistic prosperity.

2.9 The Success of the Industrial Revolution Was Based on the Principle of Humanity's Separation from Nature

Another important thing we can learn from nature is the way in which the success of the Industrial Revolution was based on our separation from that very nature. Why did the anthropocentric technology which has invited a civilizational crisis develop to the extent we see today? And, what do we have to consider today in order to change the direction of technological development away from material desires towards the encouragement of spiritual or immaterial desires? To pursue these issues, we need to take a look at the development of modern technology in Europe which led to the birth of our present underground-resource-based civilization. From the 8th until the 13th century, Europe was in what is called the medieval warm period. Agriculture spread across the Alps into

Western Europe, where technological innovation to utilize water and wind mills extensively continued. These technologies gradually became essential to the textile and steel industries and, as we shall see later, were an important factor in setting the direction towards the modern, Western European civilization in which man aims to conquer nature. The cooling of the European climate starting in the latter half of the 13th century, with a small intermission in the 16th century, reached a peak in the second half of the 17th century. Harvests in Europe around that time dropped drastically and the cooling triggered major famines. Around the year 1500, when the small ice age was at its height, the climate not only reduced agricultural productivity, but also caused the spreading of pests which, combined with the effect of numerous wars, led to sharp decreases in population. European society at that time, it is said, was on the verge of collapsing. On the other hand, a lack of labor led to the growth of livestock and dairy farming, and machines were introduced to make up for the lack of hands in agriculture and industry. Gradually, the idea that man was part of nature faded, and humans became increasingly removed from nature and, attempting to alter and reorganize nature artificially, aimed to become its conquerors. Such developments led to, and were further fueled by, Francis Bacon's concept of human dominance over nature and Rene Descartes' mechanistic worldview. This gave birth to a modern, Western philosophy justifying modern civilization and, in a concrete way, helped build a human empire enslaving nature.

The philosophy of dominance over nature advocated by Francis Bacon expresses itself in his philosophical work *Novum Organum*, where he—while acknowledging the overwhelming power of nature—at the beginning of volume 1, writes, "Nature cannot be conquered but by obeying her." Rene Descartes, by contrast, was a skeptic who based his philosophy on the statement "Cogito ergo sum"—I think, therefore I am. Nature was seen as existing in opposition to man, and thus, the better human beings understood the laws of nature, the easier it would become for them to control and manipulate nature at their will.

Such a mechanistic worldview became the most fundamental principle of modern science and technology, and as a result a convenient modern society was built, enabling the endless expansion of material desire. These ideas became so dominant across the world that even non-Western countries could not develop without adopting Western civilization. If attempts were made to reject this civilization, countries would be colonized, and, thus, regardless of whether or not there was any willingness to adopt Western thought, the world was gradually swallowed up by Western civilization. This anthropocentric worldview, however, does not allow for any restraint in relation to nature, and thus led to the creation of severe environmental constraints.

A dualistic worldview developed in order to understand the world in this way ended up creating a dichotomy between humans and the natural world. Of course, Cartesian thought did not suddenly trigger modern civilization. Descartes perfected mind-body dualism, but it is said that such dualism is found even in the works of Plato or Aristotle—major founders of Western thought—or in Hinduism.

Human dominance over nature was, thus, not just an idea arising from the works of Bacon and Descartes. The Old Testament, Genesis 1:26, states, "We will let them rule the fish, the birds, and all other living creatures." The Christian view of nature, which took shape long before the scientific revolution, looked at human beings as being in opposition to nature. Francis Bacon's thoughts on human dominance over nature derived from an objectification of nature for scientific investigation, thus creating an intrinsic opposition between human beings and nature, and Descartes took this worldview even further. When did such views of nature originally start taking shape? According to the Japanese philosopher Takeshi Umehara, they had their roots in the civilization existing in the Western part of the Eurasian continent where wheat and dairy farming formed the foundations of society. The first nation (city-state) was created by the Sumerian king Gilgamesh (Gilgamesh is regarded as the historical king of the Sumerian city-state Uruk, in modern-day Iraq, around the year 2600 BCE). According to *The Epic of Gilgamesh*, the king went on a journey to search for cedar trees not found in Mesopotamia. The cedar forest was, legend tells, protected by the guardian spirit Humbaba, but turning his back on the gods, Gilgamesh killed Humbaba and brought home cedar trees to Uruk. In this way, after unifying the city-state of Uruk, the first undertaking of

Gilgamesh was to kill the guardian of the forest, Humbaba. According to Umehara, allowing the murder of the forest's guardian spirit is a train of thought found in the Sumerian civilization—which came to form the foundation of Western civilization—and similar thinking was later reflected in the works of, for example, Plato or Descartes, gradually becoming a basic principle of Western civilization.

Did the idea of human dominance over nature then begin in this era? Apparently not; our species, homo sapiens sapiens, originated in Africa some 200,000 years ago, and some 40,000–80,000 years ago had spread to the entire African continent, in the process eradicating some 40% of large mammals such as great wild boars and wild stallions. Some 30,000 years ago, after our species had spread to Europe, 50% of great mammals such as mammoths and bison had disappeared, and 15,000 years ago the warm period following the Ulm glacial stage (the last glacial period) caused the Alaskan glaciers to melt, and with this, humanity rapidly spread southwards, within 1000 years reaching the Southern tip of South America. In this time frame, humanity caused the eradication of some 73% of herbivorous animals above 50 kg in North America, and for South America, the figure was as high as 80%. There are locations where it appears that humans, almost as part of a game, hunted and forced hundreds of mammoths over cliffs. As part of our DNA, we humans have genes of desire which enable us to conquer and dominate nature without moral restraint in order to satisfy our own greed. Apparently, a philosophy of human dominance over nature, and the existence of genes of desire, are some major characteristics of homo sapiens sapiens.

Nature technology, which harnesses the amazing powers of nature intelligently, is a process which first envisions lifestyles with a backcasting mentality, then distills the technologies required to realize these, search for hints and ideas for technologies in nature, and finally, through the filter of sustainability, designs new technologies. In lifestyle research we have learned that people strongly yearn for nature, and that numerous seeds of the new technologies we need are hidden in nature. The Spanish architect Antonio Gaudi once said, "Human beings cannot create anything, they can only discover." Nature is a textbook in which all the answers are written. Why, then, did technologies separating humans from nature come to rule the world and lead to environmental degradation? This, of course, is due to the nature of human arrogance which led us to believe we can control nature as our slave, and our predicament was caused by the illusion that underground energy and mineral resources could be a realistic solution to meet our desires. If we continue as presently, will future generations, when trying to unravel the "picture scroll" of Earth's 4.6-billion-year history, be able to find in it an era, as thin as a thread of hair, with immense impact on the future of life on the planet? This thread is the 21st century, in which humanity will have used up all underground energy and mineral resources.

Let us reconsider where we stand and what we can do. What would a technology be like that does not separate itself from nature? What would an industrial revolution look like that does not separate itself from nature? And would such technology and industry be able to avoid causing environmental degradation?

2.10 A Philosophy of Unity with Nature

With the Industrial Revolution as a major impetus, the philosophy of human dominance over nature—a trait apparently built into our genes—led to the creation of a civilization based on underground resources, which in turn caused the global environmental problems we are facing today. Can we, then, not free ourselves from this philosophy of dominance over nature? Apparently, an answer to this question may be found in Asia. In particular, the distinctive view of nature cultivated in Japan's unique environment may provide new insights. It is said that in the civilizations found in the Eastern part of the Eurasian continent, based as they were on rice farming and sericulture, anthropocentric thinking and the idea of human dominance over nature were weak. Yoshinori Yasuda calls the people who created the Western civilizations "field and dairy farmers" and those of the Eastern civilizations "rice (paddy) farmers and fishermen," the difference being the degree to which water was needed for farming. For rice paddy farming, much water is required, and thus forests that can preserve water are needed, which in turn means that the killing of the guardian spirit of the forest (as in the Gilgamesh epic) would never appear in the cultural narrative of the East. In Japan, in particular, the introduction of rice paddy farming came as late as a mere 2000 years ago (the oldest examples of such farming, possibly, date back as far as 14,000 BCE, to the Chinese Chang Jiang culture). As a result of this, people in Japan did not see themselves as special creatures, but rather as being of the same nature as all other living things. Humans, while struggling with other living things in a harsh environment, would still aim for unity with nature. Thus, it has been said that the principles of hunter-gatherer societies were still strongly present. Then, in the sixth century, Mahayana Buddhism was introduced to Japan, and as the philosophies of Saicho (founder of Tendai Buddhism) and Kukai (founder of Shingon Buddhism) merged toward the end of the Heian period (794–1185), a distinctly Japanese form of Buddhism took shape. In the Kamakura period (1185–1333), the imported Buddhism was completely Japanized and the major Buddhist philosophies had as their common strand the Tendai teachings of innate enlightenment (in Japanese, *Tendai hongakuron*), a philosophy which had at its core the idea that all living things—mountains and rivers, grasses and trees, and all the land, are imbued with the Buddhist spirit (in Japanese, *Sansen-souboku-kokudo-shitu-kai-joubutsu*). Here, we obviously do not find the idea of human, or even animal, supremacy; rather, all living things, including humans, are seen to be part of the same cycle of life. This kind of thinking is the same as found in the indigenous beliefs of the Jomon period (from around 10,000 BCE to 400 BCE) and beyond—namely that all living things are equal and share the same spirit of life. Buddhism in Japan was transformed by such strands of indigenous thought and was remolded into a uniquely Japanese form of Buddhism. During this process, the so-called manifestation theory holding that (indigenous) Shinto gods were actually manifestations of Buddha, and the syncretism of Shinto and Buddhism progressed in a multitude of ways.

It appears likely that the unique natural environment found on the Japanese archipelago helped make such views of nature even more distinct. Japan is a country in which the ground often trembles. The geographical area of Japan is only 0.25% of global land area, but the country has 7.1% of the world's volcanoes, and 20.5% of earthquakes above magnitude 6. Thus, the very ground beneath our feet trembles quite frequently and dramatically; in such a situation there is little a human being can do and little to hold on to. Also, the greenery in Japan almost violently imposes itself on human beings. The small mountainous hills found near Japanese villages were used in many intelligent ways. Fallen leaves or forest undergrowth would be used as fertilizer, and wood would be used to supply charcoal or firewood. Just in front of the house, there would be gardens; further out, fields; yet further out, the small mountainous hills, and behind these, the untouched, deep mountains—the latter a place worshipped and treasured as the residence of the gods. Even further away, these deep mountains would connect to the high peaks, which were like ladders to the heavens. The small mountainous hills served as a buffer zone between the deep mountains and populated villages—a buffer ensuring that the wildness of the deep mountains did not impose itself too strongly on village inhabitants. Actually, many efforts were constantly made to protect the mountainous hills from the threatening wilderness of the deep mountains. Grasses and weeds would have to be pulled out again and again—this is the nature of Japanese greenery and the power behind the continued existence of forests on more than 60% of the Japanese archipelago. (By comparison, the large-scale cultivation taking place around the 12th century in Europe meant that 90% of wooded land disappeared in England, 70% in Germany, and 90% in Switzerland). This is how strong and robust the greenery is in Japan. In Europe or the United States, turf would often be laid in church grounds or other sacred places. But in Japan, the green was excluded, thus creating another kind of sanctuary, where—in the case of both Shinto shrines and Buddhist temples—large, round stones would cover the ground. The green in Japan is this powerful—quite different from the idyllic image espoused in, for example, Europe. The same was the case with rivers. When heavy rain fell, the large rivers—where the dragon god was said to reside—would swell with anger and swallow up all in its surroundings. To Japanese, nature is a thing to be held in awe, but at the same time, we know it is this power which gives us life. The ideas held of nature in Japan are far removed from any thinking of human dominance.

If we can say that technology, based on the idea of human dominance over nature, has triggered the present environmental problems, then the approach needed to solve this problem must entail a new awareness of nature built into technology as well as a rethinking of the relationship between nature and human beings. It is our belief that the Japanese, with their uniquely deep view of nature, has the potential to provide solutions to this conundrum.

The historian Arnold J. Toynbee pointed to the fact that between the 16th and the 20th century all major civilizations surrendered completely to science and technology, and while the world was conquered by a materialistic paradigm, a great spiritual vacuum was created in the unifying principles of the modern world. What we must do today is not to reject technology, but to create a new kind of technology which helps people live wholesome, spiritually affluent lives in a world of environmental constraints. This is also a technology which encourages immaterial, spiritual desires, thus enabling a shift from a world of insatiable material desires based on a philosophy of human dominance over nature to one in which unity with nature is pursued and the fulfillment of our desires is realized in our inner world.

We believe that this Japanese view of nature may become a force for the creation of new forms of technology, leading to a new global standard for technology originating in Japan, which supports a civilization of life and enables new lifestyles.

In Japan, there are unique concepts of the way to understand nature. One such concept is expressed in the Shinto idea of myriads of gods residing in all elements of nature (in Japanese, *Yaoyorozu no kami*), another in the Buddhist idea, mentioned above, that all living things—mountains and rivers, grasses and trees, and all the land, are imbued with the Buddhist spirit (in Japanese, *Sansen-souboku-kokudo-shitu-kai-joubutsu*). These suggest that all things are of equal value and all are of Buddha nature or possess the divine spirit. An actual culture of living in which such veneration for nature was held while people enjoyed life—and in which people aimed for unity with nature living vigorously but with a minimal impact on the environment—was, we believe, found in the iki culture of the Eco period.

The first Shogun order to close Japan to foreigners was issued in 1633 by Iemitsu Tokugawa. According to history books, Japan's isolation from the outside world was completed with the fifth order issued in 1639 to close the country, but in reality things were not that straightforward. After this date, the policy of isolationism was enforced through incidents involving foreign ships (the Dutch *Breskens* and the English *Return*), but trade was maintained with Holland, China and Korea, and it was only around the year 1800 that all products needed could be procured domestically, finally enabling the completion of Japan's isolation. To procure all that was needed domestically for a population of some 30 million people was an immense task, and the success in doing so has been dubbed, by the economist Akira Hayami, the industrious revolution. This was a revolution in productivity which, in order to save on capital investment, used a concentrated input of labor to raise the productivity of the land. This stands in contrast to the British Industrial Revolution, which involved the heavy investment of capital in industry and the massive introduction of machines in order to reduce labor input, in the process creating the concept of mass production.

The fact that people in Edo Japan did not waste, or make light of, the limited resources available was a major factor in the success of the industrious revolution and, as a result, enabled the reduction of capital input.

To make this possible, a path of very severe labor was chosen in which even the use of horses and oxen was reduced and substituted by human power. Why, then, did such an approach not lead to a culture of human despair caused by the harsh and unhealthy conditions? We think this is because of the Japanese view of nature which holds that all elements of nature possess a divine spirit. The ability of the Japanese to empathize with all things, and their unwavering belief that all things have emotions, made it possible to succeed with an industrious revolution in which humans willingly took over the work of oxen and at the same time helped nurture an astonishing aesthetic sense, as we shall see later.

The Edo period was an era, unique in world history, which saw no war for 265 years (1603–1868), allowing culture to flourish and technologies to develop that were not for the purpose of waging war, but genuinely for the benefit of common people. Edo was thus a rare example of an era which was truly shaped by and among common people.

The sublime movements of mechanical dolls found in show tents in the city are one example of how, in the peaceful environment of Edo society, technology was popularized. Ikkansai Kunitomo, an inventor and gunsmith in late Edo, learned the mechanism of air guns through a gift presented by the Dutch government, and applied this in the making of a lamp with long durability for popular use called "the never-ending lamp." Western-style mechanical clocks introduced into Japan were not of much use in the farming practices in Japan, and were reinvented as a Japanese clock using the so-called "indefinite time method," a way of measuring time by equally dividing daytime and nighttime, as determined by the times at which dawn and dusk occur. Through the dynamic curiosity and inventiveness of the Japanese, advanced technologies imported from abroad were, one after another, transformed into items of practical use to people in their everyday lives. Technology was not the possession of a limited group of experts, but was brought to life in the everyday. This is, indeed, the essence of a cultural attitude found in Japan—the "combination of the Japanese spirit with Western knowledge"—*wakonyosai*.

Why, then, despite a widespread popularization of technology occurring earlier than in many other countries, did Japan not create a wasteful society of mass production and mass consumption? In the Edo period, the highly sophisticated view of nature held by the Japanese merged with technology, which then became available even to common people—a process of technology development and diffusion quite rare in the world. Here, skills unconsciously discovered through play and skills used to entertain other people were regarded to be valuable, and a spiritual culture was naturally cultivated in which entertainment fueling spiritual rather than material desires was seen as the preferred path to a life of enjoyment. This is where a "technology wrapped in culture" was born, and perhaps we can say that it led to an industrial revolution which incorporated a view of nature.

These currents of thinking still flow strongly in the hearts of the Japanese. At least, in the aftermath of the Great East Japan Earthquake, as the shiny makeup of financial capitalism seemed to peel off, such a philosophy was very much alive. Before the makeup is again put back on, we need to recapture the original beauty of the face of society.

2.11 Iki, the Spirit of Edo

While enjoying entertainment and developing world-class skills and technologies—as epitomized in the refinement of mechanical devices or the selective breeding of carps—no industrial revolution leading to mass production and mass consumption occurred in Japan. The culture of iki refined in the Edo period was the natural response of the Japanese to the challenge of how to create a sustainable society. Shuzou Kuki, a student of Heidegger, defined iki as, "a sophisticated, vibrant form of coquettishness." People would not boast, but subtly express themselves; they would not pay much regard to money and would not spend large amounts on things with a physical shape (architecture), but rather use the money they had on the ephemeral (Rakugo storytelling and other performances, Kabuki Theatre, Sumo, etc.); they would first of all try to ensure that their friends were having a good time, and they would experience a sense of quiet satisfaction in the simple beauty of a single flower or the freshness of a light breeze—a sensitivity expressed in Japanese as *wabi-sabi*. In the culture of iki, all things including humans were seen to be of equal value and people would enjoy their everyday while paying respect to nature, all of which was thought to be of divine or Buddha nature. This was a culture which nature had helped give birth to, and here we find a philosophy of living which, we believe, is essential to create an altruistic society.

There are four important elements in the culture of iki. The first is living in harmony with nature, enjoying life to the full without placing a large burden on the Earth. It is a culture in which satisfaction is derived not from the purchase of things, but rather from the enjoyment of various forms of entertainment—

watching performances at the street tents or going to a Kabuki show or Rakugo storyteller's performance with friends or neighbors. Secondly, the culture of iki was one which aimed not to create losers in society. Mainstream elements of culture and society were not reserved for the rich and the strong; special attention was also paid to people who would normally be seen as losers; indeed, it was a culture in which such people had the opportunity to play key roles. It was, thus, a society in which "the common Joe," even if he were out of work, could live unashamedly. The third element of iki culture was a sense of moderation—of knowing how much was enough. People would not seek to gain more than was needed in daily life, and all things would be used to the full in respect for life and nature. As best expressed with the phrase *mottainai*—all things are too precious to waste—it was believed that the gods resided in all and everything and gratitude was felt even for a single grain of rice. The fourth element of the iki culture involved the metaphoric enjoyment of nature. We could say this is the same as the wabi-sabi culture in which the greatest satisfaction is found in the simple and the quiet. Even if people had money to buy a mountain, there was no need to do so; by arranging sand in your small enclosed garden and pretending this was Mount Fuji, you had achieved the same sense of satisfaction in a metaphoric manner. This was called the culture of mitate, and is also found in, for example, the traditional tea house which is a representation of the cosmos.

To summarize, the essence of the iki culture was to aim for harmony with nature, to live and enjoy everyday life with nature as the foundation. There was a mental and spiritual freedom, and since the principle of competition was not strongly present, losers were not created. By living a life of moderation (the spirit of *mottainai*), overshoot did not occur in any of the relations humans had with their surroundings. And, finally, a cosmological worldview in which the metaphoric enjoyment of nature was practiced meant that even everyday life was connected to the wabi-sabi culture of discovering beauty in the simple and the quiet. Iki, to us, appears to be a truly unique form of culture.

2.12 The Japanese Industrial Revolution

The main industry of Edo Japan was, obviously, agriculture. Things needed in daily life, including for example energy (firewood, charcoal, wooden torches, vegetable oils, wax), were procured and made from the land. Also, since land taxes were paid in rice, society could not function without a market. Farmers did not live isolated lives of self-sufficiency tied to the land, but apparently enjoyed a great range of freedoms, and in the middle period of Edo, commerce as well as a coin-based currency were indispensable.

The Japanese population at the time was some 32–33 million—about a fourth of what it is today—and was the fifth largest in the world, surpassing both the United States and England. This large population was maintained in a country closed to the outside world, with agriculture as its main industry and all things needed manufactured domestically. There was not much leeway when it came to the ability of the environment to provide for all this, and thus, regulations on the use of forests were introduced and jointly administered by bureaucrats and common people, and wooden material was thoroughly reused and recycled. The town of Edo (Tokyo) was one of the largest cities in the world; it was extremely densely populated and the number of inhabitants may have reached some 1.3 million at the end of the Edo period. Enormous flows of food and fuel entered Edo, and human excretions and waste also increased. In contrast to the situation in European cities, though, human excrements were collected and used as fertilizer, and it was strictly forbidden to dispose of trash and waste on streets, empty land, or in the moats and rivers. As a result, Edo was kept so clean that it astonished all visitors coming from abroad.

Edward S. Morse, an American zoologist and orientalist who discovered the Omori shell mound in the town of Edo, was amazed at the level of culture found in Japan. The literacy rate of people in Edo was 98% (Hiragana, at the beginning of the 19th century), and both people's interest in culture and their level of education were the highest in the world. Towards the end of the Edo period, the school attendance of children had reached more than 80%. Not many things were needed in daily life; living was simple and prices low, and if only you had learned a craft or mastered handiwork and worked diligently, you would have enough even to enjoy a cup of sake before bedtime. There was no need to be bothered by much else in daily life, and, apparently, life in the town was a micro-cosmos on its own, there was always plenty of laughter, and even when a person had to leave this world, this was calmly accepted much in the same way as when the withered flowers of a cherry tree flutter to the ground.

Let us take one more look at the differences between the British Industrial Revolution and the Edo ditto. In England, the Industrial Revolution made it possible to manufacture goods in much larger volumes and to transport both people and goods much farther than earlier, thus leading to major changes in the social structure. What was the situation like in Japan? As we have seen, many sophisticated and precise technologies were born, but most of them did not leave the sphere of entertainment or play. Does this, then, indicate that the British Industrial Revolution was innovative whereas the Japanese was somewhat behind? We do not believe that is the case. Maybe the Japanese were simply not particularly interested in using technology to travel long distance or to produce large volumes of food or goods. Of course, being materially well off enough to live a decent life was a precondition also in Japan, but as long as people could maintain a proper everyday life, they did not yearn for more money, and, it appears, placed the highest preference on how to enjoy and have fun. This is, indeed, the way of manufacturing enabled by the industrious revolution. Against this background, technology often took the shape of skills developed unawares in a flow of cultural entertainment, and value was placed on skills that made people enjoy life. A spiritual culture with nature as its mentor was nurtured in which people enjoyed everyday life, and which fanned people's immaterial rather than material desires.

Textiles, lacquerware, craftsmanship, all forms of technology were perfected to the level of art, and it is no exaggeration to say that these craftsman skills are what came to form the backbone of Japan's future technological development. This is how mature Edo society and the technologies nurtured therein were. Play became a mainstream activity in society and contributed not to a system of mass production, but to the development of a highly sophisticated culture.

2.13 The Four Elements of Technology We Can Learn from Iki

The concept of iki, which we have tried to describe from different angles in this chapter, is a basic concept of nature technology. The Industrial Revolution triggered the development of modern technology, but the success of this enterprise was gained only by removing humans and their technologies from nature. This philosophy of human dominance over nature became a key principle of modern technology and civilization and led to an ever great burden being placed on the Earth. Reintroducing a sense of nature into technology—that is the reshaping of technology much needed today. In Japan, there was no removal from or parting with nature. Actually, it may be more correct to say that in a country such as Japan where the ground shakes, the rivers flow over at major rainfalls, and the greenery imposes itself on people, it was not possible to remove ourselves from nature. This reality was what helped create the culture of iki in Edo and a Japanese-style industrial revolution which was entirely different from the one originating in Great Britain.

As described in the above, the culture of iki, which has not lost a proper view of nature, consists of four elements: "enjoying life with nature as a foundation," "the metaphoric concept which includes a cosmological dimension," "avoiding to create losers," and "moderation—knowing how much is enough."

Transferring these ideas onto concepts of technology, we arrive at the following four elements. Technology that "realizes high-function/ultralow environmental impact with nature as a point of departure," "is simple and easy to understand," "encourages communication and community," and "inspires attachment and affection." A technology which incorporates such elements is, indeed, what we call nature technology.

The first element:

Technology which realizes high-function/ultralow environmental impact with nature as a point of departure

Nature drives perfect cycles with a minimal input of energy. Nature technology looks at such mechanisms and systems through the lens of science and redesigns technology so as to incorporate them.

(At our present level of scientific development, merely mimicking nature may often lead to a situation in which huge inputs of resources and energy are required. Therefore, it is imperative that we, upon gaining an understanding of nature's mechanisms and systems, look at these through the filter of sustainability and reconstruct—or redesign—them as new forms of technology).

The second element:

Technology which is simple and easy to understand (clear)

The iki notion of metaphoric concepts incorporating even a cosmological dimension translates into the idea of simple and easily understood (clear) technology. This is a technology that is approachable and understandable to anyone. There are no "black boxes," and since it is a technology that both adults and children can easily understand, it also becomes possible for anyone to participate in the technology. Of course, the face of the engineer behind the technology must also be visible. It must be a technology which allows us, the users, to confirm important aspects of safety or secure use, and which thus allows for a dialogue on an even footing between the user and the engineer.

The third element:

Technology which encourages communication and community

This is an element of technology important in lifestyle design. A technology around which people might gather almost as if they were enjoying each other's company at a camp fire and which allows people to connect and initiate conversations. Unfortunately, modern technologies—be it mobile phones or the internet—lack in aspects such as human warmth or sympathy, and have thus evolved along a path far removed from the original purpose of communication. The communication that naturally occurs when you go to the neighbor to borrow miso or soy sauce, or the conversation that naturally starts around a camp fire even among strangers—we need technology that can play a role similar to that of miso or soy sauce or camp fires in the community.

The fourth element:

Technology which inspires attachment and affection

The user does not get tired of a product made with robust, quality technology, but rather wants to continue repairing it for long-term usage. When such a product is used over time, a feeling of affection arises and the product becomes one for which strong attachment is felt. This is the reason why foreigners visiting Japan in the Edo or Meiji period (1868–1912) were astonished at the almost artistic level of everyday items and often said that the Japanese were surrounded by objects of art in their everyday lives. This situation existed thanks to the spirit of affection deriving from people's belief that the divine dwelt in all things and that even everyday items should therefore be used with great care over long time.

If the black boxes in technology disappear, and it is simple and easy for anyone to understand, technology will become a closer and more intimate existence in people's lives. Such technology encourages communication and community, which over time leads to a sense of attachment and affection. This is the kind of technology we are in need of today.

One should not believe the already outdated myth that the more the black boxes included in technological development, or the more complicated the technology, the stronger will be a company's ability to differentiate itself and compete with other companies. What is needed today is technology than can be understood by anyone, and the premise upon which this is created are visions of wholesome, fulfilling lifestyles even under severe environmental constraints. Differentiation in the future will derive from the ability to manufacture products on the basis of an ability to envision and make proposals for new lifestyles.

2.14 A Waterless Bath Learning from Foam

After the Great East Japan Earthquake in March 2011, water supply was severed for as long as three weeks (the author's house). This gave one of the authors of this book reason to feel deep gratitude for the blessing of having water and also spurred many other thoughts. During the weeks with no water, it was possible to live on just 2 liters of water per person per day for cooking, washing, and cleaning the body with a washing cloth. In comparison, a normal toilet uses some 6–10 liters of water in one flush, and when we think of bathing, a shower requires some 15–18 liters per minute, whereas filling the tub takes some 200–300 liters of hot water. The number of households in Japan was approximately 48.64 million in 2010 and, although population continues to decrease, is expected to continue to increase until around 2015. After that a gradual decrease is forecasted, but the total number of households will still be around 48.80 million in 2030. The reason behind this is that single-occupant households continue to increase (29.5% of the total in 2005, and expected to reach 37.5% in 2030). If we assume that all households fill one tub every day (in Japan it is a custom to take a hot tub bath every night using the same water for the entire family), the volume of water consumed just to do this will reach 14.64 million tons/day in 2030. This is equivalent of some 37,500 25 meter swimming pools. Furthermore, since it is necessary to heat this water to 40°C, tremendous amounts of not only water, but also energy will be required. It is obvious that we will not be able to supply such huge amounts of water and energy in 2030 facing severe environmental constraints.

Then, how will we be able to bathe? With a forecasting mentality, the response might be that we should reduce the number of baths taken or make the time for each shower shorter, but are there no other solutions? With such approaches, people would have to endure scarcity and would not be able to enjoy life. If we start from the premise that we must create wholesome, fulfilling lifestyles that are possible even under severe environmental constraints, what kind of bathing could we envision? A new way of bathing derived from an example of lifestyles envisioned with backcasting has come into view.

Lifestyle Example. Enjoying Shared Bathing

These days it is hard to believe, but in the past, people had the custom of taking morning showers every day, and hot tub baths were ready for use 24 hours a day in every household. One can only wonder how they were able to secure enough water and energy to enable this. In Japan, the air is moist and people are clean, so tub baths are indispensable. These days, a new trend is "community water sharing." People fill the tub at home and invite friends to come and bathe. Sometimes, even friends of friends will come, and the chat after the bath greatly helps expand the host's social circle. There are more and more events organized by members of this water-sharing community, and within the community people take turns to teach classes of, for example, sweets making, DIY, or outdoor skills.

If we think with a forecasting mentality, people have to endure worse living conditions, but with a backcasting mentality, other possible bathing options appear on the horizon. As we have touched on earlier, the structure of our desires is such that perceptions of quality of life are irreversible, and on top of this, in a hot and humid country such as Japan, taking a daily bath is a necessity. What about taking a bath every day, but without using water? A bath which cleanses the body, retains heat, and is as relaxing as possible. We do not necessarily have to use water for such a bath.

Let us, again, knock the door of nature's treasure house. How do insects manage to live comfortably? Or what about birds? Some of them use sand to bathe, others even bathe in ants—and thinking of pigs or elephants, we see that they enjoy bathing or rolling in mud. Mud bathing, sand bathing, or ant bathing may not necessarily appear attractive options to us humans. Let us try to knock that door again. The chrysalis of a spittle bug (of the Cercopidea family) is enveloped in foam. The foam protects from ultraviolet radiation, the plentiful air contained in the foam has an insulating function and reduces the effect of external differences in temperature, and thanks to this, even a weak chrysalis can maintain life in a comfortable environment. The fish called *Betta splendens* (Siamese fighting fish) lays its eggs in foam. The foam has a surface tension which enables it to bond to various things. Why do insects and animals tend to gather in the basin under a waterfall? When the water bubbles break, ultrasonic waves are generated, and, possibly, these bubbles help remove dirt. The crackling sound of bubbles breaking has what is called a $1/f$ fluctuation, widely found in nature, and may be generating a relaxation effect.

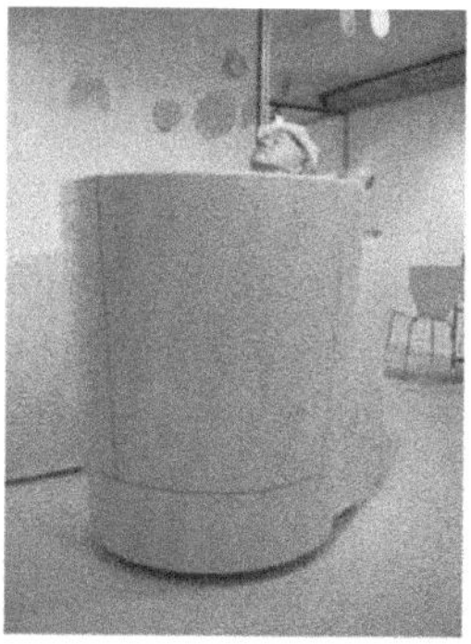

By setting a simple door, as heavy water-tightness is not necessary, it is possible having a bath with a wheel chair.
(6 liters)

Take a bath alongside of my bed, today, at the balcony tomorrow .... Water free bath is light and can be carried anywhere you want.
(3 liters)

Figure 2.13 Water free bath using bubbles.

Thinking along such lines, the possibility that bubbles or foam might lead to new ways of bathing starts to appear. By creating bubbles at a temperature of 70°C, they become able to retain enough heat to warm a person's body. When the bubbles break, ultrasonic waves are released that help remove dirt, and the bubbles should then bond with the dirt removed. If one could take a bath with soft, airy bubbles holding warm air, the sturdiness of the bath tub need not be as high as previously, and it becomes possible to achieve a reduction in the materials used. Being able to bring your portable bath to bathe anywhere you like might also create new ways of enjoying life. Today, you bathe in the bedroom; tomorrow, on the veranda. People may become able to discover new and more attractive lifestyles. What is more, since the bath tub is only filled with bubbles, a much lower level of water tightness than with conventional baths should suffice, which opens up for the possibility, for instance, of bathing in a wheelchair with a special gate attached. This could create news ways for the elderly, or people with disabilities, to enjoy taking a bath. For the elderly, in particular, possible health benefits may be realized from taking a bath in which there is no water pressure on the body. If an effective cyclical system is installed, it is more than possible to take a pleasant bubble bath with a mere 4–8 liters of hot water. Taking a bath in a tub with no water pressure feels a bit like floating in space, a strange sensation the authors of this book had never experienced before. There are still several issues to solve, such as how to make stable bubbles without the use of surfactants, but this waterless bath undoubtedly has the potential to unleash new lifestyles (Fig. 2.13).

In this way, environmental constraints are not demanding of us to endure living a harsh life; rather, they are positive constraints urging us to discover new lifestyles we had not even contemplated so far.

Acknowledgment

This work was supported by JSPS KAKENHI Grant number JP24120008 in Scientific Research on Innovative Areas: "Innovative Materials Engineering Based on Biological Diversity."

References

A Showroom of Nature's Wonders, http://nature-sr.com/

Bacon, F., and Katsura, J. (trans.). *Novumu oruganumu – shinkikan* (*Novum Organum*), Iwanami Bunko, Tokyo (1978).

Benyus, J. *Biomimicry: Innovation Inspired by Nature*, 2nd ed., William Morrow Paperbacks, New York (2002).

Descartes, R., and Tanigawa, T. (trans.). *Houhou josetsu (Discours de la méthode pour bien conduire sa raison, et chercher la vérité dans les sciences)*, Iwanami Bunko, Tokyo (2002).

Esty, D. S., and Winston, A. S. *Green to Gold*, Yale University Press, London (2006).

Forbes, P. *The Gecko's Foot: Bio-inspiration: Engineering New Materials from Nature*, W. W. Norton & Company, New York (2006).

Fukuzawa, Y. *Fukuou jiden* (Autobiography of Yukichi Fukuzawa), Keio University Press, Tokyo (2009).

Gill, V. G., and Dale, T. *Topsoil and Civilization*, University of Oklahoma Press, Norman (1974).

Hawken, P., Lovins, A. B., and Lovins, L. H. *Natural Capitalism*, 1st ed., Back Bay Books, Tokyo (2008).

Hayamizu, Y. *Rekishi jinkougaku de mita nihon* (Japan seen from the angle of historical demographics), Bungeishunju, Tokyo (2001).

Heidegger, M., Hara, T., and Watanabe, J. (transl.). *Sonzai to jikan (Sein und Zeit)*, Chuo Koronsha, Tokyo (2003).

Honda, H. *Katachi no seibutsugaku* (A biology of forms), NHK Publishing, Tokyo (2010).

Ishi, H., Yasuda, Y., and Yuasa, T. *Kankyo to bunmei no sekaishi* (A world history of the environment and civilization), Yosensha, Tokyo (2001).

Ishida, H. (Supervisor). *Sugoi shizen zukan* (Amazing picture encyclopedia of nature), PHP Publishing, Tokyo (2011).

Ishida, H. *Channeling the Forces of Nature*, Tohoku University Press, Miyagi (2009).

Ishida, H., and Furukawa, R. *Nature Technology: Creating a Fresh Approach to Technology and Lifestyle*, Springer, Japan (2013).

Ishida, H., and Shimomura, M. *Shizen ni manabu neichaa tekunoroji* (Nature technology learned from nature), Gakken Publishing, Tokyo (2011).

Ishida, H., Matsuda, M., Eguchi, E., and Nishizawa, M. *Yamori no yubi kara fushigi na teepu* (Wondrous tape from the foot of the gecko), Arisu-kan, Tokyo (2011).

Kawakatsu, H. *Bi no kuni Nihon wo tsukuru* (Creating a beautiful Japan), Nikkei Business Bunko, Tokyo (2006).

Kawakatsu, H., and Yasuda, Y. *Teki wo tsukuru bunmei – wa wo nasu bunmei* (Civilizations that create enemies: civilization that create harmony), PHP Publishing, Tokyo (2003).

Kobayashi, M. *Dekaruto tetsugaku to sono shatei* (The philosophy of Descartes and its reach), Iwanami Shoten, Tokyo (1995).

Koizumi, K. Kikou hendou to bunmei no seisui (Climate change and the rise and fall of civilizations), *Journal of Geography* (Chigaku Zasshi), **116**, 62–87 (2007).

Kuki, S. *Iki no kouzou* (The structure of iki), Iwanami Bunko, Tokyo (1998).

Morse, E. S., and Ishikawa, K. (trans.). *Nihon sono hi sono hi* (Japan day by day), Heibonsha, Tokyo (1990).

Naess, A., Saito, N., and Hiraki, T. (trans.). *Diipu ekorojii to ha nanika* (*What Is Deep Ecology?*), Bunka Shobo Hakubunsha, Tokyo (1997).

Nayagama, K. *Jikoshuuseki no shizen to kagaku* (Self-aggregating nature and science), Maruzen, Tokyo (1995).

Oppenheimer, S. *Out of Eden: The Peopling of the World*, Robinson Publishing, London (2004).

The Biomimicry Institute, http://biomimicryinstitute.org

Toynbee, A. J. *A Study of History: Abridgement of Volumes I–V*, Oxford University Press (1985).

Umehara, T., and Yasuda, Y. (eds.). *Bunmei to kankyou* (Civilization and environment), vol. 3, 9, Asakura Shoten, Tokyo (1995).

Underwood, P. *The Walking People: A Native American Oral History*, A Tribe of Two Press, San Anselmo, CA (1993).

Wade, N. *Before the Dawn: Recovering the Lost History of Our Ancestors*, reprint, Penguin, New York (2007).

Watanabe, K. *Edo to iu genkei* (The phantom image that is Edo), Genshobo, Tokyo (2008).

Watanabe, K. *Ikishi yo no omokage* (Vestiges of a fading world), Heibonsha, Tokyo (2005).

Wilson, E. O. *The Diversity of Life*, W. W. Norton & Company, New York (1993).

Yasuda, Y. *Joumon bunmei no kankyou* (The environment of the Jomon civilization), Yoshikawa Hirobumikan, Tokyo (1997).

Chapter 3

How Japan Found a Way of Life Producing Spiritual Affluence from Limited Resources

Ryuzo Furukawa
Faculty of Environmental Studies, Tokyo City University, 3-3-1 Ushikubonishi, Tsuzuki-ku, Yokohama, Kanagawa 224-8551, Japan
ryuzofkw@tcu.ac.jp

Before modernization, our way of life was relatively inconvenient. Then, in the period bracketing the Second World War, there was a transformation in how the Japanese lived. I conducted interviews with more than 500 90-year-old people called the "90-year-old hearing" about the prewar way of life in various parts of Japan. What I found was that despite the inconvenience of the prewar way of life, it offered enjoyment and other elements of spiritual affluence. Most people are perhaps skeptical that the prewar way of life could have yielded spiritual affluence, but my findings contradict that skepticism. Most elderly people around 90 years of age at the time of the interview attested that prewar life did offer spiritual affluence. After the war, as modern conveniences were introduced, the type of spiritual affluence that existed before the war was lost; at the same time, the environmental impact of Japan's way of life grew. In this chapter I will explore that transition in depth.

Lifestyle and Nature: Integrating Nature Technology to Sustainable Lifestyles
Edited by Ryuzo Furukawa

ISBN 978-981-4774-62-8 (Hardcover), 978-1-315-14687-4 (eBook)
www.panstanford.com

3.1 A Way of Life That Depends on Nature

The author has conducted over 500 interviews about the prewar way of life with elderly people, especially those around 90 years of age, from all of Japan's prefectures (Hokkaido to Okinawa). Although I started in Miyagi Prefecture and worked outward from there, and the sample size of interviews is smaller for more far-flung prefectures, I have been able to collect from elderly men and women a rich set of information about the prewar way of life associated with varying environmental conditions and primary occupations in different regions of Japan. And I continue to conduct additional interviews. It all began on a small scale in 2008, so the year 2016 marks eight years of work. Initially, in order to solicit firsthand accounts of prewar life seen through adult eyes, I sought out interviewees born around 1922, who would have been 20 or so when the Pacific War began. As I refined my methodology, and the targeted age cohort began to pass age 90, I began to refer to these as the "90-year-old hearing."

The 90-year-old hearing have yielded a treasure trove of insights. But perhaps the most important of all is that prewar living meant "a way of life dependent on nature." I'd like to explain that by citing a few excerpts from interviews.

"We pickled around 800 daikon radishes. There were three kinds of pickles we made from them: amazuke, nakazuke, and honzuke. Pickling allowed us to eat daikon all year round, even when it was not in season. Honzuke pickles were made by sprinkling salt and rice-bran over the daikon—it looked like a snowfall—plus persimmon peel for sweetness. Then we laid daikon leaves over the top and put in chili pepper: that would keep bugs away from the pickles even in summertime. Of course, we ate daikon pickles with meals, but also snacked on them with tea. In those days the only sweets we really had were hard candies that we made from rice and malt. We had biscuits, but they were really more like hard tack, serving as emergency rations." (A 90-year-old from Miyagi Prefecture)

"We stored a year's worth of rice, which was two and a half sacks per person. For our family, that meant 30 sacks or so. In addition, as a guard against bad harvest, we kept six sacks of rice up in the

rafters as 'starvation supplies' [gashikoi]. *The smoke from the hearth was always swirling around those sacks up there, so we didn't have to worry about pests getting into them."* (A 90-year-old from Miyagi Prefecture)

Preserved foods were a common practice among people living in regions with harsh winters. And they were not considered just a necessary evil. Preserved foods gave reassurance against the specter of starvation. And as people perfected clever ways of pickling and flavoring pickled daikon, the process of preserving foods became an enjoyable pastime from which people derived spiritual affluence.

In Miyagi Prefecture, the way of life in mountain communities was vastly different from that of seaside communities. Shiogama was a thriving port in Miyagi from which peddlers carried fish to towns and upland communities. In regions where salt was readily available, people preserved and ate salmon, as in the following accounts:

"We went salmon fishing in late October and November. We would take the salmon we caught and pickle them in salt in large tubs. Then we ate them until the spring equinox, which is a Buddhist holiday in Japan. Those salmon were a real blessing to us." (A 90-year-old from Miyagi Prefecture)

Meanwhile, further north in Akita Prefecture, people had similar practices (although with different names) for making preserved foods to last them through the harsh winter. Their testimonies reveal how they pursued a way of life that gave spiritual affluence despite living in a harsh climate.

"In my hometown of Nakasen, we didn't have wild vegetables from the mountains like they did over in Yuwa. Instead, we had lots of cultivated fields. And what we grew in the fields, we would preserve using various methods. We were experts in drying and smoking. In our area if you couldn't make your own smoked pickles, it was considered an embarrassment." (A 90-year-old from Akita Prefecture)

The further south one travels, the less people's way of life depended on preserved foods. Let's consider the food ways of Okinoerabushima, a subtropical island in Kagoshima Prefecture.

There, people could eat what grows naturally, even during the mild winter, so there was little need to preserve foods.

"We had no persimmons or apples. We did have mandarin oranges. We had many varieties of orange—several local, native ones, if I'm not mistaken. Otos and Tokuribus—nice fragrant types of oranges. And Banshiros. And flat lemons of course: we grew a lot of those. We were never short of fruit around here, I can tell you that. We didn't process them or anything, we just ate them off the tree. We'd slip a few in our pockets when we went out to play, and eat them whenever we felt like it. I had a friend at school who always knew when the oranges were ripe, and he would lead five or six of us, a whole group of friends, would go into the forest and find tons of them. The ones lower on the tree were tart, the ones growing higher up, which got more sun, tasted the best. So, if you couldn't climb trees, you were stuck with a bunch of tart oranges. We'd climb up and eat them while sitting in the tree, then climb down with full bellies. If we got thirsty while romping around, we'd dodge into a sugarcane field along the way and snap off a cane to lick. The kids who were best at it would dash in there and snap off a whole bunch of canes and toss them out to us. We could grab as many flat lemons and other fruit as we wanted. Nobody ever got mad about it. Same with wild strawberries. And we could pick the Banshiros that grew in the woods. They weren't anybody's orchard trees or anything. I never ate too much at once and got a tummy ache, because I could stand to wait; I knew there would always be more available next time I wanted some." (A 90-year-old from Okinoerabushima)

These accounts reveal something very interesting. In the Tohoku region, with its harsh winters, people tried to make smoked pickles as tasty as possible, and enjoyed figuring out different ways to flavor pickled daikon. On the subtropical island of Okinoerabushima, they had an abundance of fruit, yet no one gorged themselves; they appreciated the subtle differences in flavor from one fruit to the next and friends kept each other up to date about what was in season. Clearly, in different natural environments, there are different flavors to enjoy, and different knowhow that is used for activities like making preserved foods. But there is a common element: a philosophy of "making the most of what you have" that is independent of the specific natural environment. Meaning that ways of life based on a common set of values can take many specific forms of spiritually affluent lifestyle and practical culture.

Perhaps we can say that as long as this underlying value structure was in place, it ensured that only ways of life which generated a plethora of food, shelter, craftspeople, mutual aid, rituals, work, and other folkways conducive to spiritual affluence and adapted to the environment were passed on and perpetuated over the long term in Japan's geoclimatically diverse regions. According to my 90-year-old hearing, an analysis of the prewar way of life shows that typically people in the same area led a similar way of life, but a different area, even just on the other side of a mountain range, might have a different way of life. This could be considered evidence of adaptation to the natural environment over the long term. A way of life rich in spiritual affluence is strongly dependent on the natural environment of its locale.

3.2 A Way of Life Attuned to Nature

In prewar Japan, despite severe resource limitations and privation, people were able to achieve a way of life conductive to spiritual affluence by being attuned to nature.

Rather than trying to fight the climate, they generally opened and closed sliding doors to adjust the indoor temperature. Since the doors were left open during the summer, they put up netting to keep out mosquitoes and other insects. People found ways to introduce an element of fun into this form of climate control, such as releasing lightning bugs inside the netting.

And because life was conducted in sync with the sun, people generally returned home when it grew dark out in the evening. But they also sometimes frolicked in the cool of moonlit nights. The natural world provided constant opportunities for younger children to learn from older children and from adults in the community, and then pass down that knowledge in turn when they became older. It was an entirely different world and method of learning than studying inside a classroom.

Fruit trees were planted in the yard, and fruit is what many people ate as a snack. The type of fruit tree varied by region, and featured as an element of the local culinary culture. Because fruit trees also add a splash of color, they made communities look more beautiful than you can imagine. The freedom to pick and eat fruit created a rich environment for children that has largely vanished today.

Example from Aomori Prefecture

Reading the rhythms of nature

"To get to school, you either had to run up and over the mountain, which took a long time, or go around the base, which was faster. But to go around the base, you had to run about 50 meters along a cliff exposed to the waves. You had to watch the waves and time your run between them. Get caught by a wave and you'd be swept away…you were risking your life. So if you timed the waves wrong you were in big trouble. On days like today with calm winds, you didn't have to time the waves and run for it. They would only come about so high and you could take your time going around. But if you got swept away, you were a goner. Even if you thought no waves were coming, you wouldn't want to dawdle there. Every few waves the water would recede and you would have some time until the next one. You needed to spot that window and make a run for it." (Man, born 1926, Narusawa Village, Aomori Prefecture)

Examples from Miyagi Prefecture

Beating the heat

"For heat in the winter we had a kotatsu *[a low table with a blanket skirt around a brazier]. Hot water bottles and bed warmers were only for the old folks. The best way to be warm was to get straight from a hot bath into bed. During the summer we opened up the whole house, and the only thing we had to keep cool with was hand fans. We slept inside a big mosquito net. We burned* yomogi *[a kind of mugwort] like incense sticks. There was a mushroom called* haetorimodashi, *which if you put it in water, would draw flies and kill them. But it was actually an edible mushroom. As for lice, we got rid of them with hot water."* (Woman, born 1926, Sendai, Miyagi Prefecture)

Frolicking in the moonlight

"We had a grove of trees around our house as a windbreak, and people came and went freely there. There was a kind of shaded bench where the kids would gather, and the older kids would teach stuff to the younger kids. 'Know that house with the plum tree? It's easier to pick them from such-and-such direction.' We'd play together under the moonlight until it got dark, learning stuff like that from other kids." (Man, born 1927, Ozaki, Miyagi Prefecture)

The foregoing accounts come from Aomori and Miyagi Prefectures. Now let's turn to some examples from Mie and Nara Prefectures, which have a warmer climate. In seaside communities, seafood was abundant. Longtime residents report a sense that catches of seafood have fallen and the marine environment is changing. In the old days, abalone, turbo, crustaceans, and fish were all taken in abundance, so people knew how to clearly distinguish between fresh and not-fresh seafood.

Some people also ate locusts and cicadas. This was especially common in mountain communities where the diets tended to lack sufficient protein and calcium.

Examples from Mie and Nara Prefectures

Yonade

"We had a type of fishing called yonade. *We would go out on the rocks at night and find big turbos clinging to them. In five minutes you could fill a basket on your waist with turbos. During the day they hide underwater. But they come up out onto the rocks at night. You could literally fill a basket in just two or three minutes. We'd carry them back home and set them down beside the hearth, then the operculum would pop right open. You'd have to grab onto that and wind the meat out of the shell. The very bottom of the meat is black, and the tip has some sand in it, but you brush that off and then eat it. Delicious! I can't even describe the flavor. The ones they sell now are cultured and the flavor is nothing like the ones we used to catch. Wild turbo is delicious beyond words."* (Man, born 1929, Kuwana, Mie Prefecture)

Reveling in a bounty of delicious seafood

"So it was great to be a diver back then. In my time we caught a lot of abalone. Elegant abalone: that species are the big ones. We caught lots of them. The reef was shallow and we dived for tons of abalone. They were everywhere offshore. My grandmother loved the rocky foreshore. She collected nori seaweed there and shared it with us. In the old days you would see the rocks carpeted with nori and all you had to do was pick it." (Woman, born 1927, Toba, Mie Prefecture)

Fresh-picked flavors

"Of course, produce you buy at the supermarket can't compare to produce you bring in from the fields with the dirt still clinging to it. It has a whole different texture and flavor. Just one example is the cucumber: fresh picked ones have an incredible crunch and snap to them. Ones from the supermarket lose that because they were picked days before." (Woman, born 1921, Gose, Nara Prefecture)

A flowing stream feels wonderful

"We used our river water for drinking and everything else. We washed everything in the river—the water was so clean. There was a brisk current so it felt great to wash things in it. Water from a faucet is nothing but a trickle, compared to a river rushing past. There was a little pool built along it we called the dondobachi. We would cool watermelons in it. Water could pass through the gaps between the stones making up the wall of the dondobachi. In summer we would make barley tea in a teapot, then wedge a length of bamboo into the gaps in the rock and hang the teapot on it from its handle. Dangling in the dondobachi would chill the barley tea nicely, naturally. The water was so great." (Woman, born 1921, Gose, Nara Prefecture)

Okinoerabushima is an island in Kagoshima Prefecture with a warm subtropical climate, quite different from the regions in which previous interviewees live. The island has limited freshwater resources, and securing a water supply has long been a challenge of life.

In contrast, the ocean used to abound in seafood, and the island had an abundance of fruit, so food was plentiful. Unfortunately, there has been a dramatic decline in the wildlife and food supply of the island waters and forests in recent years. Degradation of the environment has reduced opportunities for people to enjoy and benefit from nature's bounty.

Interviewees spoke of young people gathering in the evening to play the shamisen and dance. The sugar shacks where sugar cane was processed into sugar were a favorite spot for young couples to spend time together.

Okinoerabushima

Island flora supplies a different diet

"On Okinoerabushima, sotestu *[sago cycads] grow everywhere in abundance. We would grind the seeds into powder and make a porridge from it. We didn't grow any rice. We also had yanabu, the seeds of the sotestu. So we cut out the pith of sotetsu and ate it; we called that* kyara. *We ate the seeds and the pith. But since yanabu contain toxins, we had to wash and dry them first."* (Woman, born 1928, Okinoerabushima, Kagoshima Prefecture)

Romping in the sugar shack

"We would play the shamisen and dance along to the music in a kind of open space, somewhere outside the village. People don't do that these days. There were sugar shacks there, a whole bunch of them. Little huts where they made sugar, next to the fields. Sugar was only milled during the winter, so until late December or so, the sugar shacks were empty. And that is where we went for dates. There would be lots of bagasse left over from pressing the sugarcane. That made it cozy. I guess we just didn't have enough beautiful girls in our village, because I used to walk all the way to distant villages, carrying my shamisen, to find beautiful girls to sing along with me as I played." (Man, born 1929, Okinoerabushima, Kagoshima Prefecture)

Our society once coexisted with nature. What was life like in a world of coexistence with forests, rivers, ocean, plants, and animals?

Nature bestows abundant blessings on people, giving rise to a willingness to share unstintingly with others and a feeling of affinity with nature as the source of these blessings. This reflects a perception that things are bestowed by nature rather than one's own belongings, expressed with these words of gratitude: "I partake of this overflowing bounty." Moreover, abundance enabled people to unreservedly accept nature's appealing and appetizing produce, and the beauty, delight, comfort, and seasonality of fresh local foods. These qualities weave a keen sensitivity to nature into people's way of life.

But nature is not always tame. From time to time, the natural world throws curves. This uncertainty breeds serenity rather than frazzled nerves. Unexpected deviation nurtures an appreciation of things at their optimum. When people craft their necessities out of what nature provides, the results are not always perfect. This spurs pursuit of higher degrees of craftsmanship and ingenuity. At times nature poses a threat to humans. When people face uncertainty over when resources or food supplies might run out, they prize them more, prolong their use as far as possible, and avoid using them wastefully. That's how I see it, anyway. When people are conscious that nature could menace their livelihoods at any moment, they are more conscientious about storing things up, helping each other, being hospitable, and being thankful than people in today's artificially stabilized way of life.

3.3 Craftsmanship as a Way of Life

After I finished conducting interviews with 90-year-olds in Okino-erabushima, we went out for a walk and he demonstrated how to build an insect cage from sotestu fronds. He snapped off some sotetsu growing along the road and in no time finished the cage, in a shape I had never seen before, made from materials that nature provides easily to hand in abundance. He said that when he was a child, he would go play outside, catching bugs and making sotetsu cages to carry them home. Of course, kids today who go out bug-catching don't know how to build their own insect cages out of natural materials. But in the past, in many parts of Japan, kids made their own playthings, mainly out of bamboo, sticks, and so on. Making things to play with was part of the fun, and an introduction to crafting things.

Making stilts and bamboo-copters

"I made my own stilts. A lot of kids knew how to make their own bamboo-copters. But nowadays you never find kids who can even make something as simple as a bamboo-copter. I would stilt-walk all the time. My dad was always telling me 'Cut it out! Those are too tall! That's dangerous!' And my stilts were tall: a whole meter!" (Man, born 1929, Kuwana, Mie Prefecture)

Making and using fish traps

"We made basket traps every day, woven out of bamboo. Then dug out pond snails out of their shells and put them in the cage as bait. Sometimes we caught catfish, and we caught tons of eels! We caught loach with the help of rice farmers. You see, the water in the rice paddies would get too hot in the sun, so every so often at night the farmers had to change the water and fill the paddies fresh. They opened the sluice gate a certain amount to drain water into the stream. That would draw lots of loach, and the farmers let us catch them there." (Man, born 1924, Nabari, Mie Prefecture)

Other playthings children made for themselves, varying from region to region, included skis, kendama, skates, sleds, and bugs nets. Sometimes they learned how to make them from siblings, sometimes from other children in the neighborhood. Ways of play were passed down within a community of children, enabling them to construct their own world of play that made it more fun to be a child.

In the prewar Japanese way of life, children had their own world. For example, in Hyogo Prefecture's Toyooka City, there was a clam fishery exclusively for children along the Maruyama River. Nowadays, that kind of activity is no longer part of children's play. But in the old days, clamming was a cherished pastime. To earn the right to take part, a kid had to be able to swim all the way across the river. Older kids would put a boat out into the river and act as lifeguards. Children understood the dangers that exist when living in coexistence with nature. If a surplus of clams were harvested, the children would give some to their neighbors. The way clamming and swimming were associated through this tradition is of great interest. By understanding how the prewar way of life incorporated a world of their own for children, we can better appreciate the enjoyment to be derived from play that includes children crafting their own playthings.

Another thing children enjoyed was simply watching craftspeople at work.

"Most of our kitchen implements were made of wood or bamboo: buckets, paddles, chopping boards, strainers. There was a bucket maker in our neighborhood. Metal pots were highly valued possessions. There was a tinker who came around with his tools to mend holes in pots. He would heat metal in a charcoal fire, pour it into a mold, and plug the hole. When I was a kid, I loved to go watch him at work whenever I had the chance." (Man, born 1924, Shiogama, Miyagi Prefecture)

Among the many household implements in everyday use in the old days: zori (straw sandals), straw boots, bamboo baskets, wooden boxes, straw mats, tatami mats, wicker trunks, pots, kettles, buckets, barrels, clay pots, bottles, trays, dishes, teapots, ceramic bowls, clothes, charcoal, bags, mosquito nets, bed warmers, lanterns, lamps, umbrellas, farm implements, and more. Even in the prewar era, people did not usually make all of their implements; they bought some artisan-made ones from stores and used them with care for many years. But people in farm villages or up in the mountains lacked a means of obtaining these products, so they supplied them for themselves out of locally available resources. For example, grandmothers would work each night weaving zori for their children. Since children walked long distances and spent a lot of time running around and playing, their sandals would wear out quickly. We cannot assert that this was a source of spiritual affluence in all cases, but it was for some households.

It was an era when crafts flourished. There were many different artisan crafts, and everyone had an opportunity to master some kind of craft. People could begin exploring which crafts they were suited to during childhood. In the hamlet of Tachihama, part of Ishinomaki City in Miyagi Prefecture, there were those who wanted to become inkstone makers, those who wanted to become fishermen, or charcoal makers, and so on. In this area it was common for people to try their hands at several crafts. According to interviews with nonagenarians from Tachihama, that freedom of opportunity gave rise to spiritual affluence.

Many people did not take up a particular craft out of an initial interest in it, but because they had a need to make a certain thing. But then, having mastered the basics of the craft, and finding more and more applications for it, it came to fascinate them.

For example, many women learned knitting at school, but found that it became a favorite pastime.

"I started knitting with chain stitch, then single crochet, double crochet, half double crochet, decrease stitching. You can see many stitches within a single piece like this. Sewing too. It was in the third or fourth grade that I started needlework, learning how to work the needle by tracing lines drawn on thin towels. We actually had sewing races but I was slow. For my sons and grandsons, I would darn the holes in their socks. Then they'd get more holes and come back again saying, "Please mend them!" I'd ask, "Don't you feel embarrassed to wear darned socks?" And they'd answer, "But they're warmer!" The fabric is actually thicker where they've been darned. In the old days, to darn socks, I'd put a burned-out lightbulb in the heel, lay a scrap of fabric over it, and sew with the lightbulb as a form. The lightbulb formed the heel shape and the toe shape. That's how I darned socks. Oh, it's been at least 20-some years since the last time. I was always happy to do it." (Woman, born 1928, Owase, Mie Prefecture)

When making things becomes an enjoyable pastime, the next step is to master a higher level of skill in a craft, which offers a growth experience. The feeling of growing accomplishment makes it more interesting and people become enthralled. Initially, they are only making things for themselves, but eventually they start making things for others. And that spurs further time spent on the craft for the sake of further increase in mastery. Over time, this process nurtures a lifelong passion for the craft, as the following interview excerpts testify.

A love of sewing

"Over the decades, I had fewer and fewer things to sew. For myself, I would sew what my sewing teacher assigned me. Then I also had piecework sewing, which earned me a little spending money.

Nothing fancy, just things people asked me to sew for them. I like sewing. Even now I keep my sewing machine out. I've made who knows how many garments—hundreds, maybe thousands—and I still sew. That bag is hand-sewn. I made ten or so of them for a kindergarten bazaar. Whether it's sewing on a machine or hand-sewing a kimono, I just love making clothes! When we lived in Manchuria, my family bought me a sewing machine. I wasn't familiar with Manchurian clothing styles, but I tried making them anyway. Then, when we came back to Japan, neighbors asked me, 'Would you sew one like that for me?'" (Woman, born 1920, Osaka)

Meticulously unraveling, washing, and reknitting wool

"In the old days, we fully appreciated the things we had. We would unravel old wool sweaters, wash the yarn, and then reknit the sweater. People did it all the time. Or maybe when you bought a sweater, you would unravel and then reknit it. Because it was much too expensive to go buy new wool yarn. They weren't importing much of it in those days. And Japan didn't produce its own." (Woman, born 1919, Ise, Mie Prefecture)

Accounts of the prewar way of life illustrate the relationship between craftsmanship and spiritually affluent lifestyle. From them we can say that it was not technological competition between regions or artisans that produced spiritual affluence, but the application of local resources and local techniques to make things that the artisans themselves needed. This led to crafts that were beautiful and characteristic of their place of origin.

3.4 Togetherness as a Way of Life

Experts say that Japan's demographic decline will continue, with a decreasing birthrate and aging population leading to a further increase in the prevalence of elderly people living alone. Being alone is no way for anyone to live. A society that provides spiritual affluence can only be established where people gather together in settings that foster spiritual affluence. Does today's Japan provide such occasions for togetherness?

Here are some comments on that subject by 90-year-olds with memories of life in prewar Japan: "Recently I've drifted apart from my neighbors and feel lonely now." "In the old days, if you wanted to get to know people, there were situations where people gathered naturally, just to chat and gossip. But those are gone now. Now you have to go out and create occasions to socialize or it doesn't happen. You have to actively seek out opportunities to get together with people. And many of us just can't manage that." What is the story here? In the old days, even in cities, neighbors had a spirit of solidarity with each other, but now they are atomized, and only come together when a conscious effort is made. The idea of bringing back that spirit of neighborly solidarity is now perceived as too much of a hassle. Why is that? That is a major question looming over gatherings to foster relational connections between people in Japan today.

(1) Spontaneous gatherings

The prewar way of life involved spontaneous gatherings with other members of the community when going to the local laundering spot or bathhouse, or simply striking up a conversation on the veranda "Engawa" of a neighbor's home. Let's look at some examples of the many occasions that used to be prevalent in communities all up and down Japan.

A laundering spot where the water is nice and clean

"In our house, until the washing machine came along, my mother did all the laundry using a washtub and washboard in the stream. We had a stream of nice clean water running by up in the woods, where everybody did their laundry. Everyone would walk up to the stream in the woods chattering up a storm; laundry time was gab time for all of them." (Man, born 1926, Toba City, Mie Prefecture)

A laundering spot at a small waterfall

"There was a small waterfall there, and it had carved out a basin in the rock. We used that basin for washing food. We had a spot for laundering, a spot for rinsing, a spot for washing. Any time you would go there you'd find five or six people there chatting: all women. The waterfall is gone now. When the piped water system was laid

in, the waterfall disappeared." (Man, born 1928, Ajigasawa, Aomori Prefecture)

Many of the places where people spontaneously gathered, like those described above, featured clear water and natural beauty, which made them pleasant places to spend time as well as serving a functional purpose. So people did not consider it burdensome to spend long hours there.

The following examples cite gathering places people would drop by casually, no invitation needed and no offense taken if no one was around.

Veranda "Engawa"

"The veranda was where the kids would play. We'd play with marbles or beanbags or other simple toys. Because all the houses had verandas in the old days you could come right up there and holler out, 'Is so-and-so home?' If she was home, she would come out and you'd talk on the veranda. Nowadays you would have to deal with all the fuss of knocking on the front door. It wasn't so stuffy back then. The adults would also hang out on the veranda drinking tea and socializing." (Woman, born 1928, Kakuda City, Miyagi Prefecture)

Morning bathing

"Not many homes had their own baths, so the custom was for all the men of the merchant families, including my father, to go take a morning bath. And as I heard it, they often grabbed a beer on their way back from the morning bath!" (Woman, born 1918, Sendai City, Miyagi Prefecture)

(2) Nature-inspired gatherings

Gatherings to gratefully cash in on blocks of time "donated" by mother nature were boisterous occasions. When rain or other inclement weather prevented people from working in the fields, it meant free time. That free time was appreciated as a chance to rest from the tiring farm work and get together with neighbors for a good time. In an era when many houses lacked their own bath, people would bathe at the home of a neighbor, with much the same attitude of gratitude that they took what nature provided. Also, the forest within the precincts of the local shrine was considered a getaway shared by the whole community where one could immerse one's self in a tranquil natural setting. In contemporary society, where it is no longer necessary to coexist with nature, such occasions for socializing that fostered spiritual affluence have completely vanished.

Bounty of the sea

"You know, whenever a fisherman caught a sea turtle, they would make a fire on the beach, and someone would bring a pot and pot stand from their hearth. They would cook up the turtle right on the beach. We'd all feast on boiled turtle meat with a splash of soy sauce. Anyone up on the cliff who saw that would think, 'Ah, they caught another turtle! Maybe they will cook some for me too!' and go down to the beach with their chopsticks." (Woman, born 1928, Wakayama Prefecture)

Here are some of the many other occasions for spontaneous gatherings to enjoy the magnanimity of nature.

Weaving on rainy days

"When it rained the women used to get together and weave zori (straw sandals) or straw mats." (Woman, born 1926, Toba City, Mie Prefecture)

Neighborly bathing

"Back then, we were the only family in the neighborhood that had a bath at home. So every evening our neighbors would come by to use it. And afterward five or six people would hang out on the veranda cooling off and gossiping over a cup of tea." (Man, born 1925, Ozaki City, Miyagi Prefecture)

Shrine grove as a getaway

"During the summer all the kids in the neighborhood would gather there. We didn't have any toys. And the shrine grove was the only place with enough space to play tag. Pretty much every community had a shrine with wooded grounds." (Man, born 1929, Yamatokoriyama City, Miyagi Prefecture)

(3) Gatherings for collaborative work

Part of living in harmony with nature meant that people carried out work collaboratively when that approach made sense, like rethatching of roofs and infrastructure maintenance. Moreover, there were community practices and rules to ensure the completion of the necessary farm tasks in each season. These longstanding mutual aid arrangements went by different names (*yui*, *ko*, etc.) in different regions. They created occasions for people to gather, fostering spiritual affluence. They prevented neighbors from growing distant towards each other. But in today's world, although some places have kept up these traditions, the mechanization of farm work and the labor shortage caused by the aging of rural populations have steadily eroded the logic of collaborative work. The systems and practices have been simplified and are ceasing to exist as a means of mutual aid and an occasion to gather that fosters spiritual affluence.

Rethatching a roof

"All villages had neighborhood associations, and they would take part in occasions such as setting up for a festival. There was also, until they went off to war, a youth association which carried out all kinds of tasks. They handled rethatching, for instance. All the houses in our village had thatched roofs, so villagers would help each other do that in turn. And when someone passed away in the old days, everyone pitched in to dig the grave, because we didn't cremate them like nowadays; we buried them. And at rural funerals, relatives would still wear something like a rope around their waists and carry special funeral banners in the procession." (Woman, born 1923, Akita City, Akita Prefecture)

Building a school

"Our schools were all destroyed in the Ise typhoon of 1959. To raise money, we undertook collaborative work to harvest hijiki and arame seaweed. Not wakame seaweed, though—people still harvested that for themselves. But we all worked together harvesting hijiki and arame to raise the funds for rebuilding the schools. Despite that, we later sold the forest behind the village on Mt. Zige for more funds." (Man, born 1929, Toba City, Mie Prefecture)

Road maintenance and community hall construction

"When the roads needed repair, we used collaborative work. I suppose the road was collective property. In any case, everyone turned out to help fix it, or to build any public facilities such as a school, or the community hall. Everyone pitched in." (Man, born 1931, Okinoerabushima, Kagoshima Prefecture)

Rice planting and harvesting and other collaborative farm work

"In some areas there were associations called 'Yui' that mobilized large numbers of people for rice planting and harvesting, and to provide meals for them all. There were 'snack breaks' at 10 a.m. and 3 p.m. during which fruit and other seasonal produce would be served, which we called 'smoking breaks'. When all the work was done, we would pound rice to make 'on-the-spot' mochi for everyone to eat. Community ties were strong back in those days." (Man, born 1922, Ishimaki, Miyagi Prefecture)

The ko association

"The young men made up an association called the shojinko: they would take turns staying in a house called the shojinyado in groups of five or so, cooking their own meals." (Woman, born 1914, Kakuda, Miyagi Prefecture)

"There were ko associations for women, like the Kannon-ko and the Baika-ko for those who observed Soto sect Buddhism. The associations would take trips together sometimes, and those were fun." (Woman, born 1923, Onagawacho, Miyagi Prefecture)

Village gatherings

"As for occasions where only women gathered, there was the 'okosama' tradition. Okosama was when all the men would go out first thing in the morning to the dojo with just rice to eat, while the women made a delicious local food called warabi-zai, with bracken, taro and potato and so on. We did that about once a month." (Woman, born 1920, Onogun, Fukui Prefecture)

Although they are not formal systems and practices, the following examples functioned as mutual aid among neighbors who had a spirit of solidarity. These tasks could be done solo, but the work went faster when a group of people assembled so they could enjoy chitchat and camaraderie while doing it. When people gather with a spirit of togetherness, it becomes a vehicle for spiritual affluence.

Making futons

"My mother was skilled at making futons. Ladies from our neighborhood would come and help her. After stuffing them with cotton, they would work together to pull and stretch them out. I helped from when I was little." (Woman, born 1925, Sendai, Miyagi Prefecture)

Making miso

"Seven or eight ladies would gather at someone's house to make miso. Sometimes they would also go to a hot spring to soak away the fatigue of farm work." (Woman, born 1917, Sakurajima, Kagoshima Prefecture)

In today's society, it's easy to say, "People should help each other more!" but unless the relationships have already been established, people tend to see mutual aid activities as a pain in the neck. They prefer to repay favors with money or things, sending gifts rather than getting together to do something. This is a superficial way of handling obligations and does little to bolster relationships that have become attenuated at a fundamental level.

3.5 A Way of Life in Which People Fulfill Responsibilities

Now let's turn our attention to roles within the local community. Following are a number of interview extracts describing scenarios from daily life where there are set responsibilities that need to be filled. They make clear the existence of important responsibilities that were carried out on behalf of the whole community involving festivals and other rituals, education, funerals, circulating information, social activities, and passing on traditions. Among these, festivals and rituals remain a part of community life today, but the responsibilities associated with education, funerals, circulating information, social activities, and passing on traditions have lapsed.

(1) Responsibilities in the local community

Festivals

"I donated a banner to commemorate being named as one of the leaders of the festival committee. In my hometown of Tsuchizaki there are a lot of people who want to have an official role at Shinmeisha shrine, because you get to wear a 'kamishimo' [traditional Japanese formalwear]. The head of the district council also serves as the district festival representative, so only one person at a time could do it. But when I lived in the Tomaecho district, as many as 10 or 20 could do it. My grandchildren still joke, "Grandpa looks like he's a young groom getting married! [because a kamishimo is also worn by a groom] Which is why I love doing it." (Man, born 1923, Akita City, Akita Prefecture)

On the other hand, an educational system based on local people training and teaching local children has faded away. In the old days, a neighbor wouldn't hesitate to speak up and scold a misbehaving child, but that rarely happens nowadays. However, there are some outlying islands of Japan where unique customs are maintained.

For example, Toshijima, which is part of Toba City in Mie Prefecture, maintains to this day its 'neyako' system dating back to the Edo period. Upon graduating from middle school, boys are sent to lodge with an adult—the neya-oya—who will oversee them. This living arrangement is called neyako. Boys do not live the entirety of daily life at the neyako; they go home for meals and return to the neyako after dinner. In the old days they would spend every night at the neyako. The purpose of this arrangement was to learn the island's fishing trade. Young men of the neyako also played an important role in local festivals. In addition, once a year the neya-oya treats all of the boys to a meal. These practices continue until a boy moves out of the neyako. But the bond between neya-oya and neyako remains strong and lifelong, much like a parent and child.

A surrogate father

"Our way of life on Toshijima has featured the neyako system going back to the Edo period. When they graduate from middle school, all boys move into one house together, where they sleep every night. A fisherman serves as a surrogate father to them. Usually a guy will bunk

in the neyako for about 10 years. People say the bond between a man and his neya-oya is stronger than with his own parents. And it's true: a man who might not return home at the death of one his own parent would come running if anything happened to his neya-oya." (Woman, born 1918, Toshijima, Toba City, Mie Prefecture)

Systems of surrogate parenting in which a non-relative takes oversight of a young man's life existed in fishing, artisan crafts, and mining. The purpose was to give this surrogate parental authority to supervise the youth and make sure he diligently learns the trade. Here are some concrete examples.

The oyakata

"A fishing family can make tons of money when the fishing is good, but can be impoverished by a run of bad catches... to the point of not even being able to afford a doctor for a sick child. You see, fishermen are notoriously bad at saving; money runs right through their fingers. So when bad times hit, a fisherman goes to see his oyakata [a respected benefactor of means], who takes care of them. The oyakata will come up with the money for a doctor when a down-on-his-luck fisherman's child gets sick." (Man, born 1929, Shiogama, Aomori Prefecture)

In many areas, when funerals are held, local people divide up the responsibilities for arranging it. In the old days people were not cremated, but interred, so digging a grave was necessary. Many interviewees stated that grave digging was the most demanding responsibility. In order to ensure that such demanding responsibilities be fulfilled, there were longstanding, established protocols.

Funerals

"Because we lived out in the country, a lot of things were based on neighborly relations, regardless of whether you were blood relations or not. When someone passed away, since there was no crematorium, we carried out the funerals in our own way. The first thing for women to do was make rice-flour dumplings called shiratama. They were made without katakuriko [potato starch] and had a smooth texture. When someone died in our hamlet, there were three tasks to be done. The first role was to inform surrounding hamlets, the second was to go buy the needed supplies and the third was to dig the grave

for interment. These responsibilities didn't fall only on the family; everyone pitched in. Our hamlet of Tachihama was divided into three groups that carried out these three tasks." (Man, born 1928, and woman, born 1918, Ogatsucho, Ishimaki City, Miyagi Prefecture)

There were also numerous people with important responsibilities in making village life viable.

First, there were the peddlers who came through the village at the same time every year, or as often as every week, to sell goods and foods. Because these peddlers walked around a wide territory hawking their wares, they possessed a great deal of information about the world outside the village. The main purpose of peddlers was to bring in goods, but they also were critical in creating a conduit for information in an age before the internet or even television made information so readily accessible. The role played by peddlers was to connect the village with the outside world.

Gleaning information from peddlers

"The peddlers were real chatterboxes. They used to stay in the flophouse." (Man, born 1925, Iyo City, Ehime Prefecture)

And school, as a public institution, was a setting in which the elementary school students had many important responsibilities. Students would take turns bringing vegetables from home to be added to the miso soup served at lunch; it's a heartwarming picture. Aside from routine tasks like this, there are examples of elementary school children doing their part for the community while having fun, such as collecting locusts to sell and using the proceeds to help fund the rebuilding of destroyed school buildings.

Contributing to the public good

"Elementary school kids would take turns bringing vegetables to use in making miso soup for lunch." (Man, born 1928, and woman, born 1929, Meiwamachi, Mie Prefecture)

(2) Responsibilities of children in the family

In daily life, children were given their first chores around the house at an early age. Fetching water, gathering pine needles for the fire, heating the bathwater, scrubbing the floorboards, opening and closing the rain shutters, looking after younger siblings, and many more. In some areas they even helped in specialized tasks like drying wild chestnuts and boiling seawater to obtain salt. Since a way of life coexisting with nature made for busy days, any contribution children could make was expected of them by parents. Household chores were undoubtedly a heavy responsibility for children. But according to the recollections of nonagenarians, chores around the house were not something they were called on to do only on certain days. They were a daily routine. In some cases, helping out the family was even prioritized above attending school. People have both grim and pleasant memories of chores.

"Kids were always assigned chores of various kinds. Handling certain responsibilities was ingrained in everyday life. Little kids would open and close the rain shutters, and when they got a little older they would be assigned to heating the bath and so on." (Man, born 1928, Yamamotocho, Miyagi Prefecture)

"People had a lot of siblings back then, so when a kid hit third or fourth grade, they would start carrying their little brothers or sisters along to school with them. If the baby cried, the older brother or sister would take it out into the schoolyard to change its diaper and get it settled down. Daily responsibilities came first and studying came second." (Woman, born 1928, Shiroishi City, Miyagi Prefecture)

"Our house had a wood-burning stove, a small one. We burned firewood to heat the bath and to cook. We had a wooden tub for the bath with a stovepipe that we used to drop in wood from above. If we were out of firewood, we had to go into the forest and cut some, haul it back, chop it up, and then build the fire. Your friends and local children would help haul the firewood, which I actually enjoyed doing myself when I was little. My husband continued using that wood-burning stove after he became a police officer." (Woman, born 1922, Shiroishi City, Animachi, Akita Prefecture)

"When I was in elementary school, one task kids often got assigned was boiling water to get salt. That was kid's work in those days. Start the fire, including gathering brush for fuel if necessary. The forest was government land, so we were allowed to take firewood from fallen and

dead trees that were going to be left to rot otherwise. We used that to boil seawater. I'd fetch seawater and pour it into a big pot, then boil it until the water evaporated and the salt was left. With a big pot, in two days of boiling, we ended up with 20 or 30 liters of salt to take back to the farmhouse. In those days, if you lived on a farm and didn't have salt for pickling, it was a major problem. You could trade salt for rice at about a one-to-one ratio by volume back then." (Man, born 1932, Shiroishi City, Ogatsucho, Ishimaki City, Miyagi Prefecture)

(3) The responsibilities of adults in the family

Of course, children were not the only ones with responsibilities around the house. As boys and girls grew up into fathers and mothers, and even once they progressed to old age, they always had tasks to perform. A typical pattern was for the mother to work inside the home and the father to work outside the home and on things that involved the neighborhood.

"In his twilight years, the patriarch of the family will lay out a woven mat on the floor and make a sacred rope called a toshinawa. Because it's a sacred thing, he first bathes, then fills a bowl with water and rinses his hands. Usually ropes are made in a vertical position. But this rope is long enough to go all the way around the perimeter of a 12 tatami-mat room. It's called a 'hacchojime' and has seven hanging straw tassels woven into it, spaced every meter or so. Between the tassels are sacred paper ornaments called 'heisoku' and green sprigs of pine. Everyone was forbidden to enter the room while the sacred rope was being made." (Woman, born 1914, Marumorimachi, Miyagi Prefecture)

"The elderly members of the family sat in the most honored place beside the hearth, on 'zabuton' cushions. The old folks would let us, the kids, sit in their laps so we could get warm from the fire. Because there wasn't enough room for everyone to have a seat next to the fire. And my mother, who had married into the family, sat in the place from which wood could be added to the fire, called the 'kizuri'. Mother's responsibility was to feed the fire and fetch more firewood if needed. I was allowed to sit there only a handful of times each year. My mother did piecework as a seamstress late into the night, all year

round. Because if she didn't, she would have had to take a job outside the house making straw goods." (Woman, born 1925, Sendai, Miyagi Prefecture)

"Year round, we got up at 5 a.m. No one had municipal water service, so we had to use the well. It was the wife's responsibility to take embers from the hearth and bank them in a container for the night, then get up before anyone else in the morning, and get the fire going in the hearth. Plus clear out the ashes." (Woman, born 1923, Akita City, Akita Prefecture)

The responsibilities of children changed as they grew older. Ultimately, upon reaching adulthood, they undertook all the responsibilities that come with it. In other words, a way of life based on fulfilling one's responsibilities entailed a system of passing on the necessary skills from generation to generation. Each time a child was given new tasks to perform, he or she was one step closer to being capable of self-sufficiency. This gave people a tangible sense of growing up.

And the most important benefit of all was that this system produced spiritual affluence. For example, tasks that only a child's small hands could do were entrusted completely to them by parents. This trust and their approbation given after completion of the task enabled children to experience a greater sense of personal growth. Ninety-year-old people remember this source of spiritual affluence in their own lives.

Being entrusted with a task no one else could do

"My family were farmers, and our farm village was really struggling. So kids were needed to help out in any way they could. Our village hadn't been electrified then so we depended on oil lamps. At night, I polished the hurricane lamps. If an adult tried to polish the inside of the glass using a stick or something, it would probably break the glass. But a kid's hand was small enough to fit inside. I'd wrap a cloth around my hand and wipe the inside. That was one task that kids had." (Man, born 1919, Akita City, Akita Prefecture)

Letting the kids handle it

"The horse wouldn't obey me. So when it was cold out my feet would hurt and go numb. All the work was completely left to the

kids as young as third or fourth graders, so the smallest horse was used. The adults just left it to us kids when they were busy doing something else." (Man, born 1926, Narusawa Village, Aomori Prefecture)

Pleased parents

"We played at being soldiers, out in the woods. Then we'd gather some branches for firewood and come home. That was our playtime. We'd drag three or four branches back and our parents would be pleased with us. Making a game out of gathering brush to feed our family's fire." (Man, born 1918, Gose City, Nara Prefecture)

Perhaps we in the modern world need to reassess our current way of life and recognize the importance of being given responsibilities in our communities and families, and fulfilling those responsibilities as a way of embracing spiritual affluence.

3.6 Living as Part of a Natural Cycle

The elements of a way of life that provides spiritual affluence are integrated within a natural cycle. So, if the natural cycle breaks down, it portends the possible loss of spiritual affluence.

(1) Using forest resources for fuel

People have to eat to live. Which means they need fuel to cook food. In the prewar way of life, fuel came from resources found in the forest or the ocean. The ash left from burning fuel was used to fertilize crops, as were human and animal waste. All these resources were available within the sphere of activity of the individual. Many clever methods were developed for improved collection, processing, drying, and efficient combustion of fuel resources. The effort involved in using such resources may not itself be enjoyable, but there are many examples of the interaction with other people in the process creating spiritual affluence.

For example, as the next interview states, collecting fuel was a thankless task if you had to do it all by yourself, but when a group of people were working together to do it, the result was spiritual affluence.

Cutting firewood in the forest for fuel

"A number of people owned chunks of forest on one mountain and they put up stakes to define their properties, so they could cut the firewood they needed from their own land and haul it back home on a truck. My father often took me along into the woods. He would cut down a tree and made a see-saw for me out of it. I would play on my seesaw while I waited for him to finish the work. There were other people in the woods cutting down trees to talk to, which made it fun. All that was back before I entered elementary school." (Woman, born 1935, Natori City, Miyagi Prefecture)

Furthermore, the need for ingenuity in burning the fuel as efficiently as possible also produced spiritual affluence. Different areas used different resources for fuel, and faced different challenges in drying it, so they had to come up with local solutions. Certain regionally distinctive fuels went by vernacular terms like *saruke* in Aomori and *karakeshi* in Osaka.

"We had to have a fire in the hearth, but there wasn't much firewood available so we used saruke, which were clumps of decomposed grassroots. A group of three or four people would dig down deep, four or five meters, and cut out blocks, about 20 centimeters a side, of mud thick with roots. Once that stuff dried out, it burned great. We needed it because there was no forest accessible to us, which meant no firewood." (Man, born 1926, Narusawa Village, Aomori Prefecture)

A full year's supply of firewood from the forest, in many cases, had to be stored up in a shed or under the eaves in order to last through the winter. There was a one-year cycle: people had to go into the forest to chop wood, then dry the timber for a year, and burn it as fuel the following year, turning it into ash. Every household estimated how much firewood it would need for the year, and built a shed to hold that amount. They didn't chop any more firewood than needed in order to avoid unnecessary labor.

(2) By-products recycled for washing, dyeing, or fertilizer

Wood burned as fuel becomes ash; when ash is soaked in water, lye forms as a supernatant fluid. Lye was used for cleaning, laundering, dyeing, and fertilizing crops. In some areas people built huts specifically for using lye. Having the capability to scrub things clean can be an element of spiritual affluence.

Using lye to scour pots and floorboards

"Scrubbing pots with lye really gets the gunk off. We didn't have scouring brushes back then. We'd ball up some straw and put lye on it, then scrub. My mother would tell me, 'I washed the pot but there's still a greasy film from cooking fish. Take some lye, go out to the well and scour the pot.' And rice bran was very effective for scrubbing bowls. And floors: we'd wrap some rice bran in a cloth and scrub them until they gleamed, too." (Woman, born 1927, Shiroishi City, Miyagi Prefecture)

Leftover food, animal dung, and human waste were all assiduously collected and provided crucial sources of fertilizer for the fields. Crops grow in the soil and people eat the crops. Their excrement is returned to the soil as fertilizer. This constitutes one small cycle in and of itself, of materials circulating with virtually no waste. An environment without a single thing wasted, better yet, an environment in which not a single thing is capable of being wasted, clearly exemplified the values of prewar Japan's way of life.

Food scraps used as fertilizer

"We had a place on our property where we disposed of food scraps. They were then used as fertilizer in the fields. It was a large receptacle strictly for uneaten food and vegetable trimmings and used straw and so on. Scrap metal and other trash like that was kept separate." (Woman, born 1923, Akita City, Akita Prefecture)

Night soil for fertilizer

"We had an outhouse with a pit latrine. It was horrible when you had to go in the middle of the night during cold weather. The excrement was called night soil. It was used to fertilize the fields or mixed with straw to compost it. It helped grow all kinds of crops." (Man, born 1920, Noshiro City, Akita Prefecture)

A way of life with no waste, where resources are used in a way that recirculates them into nature, has ripple effects on other aspects of life. Other materials used for fertilizer included the water left over from boiling soybeans, washing funori seaweed, and taking baths. Rice bran and rice husks were burned for fuel, and the ash used as fertilizer. Rice straw was used for composting and to line the stalls of horse stables. Water used to wash rice was then reused to wash dishes. Seaweed was used as fertilizer. Fish bones and sea urchin shells, starfish, and other marine life were used as a soil amendment for potted plants.

The prewar way of life was not a conscious everyday effort by people to maintain a natural cycle in their community. It was simply the thoughtful outworking of pragmatic principles like "waste not, want not" and "always keep material circulating" rather than letting it accumulate and stagnate. The feeling of fulfillment engendered by this philosophy birthed a local natural cycle and spiritual affluence.

Using water from boiling soybeans and washing funori for washing; dyeing knitting wool with chestnuts

"We made miso and soy sauce at home. For miso, our local tofu seller would boil the soybeans for us in his big pot. Later, it became possible to boil them at home. The water left after boiling the beans was used to wash clothes and even hair. Silks were washed with funori in order to give the fabric a gloss. We also washed our hair with funori. For laundry we sometimes used a small amount of lye. Each family washed their own small clothes. My grandmother taught us how to wash kimonos. Align the collar, align the hems, wash the hems, wash the collar, and then fold the sleeves and wash them. That was the easiest way to wash them. We had a walnut tree. Grandmother taught us how to use walnuts to dye yarn for knitting sweaters. Green walnuts would dye yarn a greenish-brown. Black walnuts would dye it a brownish-black." (Woman, born 1924, Sendai, Miyagi Prefecture)

Mixing used bathwater into fertilizer

"Baths were generally outdoors back then. Under the tub there was a receptacle to retain used bathwater. We'd divert that water to the vegetable patch or rice paddy when needed. Often we used it to water the barley, which we grew during the winter. In wintertime we diluted

liquid manure or night soil with the used bathwater. That's why the bath was built a little bit elevated. It had slatted floors made out of a durable wood: temple juniper. You could see down through the slats under the bath. Sometimes the slats would break though. There was a changing room and then the bath beyond that. And a place where the used bathwater was collected underneath." (Man, born 1924, Higashi-Hiroshima City, Hiroshima Prefecture)

Turning old straw sandals (zori) into fertilizer.

"We didn't have enough straw, so we used zori that were half-sized. And when they got old and wore out, we turned them into fertilizer for the fields." (Man, born 1929, Iyo City, Ehime Prefecture)

(3) Repurposing things to find continuing uses for them

A philosophy of getting the maximum usage out of everything useful and never wasting anything permeated every aspect of life. People would pick out the seams of old garments and sew the fabric into new garments. When fabric became too worn to use for clothes, it was knitted into *tabi* (split-toed socks). Cotton kimonos and other garments that could no longer be worn were remade into diapers. It was a way of life based on finding ways to continue using material by changing its form and application.

As children watched their mothers diligently finding other uses for worn-out clothes and *yukata* (cotton kimonos), they naturally absorbed that philosophy. In a world where it was taken for granted that everyone would squeeze every bit of usefulness out of things, people felt ashamed to discard anything that could still be useful. A mother hand-making clothes is an act that inspires a sentimental attachment to and due appreciation for those clothes among her family, both children and adults. An environment in which material things are limited regulates people's behaviors and values. These values are at the root of the prewar way of life.

"I wore a kappogi *and* monpe *[smock and pantaloons] that I had sewed myself. Cotton clothes that became too worn-out were turned into diapers. We got seven diapers out of one yukata. We did that work by night. In the old days when you got pregnant, you hid your belly so*

the neighbors wouldn't find out, as though you had done something wrong. After giving birth, mothers carried the baby on their backs and went on with daily work as before." (Woman, born 1922, Sendai, Miyagi Prefecture)

"In the old days it was typical to take worn-out clothes and turn them into washrags, or gather up scraps of tattered fabric and make them into work clothes. In fact there was just enough fabric in one kimono to make one pair of monpe and one kappogi. So we made a lot of those, believe me. That was common practice for the time." (Woman, born 1919, Nabari City, Mie Prefecture)

"In the old days, no matter how tattered and filthy a piece of clothing might become, the fabric would never be thrown away. So in a back closet we always had a big stash of fabric scraps awaiting some use. Women wore a kind of underskirt called a koshimaki, *an old-fashioned type of underwear. Even if the knees of it wore out, the upper portion that wraps around the waist would be saved and remade into work clothes for grubby tasks like charcoal making. Sewing patchwork fabric is the kind of work that women did at night. Women could sew anything back them. Even embroider crests on things. And with no sewing machine, all by hand."* (Woman, born 1923, Marumorimachi, Miyagi Prefecture)

Decorating the coal scuttle with pictures from books

"We would paste newspaper over boxes, like the coal scuttle, and then on top of the newspaper we would paste attractive illustrations from books and magazines, as a substitute for actual wrapping paper. My mother made boxes covered in pages from weekly magazines and so on, and stacked the boxes up two or three high, using them like we would use shelves today." (Woman, born 1928, Owase City, Mie Prefecture)

One thing we can say is that the prewar way of life was not a conscious intention to create a local natural cycle on an everyday basis, but it was a conscientious outworking of principles of everyday wisdom like "waste not, want not" and "always keep material circulating." Materials with long cycle times required persistent effort to keep things in use for a long time. Knowhow about circulating material without waste, the sense of fulfillment it engendered, a dogged determination to squeeze every last use value out of materials, and sentimental attachment to one's belongings produced spiritual affluence. And things which grew in the soil were returned to the soil. These simple yet vital practices of communities facing constraints in prewar rural Japan are what made it a natural cycle society.

3.7 The Disappearance of Spiritual Affluence

(1) Why is spiritual affluence disappearing?

Prewar life in Japan was defined by a natural cycle of local procurement for local consumption. For example, as shown in Fig. 3.1, in Tanaka Village (now part of the city of Sendai), when a household made miso, they had the soybeans to be used boiled by the local tofu maker in a large cauldron. The rest of the miso-making process was then completed at home. The water left after boiling the soybeans was not thrown away, but used for washing clothes and hair. People didn't just eat funori seaweed, they washed clothes—and hair—with it because it imparted gloss. Households would cultivate chestnut trees, use them to dye yarn, and then knit sweaters. In this system everything that came from nature was put to multiple uses, rather than being discarded after serving just one purpose. Because travel was not as easy as today, fewer goods were shipped in from elsewhere, and people relied mainly on locally procured goods for consumption.

Based on what a 90-year-old from Sendai (Tanaka Village) said

The natural cycle of local procurement for local consumption created spiritual affluence. Relationships of trust built over decades were the framework for the tofu maker's provision of water left from boiling soybeans to households in the community. The kind of relationship that empowers people to go and request this favor from a local business cannot be built overnight. It is not difficult to imagine the give-and-take, favor-for-favor relationships between community members. A well-known example of the same phenomenon is the borrowing and lending of soy sauce.

According to a 90-year-old in Naruko, a village in Miyagi Prefecture, local people did borrow and lend soy sauce in the old

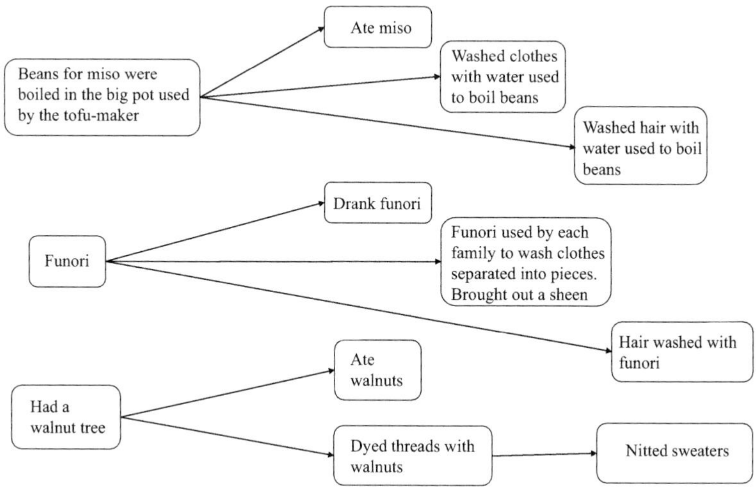

Figure 3.1 Recycled resources in everyday life.

days, but it only worked because lenders were considerate of the feelings of borrowers. People only become this considerate of their neighbors' spiritual affluence if they have lived in the same vicinity from childhood and know each other's merits and foibles inside out. Only within that framework can the lender show the needed magnanimity toward the borrower. As with neighbors within a community, the same applies to relationships within the family. A natural cycle of local procurement for local consumption does not happen automatically; it depends on people's conduct, wisdom, and considerateness. And those qualities are what bestowed spiritual affluence on these communities.

But as technologies that make daily life easier (washing machines, TVs, refrigerators, air conditioners, etc.), transport technologies (automobiles, trains, etc.), information technologies (telecom, computers, the internet, etc.) were adopted into people's way of life, they encroached on the natural cycle, disrupting the local natural cycle, and causing the collapse of the system as it had previously existed. Concurrently, the spiritual affluence derived from the circulation of locally produced/locally consumed resources has been lost, people's connection with the natural environment where they live has been diminished and the spiritual affluence experienced through depending closely on nature is disappearing. Local venues and occasions for people to gather have vanished one by one, ties between neighbors are thinning, and many communities are on the brink of utter collapse.

For example, once it became possible to easily buy soy sauce at a local store, there was no longer any need to borrow soy sauce from neighbors, inevitably leading to the fading of a share-and-share-alike relationship between people. Even a small measure of added convenience is sufficient to exert an attenuating force on community ties. People no longer gather around the local well once homes are connected to municipal water service and there is no longer a need to draw water; the local natural cycle for water has been eliminated. Rethatching of roofs and the natural cycle for thatch associated with it is eliminated when more convenient roofing materials become available; the collaborative work of thatching is also eliminated. The same goes for systems of sharing information: in the past the father of the household would bring information from the outside world, and the family would gather around him to hear it recounted. Now mobile phones, the internet, and TV have become available to all members of the family, diminishing the importance of "family time."

We used to get our water for drinking, bathing, and washing from springs and wells near our homes. We used to grow our own food or catch fish from the sea or freshwater to cook and eat. We gathered firewood from the forest to supply warmth in the cold of winter. It was a self-sufficient way of life. Our mode of transport was our own two legs. Our homes were enveloped by the natural world in immediate proximity, so we both frolicked in nature's benevolence, and dealt with nature's fury when a storm came along. In practically every regard, at a fundamental level, it was a self-sufficient way of life.

Between roughly 1920 and 1940, Japan was already orienting around big cities, and the almost completely self-sufficient way of life was gradually collapsing. After the war, the rest of the country slowly urbanized also. A dependent way of life took hold. We adopted convenient infrastructure such as municipal water supplies and convenient appliances like the washing machine, but in exchange we abandoned the well as a gathering place for conversation among neighbors. We gained access to the convenience of restaurant dining but lost the pleasure of eating food we had grown and cooked ourselves. We gained the automobile as a convenient means of transport but lost the time spent in touch with nature as we walked everywhere.

We gained the convenience of comfortable living spaces thanks to air conditioning and efficient insulation, but in the process sealed ourselves off from the fluctuations of heat and cold, the natural scents and sounds, which enabled us to relish the changing seasons. In all these things we gained convenience and a certain kind of spiritual affluence that comes with it, but lost in exchange another kind of spiritual affluence. This has been the pattern: the advent of convenience always replaces reliance on traditional practices with reliance on goods and services, and therefore the skills that people had in the age of self-reliance are lost. We have come to accept of playing the "game of life" according to a hint book, and lost the joy of discovering our own solutions. In short, we have lost the skill of pursuing spiritual affluence.

Moreover, we have acquired these conveniences by extracting the Earth's resources in a way which is causing global environmental problems. If the Earth's resources were unlimited, this trajectory for our society would not entail such serious problems. But they are not unlimited. So, unless we make a course correction, we will reach a point where continuation of our present way of life will become impossible. We now face a situation in which the proposition that our embrace of convenience is justified because it fosters a different kind of spiritual affluence of its own can no longer hold. The environmental constraints that civilization will butt up against in the years ahead demand a blueprint for a new way of life that can deliver spiritual affluence in a different form.

(2) 44 values from a disappearing way of life

Up to this point we have related numerous passages from our 90-year-olds who exemplify these values. Forty interviews with 90-year-olds in Miyagi Prefecture were transcribed. Then four scholars (including environmental science researchers and folklorists in Miyagi Prefecture) spent a year reviewing these records repeatedly.

This process identified 70 or so concepts associated with practices now disappearing from contemporary life. Adding in those items distilled from interview data with 90-year-old people outside Miyagi Prefecture, we arrived at a total of 44 items, shown in Fig. 3.2.

Although our society once had a way of life characterized by spiritual affluence born of self-sufficiency, the pursuit of convenience has removed the "safety rail" of constraints that keep society from rolling past the point of no return, as shown in Fig. 3.3, with the slope trending to the bottom right. As a result, the 44 values in Fig. 3.2, which exemplify spiritual affluence, have been lost. Many companies compete fiercely to eke out tiny margins of further convenience in their offerings to a marketplace in which an almost perfect degree of convenience is already available. It is a matter for serious consideration whether such marginal improvements in convenience translate into any significant gains in spiritual affluence.

But our society is not uniformly rolling downhill on the slippery slope of convenience; there are individuals who have reversed the course back uphill, such as those who find it wasteful to discard disposal plastic bottles after consuming beverages, so they have purchased their own reusable bottles to fill with a beverage at home and carry with them. People who have switched over to a personal reusable bottle seem unlikely to go back. Some say that after a while they begin to feel choosy about the type of tea they carry in it. Once a person starts to climb back up the convenience slope, over time they pass through stages of value transformation that dispel the desire to slide back down the slope again. But such value transformation takes time. That is why the "safety rail" is needed. Of course, it takes some sort of opportunity to start someone on the journey back up the slope, and the energy to climb. If these conditions are met, the downslide to convenience can be reversed and the individual can move from dependence back toward self-sufficiency. So far, spontaneous initiative has driven individuals who ascend the slope, but to spur a broader trend in reversing course, additional factors must be brought to bear. What those factors should be is a necessary research topic.

Ascending back up the slope of convenience does not mean returning to the prewar way of life. It means obtaining a new form

1. Living in tune with nature
2. Nature knowhow
3. Food from mountains, river, ocean
4. Getting your own food
5. Expending time and effort on preserved foods
6. Simple everyday meals
7. Special days, special food
8. Playing to your heart's content in the hills
9. Making good use of water
10. Fuel from local mountains and woods
11. A hearth at the heart of the home
12. Gratitude for nature
13. Trees at home supporting everyday life
14. A home that reflects the way you live
15. A one-year stock
16. Making everything by hand
17. Mending and careful use
18. Using things until they're useless
19. Building on ingenuity
20. Creatures all around
21. Songs at the heart of everyday life
22. Ways to help each other
23. Willingness to share
24. Fun times together
25. Giving guests a good welcome
26. A place to run into people
27. The fun of festivals and markets
28. Upholding customs
29. Life and death up close
30. Lots of people living together
31. Feelings for family
32. Roles for everyone
33. Kids working too
34. Learning through living together
35. One life, many livelihoods
36. Trading value without money
37. Towns linked with villages
38. Small shops, big bustle
39. Selling on foot and by weight
40. Walking everywhere
41. Little luxuries
42. Interesting stories
43. Just right
44. Feeling glad to be alive

Figure 3.2 44 values from a disappearing way of life.

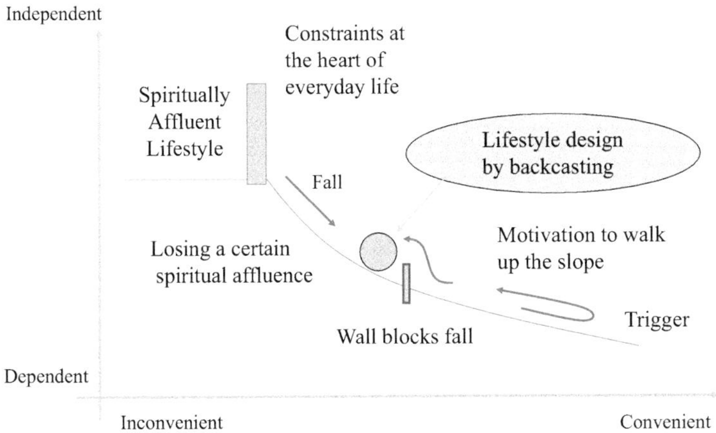

Figure 3.3 The slope of convenience.

of spiritual affluence by pursuing self-sufficiency. A different kind of self-sufficiency from the prewar way of life. A way of life that depends on the natural environment so that future environmental constraints will become a positive rather than negative factor. One that stakes out the path to a new and positive, self-sufficient, and spiritual affluence-fostering way of life.

Ascending back up the slope of convenience also does not mean foregoing all modern conveniences. We must choose which conveniences should be carried forward and which can be left behind. The process of ascending back up the downslope of convenience will generate new business opportunities and new markets in the quest for an ultimately sustainable, low-environmental-impact way of life that offers spiritual affluence.

Acknowledgment

This work was supported by JSPS KAKENHI Grant number JP24120008 in Scientific Research on Innovative Areas: "Innovative Materials Engineering Based on Biological Diversity."

References

This paper is re-written based on the following paper.

Furukawa, R. Dai ikkai kurashikata wo minaosu, PEN (Public Engagement with Nano-based Emerging Technologies) Newsletter, July 2014, **5**(4), 27–33 (2014).

Furukawa, R. Dai nikai kurashikata wo minaosu, PEN (Public Engagement with Nano-based Emerging Technologies) Newsletter, August 2014, **5**(5), 22–26 (2014).

Furukawa, R. Dai sankai kurashikata wo minaosu, PEN (Public Engagement with Nano-based Emerging Technologies) Newsletter, September 2014, **5**(6), 27–31 (2014).

Furukawa, R. Dai yonkai kurashikata wo minaosu, PEN (Public Engagement with Nano-based Emerging Technologies) Newsletter, October 2014, **5**(7), 19–24 (2014).

Furukawa, R. Dai gokai kurashikata wo minaosu, PEN (Public Engagement with Nano-based Emerging Technologies) Newsletter, November 2014, **5**(8), 32–37 (2014).

Furukawa, R. Dai rokukkai kurashikata wo minaosu, PEN (Public Engagement with Nano-based Emerging Technologies) Newsletter, December 2014, **5**(9), 19–25 (2014).

Furukawa, R. Dai nanakai kurashikata wo minaosu, PEN (Public Engagement with Nano-based Emerging Technologies) Newsletter, January 2015, **5**(10), 20–26 (2015).

Furukawa, R. Dai hachikai kurashikata wo minaosu, PEN (Public Engagement with Nano-based Emerging Technologies) Newsletter, February 2015, **5**(11), 28–34 (2015).

Furukawa, R. Dai kyukai kurashikata wo minaosu, PEN (Public Engagement with Nano-based Emerging Technologies) Newsletter, March 2015, **5**(12), 34–41 (2015).

Chapter 4

Lifestyle Design by Backcasting

Ryuzo Furukawa
Faculty of Environmental Studies, Tokyo City University, 3-3-1 Ushikubonishi, Tsuzuki-ku, Yokohama, Kanagawa 224-8551, Japan
ryuzofkw@tcu.ac.jp

In 2030, Japan will be faced with severe environmental constraints. Energy and resources will become expensive. There will be greater worries about whether the country can secure enough food for the population, and any undertaking or initiative is likely to incur excessive costs. We may become a society that is simply forced to cut back on waste and live austerely. That being the case, we should not try to maintain current lifestyles or make do with partial optimization. We need to come up with totally new lifestyles that are wholly optimized and give spiritual affluence to us despite such constraints.

Taking the above as a starting point, in this chapter I would like to discuss methodologies to transform lifestyles that I have examined, as well as various proposals about how to put these ideas into practice, and the procedures for doing so.

4.1 The Necessity of Lifestyle Innovation

Destruction of the environment continues. In the future, we can be expected to experience severe environmental constraints. As a

Lifestyle and Nature: Integrating Nature Technology to Sustainable Lifestyles
Edited by Ryuzo Furukawa

ISBN 978-981-4774-62-8 (Hardcover), 978-1-315-14687-4 (eBook)
www.panstanford.com

result of these constraints, it is said that we will no longer be able to maintain a lifestyle that places the kind of burden on the environment that is evident now. Given this, what kind of lifestyle should we try to create for ourselves?

Even though we have lost many things that have been spiritually affluent in day-to-day life, we do not try to meaningfully look at or redesign our lifestyle. From our grandparents we hear stories of past lifestyles that were spiritually affluent in spite of inconveniences, and yet we do not think to change our current lifestyle. Environmental constraints have eased because of the adoption of new technologies. Also, new products and services, as well as amenities and services offered by local municipalities, contribute to a strong sense of satisfaction, and as a result we are actually foregoing opportunities to nurture our own sense of affluence. In fact, we are losing the ability to do so. Through a breakdown of the traditional practice of using and recycling materials, on which nature and the environment depend, we have lost the source of such spiritual affluence. The introduction of convenient goods and services has replaced self-sufficient lifestyles with lifestyles of dependency, and this too is causing us to drift away from true spiritual affluence. Because we have shifted from a rural lifestyle to a mainly urban one, we are severing our ties to conditions that generate a spiritually affluent lifestyle. Lifestyles now tend to be distanced from nature. Can we bring these values back? Or can we create spiritually affluent new values above and beyond those we have lost? As environmental constraints continue to become more and more severe, it is necessary to reassess the way we live.

Our lives, moreover, are moving only toward a partial optimization, not a complete optimization. Various low environmental impact products such as energy-saving air conditioners, energy-saving refrigerators, and eco-friendly cars are available, but product optimization does not take into account the big picture. This is not the path to lifestyle transformation. Although efforts are being made to resolve the situation with technology, there is no attempt to transform the more important big-picture concept of lifestyle itself. In order to achieve a lifestyle that has a lower environmental impact, we need to aim for an entirely new lifestyle where all aspects are optimal. We need a 360-degree optimization of our lives.

4.2 How to Design Lifestyles by Backcasting

We are now on a direct path toward the loss of spiritual affluence. We are getting out of touch with the natural environment that generates spiritual affluence, moving from an independent lifestyle where we engage physically with our environment, such as by growing our own vegetables, toward a dependent lifestyle where we don't engage physically. Instead, we pay a fee to obtain commercial goods and services for everyday life. The priority is on convenience. The result is that we are gradually losing values that are important to us and we're not even aware of it.

We are also heading toward a future with increasingly severe environmental constraints. As discussed in Chapter 1, the shift to a lifestyle based on exploiting underground resources and expanding human activity has created seven risks. These risks are growing rather than diminishing, and may result in a future with even tougher environmental constraints.

Faced with these two alarming scenarios, what should we do to realize a spiritually affluent lifestyle despite severe environmental constraints? Backcasting, which was devised in the 1970s to evaluate energy policy in the UK [1], represents one possible method. Backcasting involves looking back at the present from a point in the future. This makes problems that are likely to arise visible, and then solutions are considered. My research group has applied this method and through it has developed lifestyle design methods in order to come up with solutions (Fig. 4.1) [2]. The backcasting enables us to visualize spiritual affluence lifestyles under environmental constraints. The key is to focus on lifestyle, the basic pattern of everyday life. Backcasting does not envision future lifestyles as extensions or continuations of current ways of life. It is a method of innovation that can posit a new and spiritually affluent lifestyle. The method involves five steps.

The first step is to use trusted national data across seven environmental risk areas, namely population, energy, resources,

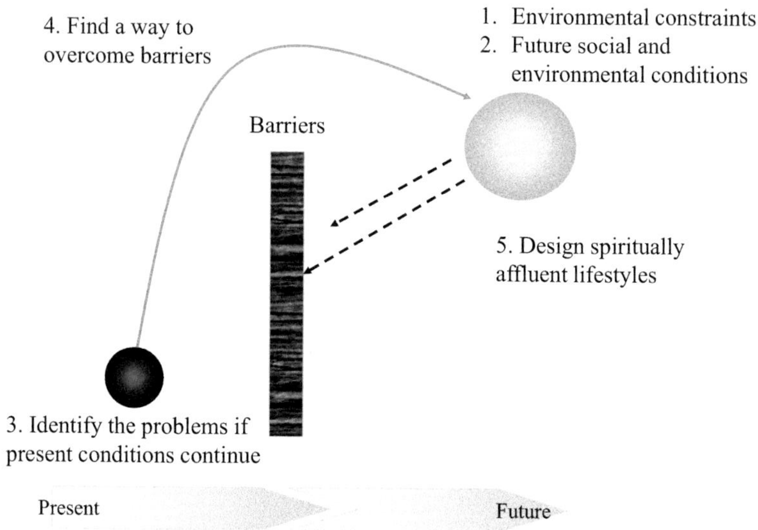

Figure 4.1 Lifestyle design by backcasting.

climate change, water, food, and biodiversity, and set the conditions of future environmental constraints. For example, establish a time frame and status by saying something like "The Environmental Constraints of Japan in 2030" and then, as far as possible, use quantitative data to picture social and environmental conditions at that time. The second step is to discuss what kind of social and environmental conditions might occur in the future under these environmental constraints. This is not forecasting. The purpose is to discuss what factors might bring about social change and in what way. It is impossible to visualize the future on the basis of quantitative data alone (the first step). The third step is to look critically at current lifestyles and innovations and pinpoint those that will become problematic under future environmental constraints. The fourth step is to think of ways to address those problems. The fifth and final step is to design a new and spiritually affluent lifestyle that incorporates those countermeasures. The lifestyles visualized here are therefore solutions, not predictions or forecasts, and it is called lifestyle innovation because it creates new values. I define "lifestyle" here as concepts held in common and shared in many day-to-day scenarios.

I design around a thousand lifestyles a year together with companies and local municipalities. The grand total so far exceeds five thousand. Soon after I started developing this methodology for lifestyle design, I ran into the problem of not being able to design lifestyles when trying to visualize them. The first time this happened was when I had designed roughly 30 lifestyles. It was easy to design the first ten or so lifestyles because I had envisioned lifestyles based on the concept of shared accommodation, local (food) production and consumption, and home power generation. I was able to design about 30 lifestyles by changing the structure of the family, introducing seasonal variations, and changing the environment (urban/rural) and life stages. However, at that point I realized I actually had not mastered the methodology of lifestyle design by backcasting. I hit a mental block and was unable to come up with any more lifestyles. Then one day, it suddenly dawned on me that I had been thinking *only* about the idea of affluence. I was not trying to come up with solutions to problems or challenges. Identifying problems or challenges is key to lifestyle design. I realized I had forgotten the basic process of backcasting (Fig. 4.1). Lifestyle design is not about changing the structure of the family and imagining different lifestyles and day-to-day scenarios. It once again became clear that there is no lifestyle design without backcasting, and identifying the challenges that are likely to occur in the future. In the process of churning out lifestyle designs, I was unwittingly focusing on forecasting. In other words, I didn't have a clear idea of the environmental constraints. I realized that if I did not discover and identify the challenges first, and then visualize lifestyles that solve those challenges, I would not be able to design more than 30 lifestyles.

Once I did get past this barrier, I was able to continue identifying challenges using backcasting. However, after I had designed around a hundred lifestyles, I hit another wall. I was able to identify various challenges with backcasting, but I was not able to visualize spiritually affluent lifestyles that would address those challenges. Now I began to wonder what constituted "spiritual affluence." I could identify challenges, but the solutions were *not* spiritually affluent. Overcoming this obstacle involved substantial trial and error. I realized spiritual affluence requires some kind of scheme from which it is created. I was wrong in thinking that if I used words such as "fun" or "beautiful," then it would be spiritually affluent. This is because words like "fun" and "beautiful" merely express isolated features of the concept. People can experience excitement or thrills without using expressions such as "enjoyment." So how is spiritual affluence generated? I analyzed the lifestyle ideas I had visualized up to that point and began to compare and analyze those that seemed spiritually affluent and those that did not.

And I made a very important discovery: spiritual affluence and constraints are closely related. They are two sides of the same phenomenon. I realized that spiritual affluence only arises in the presence of some sort of constraint. But I also came to see that we can feel spiritual affluence regardless of constraints. For example, we know that in the prewar lifestyles outlined in Chapter 3, spiritual affluence within constraints was dominant. Now, the opposite is the case. Spiritual affluence regardless of constraints is dominant.

Thanks to this realization, I was able to go beyond just a hundred lifestyles. Having overcome that barrier, I was able to visualize 200 lifestyles and more. Moreover, I realized that setting environmental constraints in the beginning and debating whether they were too severe or too optimistic was from the outset the wrong approach. This was not the right discussion. If you want to visualize spiritually affluent lifestyles that are a dramatic departure from current thinking, you must create and define spiritual affluence *after satisfying* all the original environmental constraints and assume even harsher constraints. This will allow lifestyles designs that are fresh and highly innovative. This is because if your constraints are severe, you'll come up with a new kind of spiritual affluence that does not exist today. And while it might be obvious, you will no longer feel those environmental constraints.

4.3 Lifestyle Design Case Studies

So what kind of lifestyles can be designed by backcasting? I would like to introduce some of the lifestyles designed by participants in six workshops organized in collaboration with local municipalities and businesses over six months. Rather than pointing the participants in the direction of a certain lifestyle or value, the focus was on training them in the lifestyle designing method. Each lifestyle design tells a story from a vantage point in the year 2030.

A Way of Life in Which Food Is Selected Based on Energy Consumption

A labeling system for local foods has been introduced, with "food mileage" and CO_2 *emissions labels attached to local produce. When buying vegetables at the supermarket, an increasing number of customers are selecting vegetables based on environmental impact as well as place of production, price, and fertilizer use. Locally produced vegetables are fresher, taste better, and come from a familiar source. What's more, they don't consume much energy because they don't need to be transported over long distances. Choosing vegetables with low environmental impact has become a growing trend. When distributed outside the prefecture, the locally produced vegetables are attracting attention because of the special labeling that displays the environmental impact. The local label has led to added value. As a result, the number of farmers who manage their businesses with consideration for their environmental impact has increased.*

A Way of Life in Which Agriculture Thrives with Help from Supporters

An agricultural supporter is an individual who invests in local agriculture and buys directly from farmers. This way, producers can set their own prices and consumers can get fresh vegetables without having to pay an intermediary, thereby creating a "fair trade" market. Even if there is a slight price premium, local produce is consumed locally and farmers can produce food with peace of mind.

Consumer interest in local agriculture also boosts farmers' motivation. As food information is increasingly shared among farmers, food tastings are held, allowing supporters to compare and enjoy the regional differences in food.

The rural-urban cooperation that comes out of these kinds of relationships is effective in disaster preparedness. In times of disaster, people support each other in terms of food, clothing, and shelter.

A Way of Life in Which Regional Specialty Foods Sold by Weight are Enjoyed

A growing number of stores sell products by weight and without excessive packaging. Customers bring along their own containers or bottles, which they can fill with the products they buy. There is no wasteful packaging, and quality can be maintained while cutting costs. Containers or baskets are kept at such shops, but nearly all shoppers bring their own containers or baskets. Each shopper buys only the amount they need that day, whether they are living alone or have a big family. As a result, far less food goes to waste. Besides items such as meat and fish sold at local large-chain supermarkets, there's a special focus on products direct from local areas, sold by weight. Although these products are generally not well known, customers now have more opportunities to buy them. Also, more people bring empty bottles from home and fill them with drinks blended to taste.

A Way of Life that Supports Child Raising by Allowing Parents to Work at Home for a Certain Number of Hours per Week

Today the children are all smiles from the moment they've woke up. That's because I [parent] will be at home this afternoon, waiting for them. In my company, we have the choice of working at home, and this is calculated in hourly units. Today I'm working from home in the afternoon. I schedule meetings, coordinate with associates, and do other work-related communication at the office, but I compile documents and reports at home. Due to the nature of my work, I can't work from home every day, but a couple of times a week I'm home by the time my children return from school. When my children come home, I stop working and we sit and have a snack together as they tell me about their day. When we finish eating, the children start doing their homework and I start work again. My children started to tackle their homework more positively after they saw me working from home. It was a major change. When I asked them about it, they laughed and said, "We're working, just like you!" I pay close attention to how I manage my time and work schedule, and I intend to continue to use this system.

A Way of Life in Which Empty Classrooms in Closed Schools Are Used to Bring People Together

In communities with a decreasing birthrate and aging population, especially areas considered "remote," junior high schools have been shut down and elementary schools with only a small number of children are being integrated. Such elementary schools are faced with the problem of providing air-conditioning, water, and kitchen and maintenance equipment for the whole school when there are very few classes. Meanwhile, in many companies and city government offices, the development of information and communication technology has made it possible to work just about anywhere. South of the Kanto region, where the summers are very hot, empty classrooms in schools that are near closure provide office space from May to October. City government employees engaged in agricultural and forestry work, social welfare services, or cooperation with local citizens can use the empty classrooms as workspaces. Local residents can gather in the cafeteria. There is exercise equipment and the like for seniors and others to keep them healthy and independent. For children, the school is not just a place to study reading and writing and so on, but also a place for real-life studies, learning about work, and social interaction. The empty classrooms that were previously used as storage space have become an important gathering place for people to meet, talk, and plan various activities. This evening, I will be back at school again, drink in hand!

A Way of Life Enjoying Local Food Night Markets

It used to be hard to buy local food on the way home from work. The only place open was the supermarket. However, with growing demand from people who wanted to buy and eat locally produced food, night markets have appeared. People can now buy food harvested that day on their way home from work. Partially prepared foods that have passed food hygiene standards are also available. Stores make use of collection services to procure freshly harvested vegetables from farms and home gardens in towns or villages, which are then sold that very evening. The vegetables are stored in the shade and are preserved using natural technologies, enabling people to buy fresh local produce. This is very popular with working women. From time to time I buy fish as well as vegetables. I'm happy because the food is fresh and ready to eat.

These days, the increase in the number of senior citizens living alone has given rise to a new type of restaurant within the local food night markets. These restaurants cater to the elderly by preparing the ingredients that they bring in. The service is proving to be very popular, not just among seniors who have relocated from the city back to rural areas, but among people who have moved from the city and are still new to vegetable gardening or cooking. It's a convenient way to enjoy home cooking.

Organic Agriculture: Making the Most of Living Creatures and Plants

Toyooka City in Hyogo Prefecture relies on rice farming. Artificial chemicals were once used in abundance to ensure yields and reduce the workload, but this drove the local stork population to near extinction. Today, instead of dumping artificial chemicals into the rice paddies, the city makes the most of the natural food chain in which spiders eat the pests known as plant hoppers (delciphadae): frogs feed on the spiders, snakes feed on the frogs, and the snakes provide food for the storks. Instead of using pesticides, plants such as the Dalmatian chrysanthemum, *which repel or kill pests, were planted around the fields, and the crops were sprayed with plant juice. The residents of Toyooka created an environment that sustainably produces rice surrounded by many living creatures, allowing them to coexist with the stork. Supporters (direct buyers of local produce) who are in tune with the natural environment have helped stimulate the economy, which in turn has made residents aware of the need to preserve a rich natural environment for future generations. It's a win-win situation for all. Other cities are starting to make similar commitments. Toyooka City's lifestyle comes from acknowledging that human beings are part of nature, living their day-to-day lives with consideration for other living creatures. Toyooka City's rice farming uses the strength of living animals and plants instead of agricultural chemicals.*

A Way of Life in Which People Provide Care Naturally

In home economics classes, upper-grade elementary school students learn not only cooking and sewing, but also the basics of providing care—how to give a shoulder rub, how to assist a person with walking, and how to help a person sit up in bed. Elementary school children immediately want to put what they have learned into practice. They go home and try out their newly acquired techniques on their parents and grandparents. It's nice to have their mothers thank them for a massage. But the children are surprised at how stiff their parents' shoulders are and feel they should help a little more. The children are also given a chance to experience caregiving at a nursing facility as part of their social studies curriculum. If they pass a simple practical skills test, they can receive an "Elementary School Caregivers Certificate." Because they begin at a young age, they are able to do it without hesitation when the need actually arises. And when they become parents themselves and are asked by their own children of elementary school age to be given a chance to practice caregiving, they are more than willing to offer their children the same experience.

A Way of Life in Which the Maintenance of Unstaffed Train Stations Is Passed through the Generations

Every year, just before graduation, high school students who use an unstaffed train station to commute to and from school work repair and clean the station as a way of saying thank you. This is a custom that has continued for a long time. Past improvements included putting handrails around the station, getting rid of steps, and adding slopes for wheelchair access. Recently, solar panels have been installed, along with wireless LAN for internet access. There was even a year when students not only made improvements, but turned the walls of the station into a mini art gallery. Old photos have been placed on the walls to show how the city has changed over the years. Because of this, the space in the station has become not just a waiting area, but a gathering place for local residents. When they come home to the area, former students are intrigued by developments and always stop by to have a look. They are happy to see current high school students using the wireless LAN they installed. The people who use the station worked hard to repair and improve it, so they make sure to keep it clean. Unstaffed train stations used to be dirty, but times have changed. These stations are actually cleaner than stations with staff.

4.4 Lifestyle Evaluation

We define lifestyle here as a shared concept for a way of living seen in various aspects of day-to-day life, and lifestyle scene or scenarios can be expressed by human action and values under certain conditions. For example, suppose the lifestyle concept is a "way of living that matches the season." The day-to-day lifestyle scenarios in Hokkaido and Okinawa would be different, but the "lifestyle," the core concept, would be the same.

We found that an evaluation grid method is one of the methodology to evaluate a lifestyle. The grid features many elements within each lifestyle, such as the beauty of the landscape. In this section, we will introduce the factors we used to evaluate the lifestyles using the evaluation grid method and the KJ method.

(1) Evaluation grid method

The evaluation grid method is based on personal construct theory. It is a method developed by Junichiro Sanui that expanded on the interview-style "repertory grid method" in order to know a person's original structure of perception [3]. Personal construct theory was a theory devised by the clinical ethicist George Kelly in 1955 [4]. "Construct" includes the concept of adjectival opposition, such as "the window is large/small" or "the room is bright/dark," and a cause-effect relationship between the components, such as "large window, bright room." As a whole, these structures of perception are called a "construct system." The repertory grid method is an interview-based investigative method that presents several elements and has the subject compare them and freely respond to their similarities and differences in order to know the individual's construct system. The evaluation grid method applies the repertory grid method and extracts only those parts about the construct system that are related to the lifestyles. This is done by having the subject judge which elements are superior or inferior and focusing on the decision criteria. The steps for this method are as follows:

Step 1. Create the elements
Step 2. Select the evaluation factors
Step 3. Apply laddering

Laddering, developed by Dennis Hinkle, is a technique for eliciting the upper and lower constructs of a given construct [5]. The idea of "laddering up" is asking more deeply about the reasons the test taker gives, and "laddering down" is asking about the concrete conditions surrounding the reasons given by the test taker.

(2) Creating the samples

We created different lifestyle scenes in a various theme as samples of the lifestyle scenes before the Second World War (prewar), lifestyle scenes of present day, and future lifestyle scenarios in 2030. The lifestyle scene or scenarios are classified in to 12 types (Table 4.1). The 11 classifications of the lifestyle scenes and scenarios were based on the *White Paper on the National Lifestyle Fiscal Year 1990* (Chapter 2, Section 1.2: Influence of technology in everyday life) as a reference. We added one classification to it, "interaction," which is based on the social interaction of local residents, because the results of the 90-year-old hearing (Chapter 3) revealed "interaction" to be

Table 4.1 12 categories of lifestyle scenes or scenarios used for evaluation grid method [6]

12 categories
1. Accommodation
2. Clothing
3. Food
4. Work
5. Growing
6. Leisure
7. Learning
8. Repairing
9. Shopping
10. Transport
11. Communication
12. Interaction

what they felt was missing most from their communities. Then we used those 12 classifications for the evaluation grid samples.

To create the prewar and present-day lifestyles samples, we extracted elements from books and literature, but to create the 2030 lifestyle, we used backcasted lifestyles designed by local government officials and companies. Also, when describing the lifestyles, we avoided using patently positive or negative expressions and tried to convey only the facts objectively. We limited this to a volume of about 300 Japanese characters (a relatively short paragraph; up to about 150 words in English). Below is an example.

Prewar lifestyle scene sample: The concepts of special days such as festivals and ceremonies and ordinary days (time) and public and private, front and back (space), are considered important. Sliding screens or wooden panels separate each room, but for special ceremonial occasions such as coming-of-age celebrations, weddings, or funerals or festivals, the screens and wooden panels are removed to create one large room. As society is somewhat regimented, there are rules governing how to use the room during ceremonies and in everyday life. The "front" of the house is only for important guests and the head of the household, while the living room and bedroom toward the back of the house are used by the family for daily life.

Present-day lifestyle scene sample: The xLDK, an apartment with an all-in-one living-dining-kitchen, x bedrooms, and a corridor running from the entryway through the residence, is standard. It is a space of "public and private separation," where the children's room, bedrooms, and personal spaces are separate from the shared spaces of the living room and dining area. Living spaces have become Westernized. What was once a tatami floor is now wood. Flexible partitioning screens and sliding panels have become fixed walls and doors, and the use of chairs has become the norm.

Future lifestyle scenario in 2030: The garden is a space for eating, enjoying time with the family, and cooking. When sleeping in the garden, a mosquito net is used to keep out bugs and block UV rays. The house is a simple structure just to provide shelter from storms and the cold. The bathroom and washing machines, kitchen, and other aspects of household infrastructure are installed underground. Above ground there is a large garden. In the center of the garden is a cottage on a raised platform.

(3) Investigation

Before we started the actual investigation, we conducted a pre-investigation and confirmed the need for improvements in the procedure by (i) making sure there was less bias in the categories of any of the 36 day-to-day lifestyle scene or scenarios (12 categories * 3 (prewar, present-day, 2030)) that we would use in the evaluation grid method, and (ii) making sure the steps in the procedure were accurately communicated to the test takers. After these adjustments, we created 12 categories (three scenarios in each category) and conducted the investigation with 60 individuals. We conducted the investigation with the test takers selecting six from among the 12 categories of day-to-day scenarios (the first investigation was from September 2014 to January 2015, 33 students, 6 adults; the second investigation was from December 2015 to July 2016, 21 adults; the total was 60 people).

We showed 12 day-to-day lifestyle scenarios by category to the test takers and asked them which they thought was the most spiritually affluent lifestyle: prewar, present-day, or 2030. We had them rank each and asked the reasons and factors in making their decision.

Then, we used the KJ method to summarize the results. The KJ method (also known as the affinity diagram method) [7], devised by Jiro Kawakita, is a way to summarize large amounts of qualitative data obtained through interview-type investigation, research, and brainstorming. Data is recorded on cards, one idea per card. The cards that have similarities are put together in groups, and each group is given a heading or title. In applying this method, we incorporated all the positive reasons given for each idea, gleaned from the "laddering up" process, which we then broke down into units of meaning, ending up with 6397 word phrases. We categorized these under similar headings and made sure the meanings of the phrases were consistent with each heading. If necessary, we combined some headings, got rid of others, and made

1. Freedom from waste
2. Takes no trouble
3. Costs no money
4. Takes no time
5. Efficiency
6. Easy to do
7. Opportunity to hone skills
8. Novelty
9. Qualitative improvement
10. Access to desired items/information
11. Making and looking after things
12. Attachment to objects
13. Significance accorded to possessions
14. Appetizing diet
15. Alignment with natural changes
16. Connection to nature
17. Sense of respect
18. Personal roots
19. Family ties
20. Links to the past
21. Proximity to children
22. Environmental benefits
23. Cleanliness
24. Healthfulness
25. Regularity
26. Emphasis on manners
27. Continuity
28. Preservation of culture
29. Human interaction
30. Mutual assistance
31. Sharing
32. Consideration for others
33. Benefit to others
34. Sharing of feelings
35. Sharing of ideas and information
36. Climate of tolerance
37. Appreciation from others
38. Opportunity to be oneself
39. Clear roles
40. Sense of independence
41. Making the most of knowhow
42. Self-reliance
43. Embracing a challenge
44. Positive rivalry
45. Sense of accomplishment
46. Opportunity to build confidence
47. Sense of fulfillment
48. Flexibility and a change of pace
49. Stimulation
50. Varied experiences
51. Opportunity to broaden outlook
52. Opportunity to extend knowledge
53. Opportunity to grow as a person
54. Educational benefits
55. Chance to broaden opportunities
56. Personal freedom
57. Breathing space
58. Trend consciousness
59. Compatibility with personal preference
60. Sense of luxury
61. Sense of privilege
62. Life security
63. Privacy
64. Equality
65. Peace of mind
66. Sense of balance
67. Pleasure
68. Enjoyment
69. Excitement
70. Sense of vigor

Figure 4.2 70 lifestyle evaluation factors.

improvements. The expressions used when creating the headings did not include probabilities such as "can" or "should be able to" but just the basic form of the verb, and we standardized in this way. As a result, we extracted 70 kinds of evaluation factors (Fig. 4.2) [8]. If we look at the cumulatively derived response to headings, we see that a sample of five test takers covers 81.4% of the headings, or 57 out of the total 70; a sample of 10 test takers covers 94.3%, or 66 headings; and a sample of 20 test takers covers 100% of the 70 headings, so we can say the sample number is sufficient for this research.

(4) Degree of lifestyle evaluation factor at present day

In order to identify what elements of value were regarded as important criteria both for feeling a spiritual affluence from the lifestyle (Fig. 4.3), and for having a desire to implement this lifestyle (Fig. 4.4), we conducted a quantitative online survey (from January 19 to 21, 2015; 2000 individuals from Rakuten's Volunteer Survey Group [men/women divided into equal numbers according to age, i.e., 20s, 30s, 40s, 50s, and 60+] and household income [under 3M yen annually, 3M–5M, 5M–7M, 7M–10M, 10M–15M, and over 15M]).

We found that the criteria for feeling spiritual affluence and for having a desire to implement a lifestyle did not correspond, and it became clear that the degree of relevance across the 70 lifestyle evaluation factors was different. A factor analysis of this data revealed that the evaluation for determining both the existence of spiritual affluence and a desire to implement a lifestyle were based on five factors: "connection to people," "stability," "self-realization," "convenience," and "luxury." The factor "attachment to material things" was only present among the criteria for feeling spiritual affluence. Furthermore, it became clear that the most relevant criteria for feeling spiritual affluence included "contact with people, things, nature, and food" and "psychological satisfaction." Among the criteria for desire to implement a lifestyle, the most relevant factors included "technological advancement and the guarantee of a stable life," as well as "satisfying a desire for material things."

4.5 Lifestyle Factors Desired by Japanese Local Area Under the Environmental Constraints of the Year 2030

When we use these lifestyle evaluation factors, we can quantitatively evaluate the lifestyles that have been expressed in words.

For example, let's try a comparative analysis of the spiritually affluent lifestyle ideas visualized by backcasting by the local government employee of Toyooka City, Kitakami City, Akita City, and the employees of companies that have their main offices in the

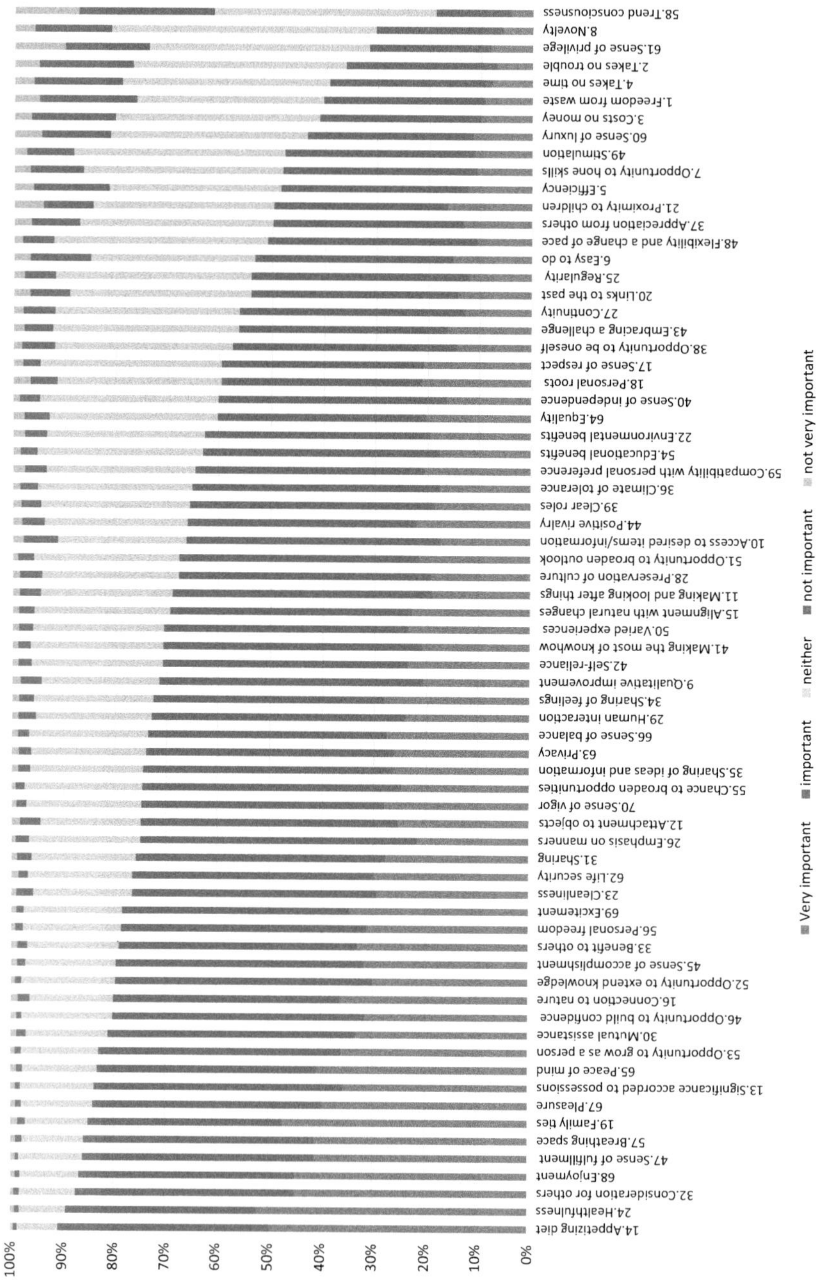

Figure 4.3 Importance of lifestyle evaluation factors when you feel it is spiritually affluent or not.

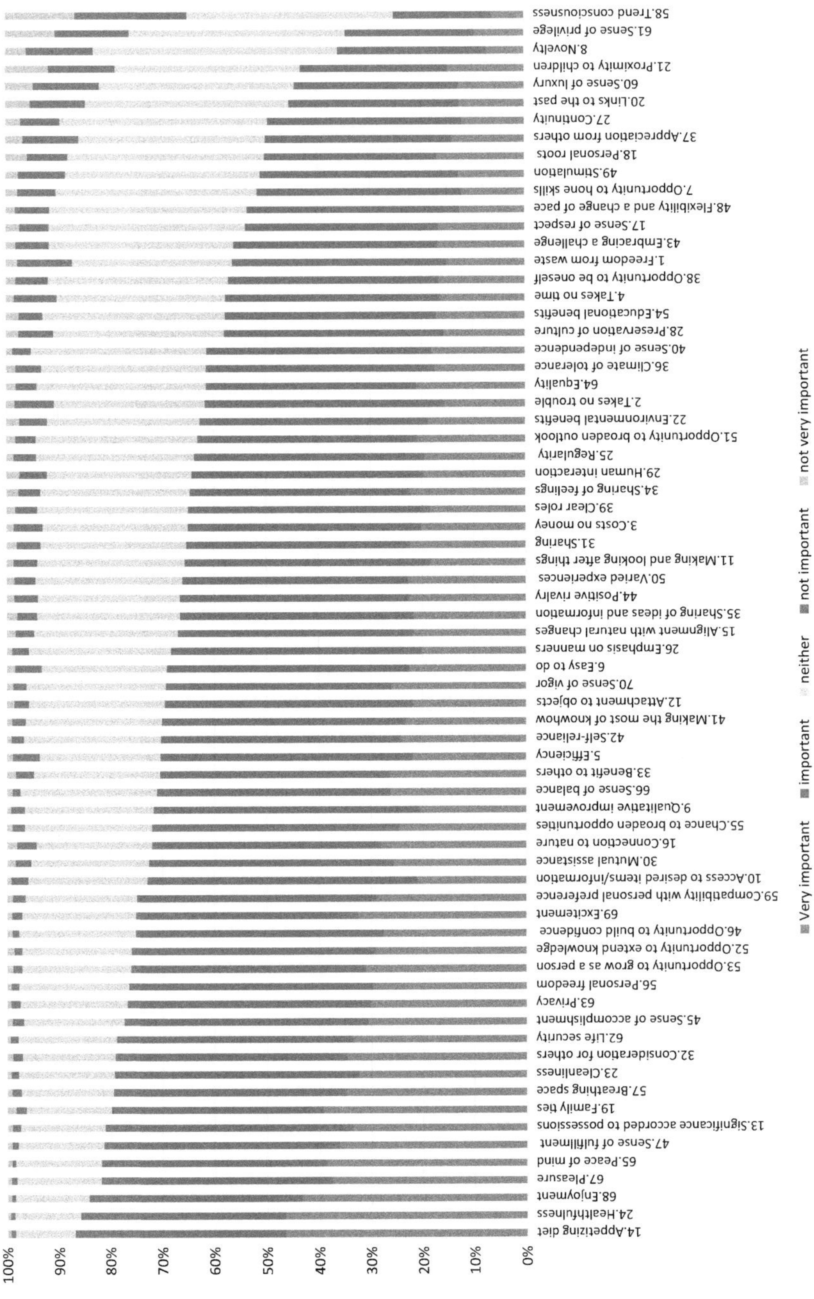

Figure 4.4 Importance of lifestyle evaluation factors when you have a desire to implement the lifestyle.

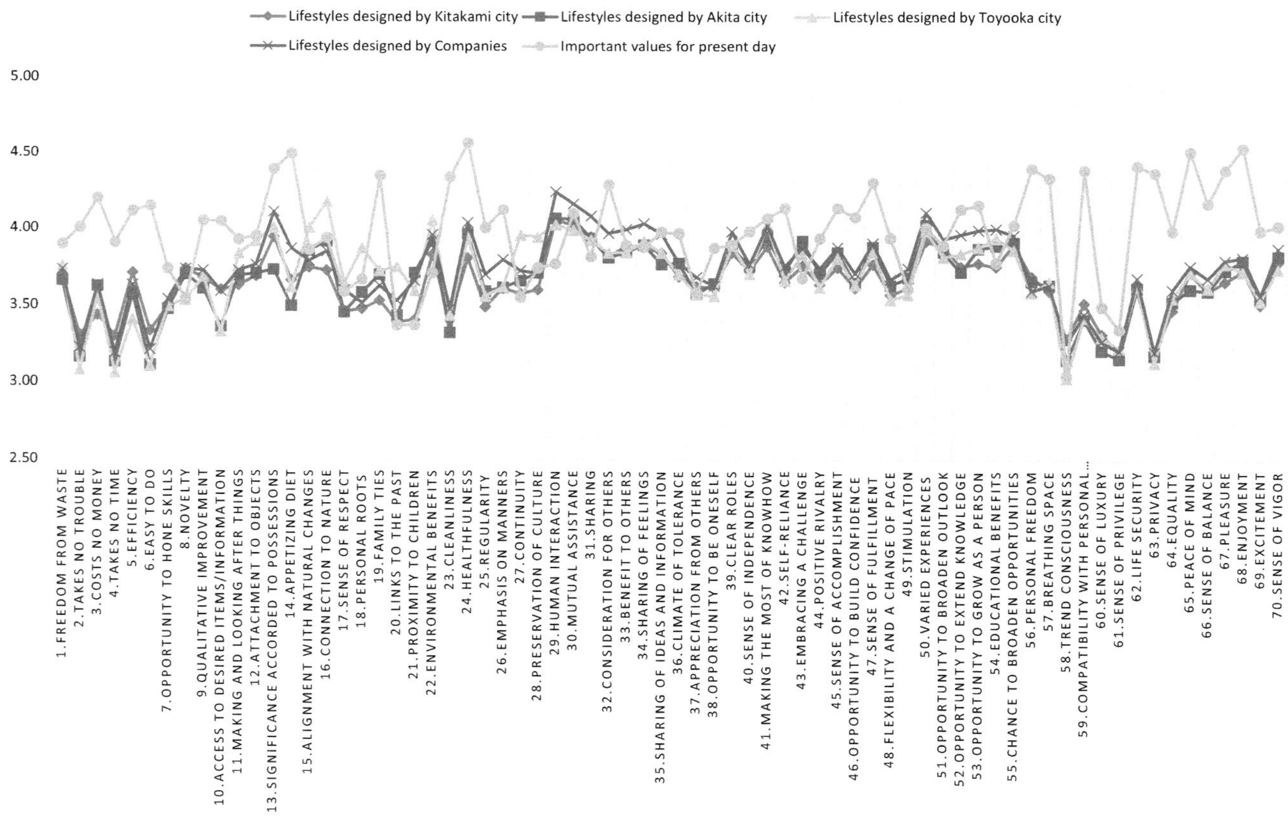

Figure 4.5 The structure of lifestyles designed by employees of local government and companies using backcasting under environmental constraints in 2030.

Kansai region. We selected 10 case studies from each region, made by them and presented as a spiritually affluent lifestyle. We then asked the test takers to determine whether or not the 70 lifestyle values were included in the lifestyles. The results were analyzed quantitatively ($N = 440$ for each lifestyle). The Fig. 4.5 compares the gap in values between lifestyles visualized in each region and those values that are currently seen as important. The result was that values currently considered as important—convenience, privacy, trendiness, a feeling of luxury—were not considered to be that important. Rather, emphasis was placed on social interaction and factors related to nature. There was comparatively little regional difference with regard to modern-day values. Clearly, the structure of the values under severe environmental constraints was considerably different from that of the values desired by people today. Since people in each region envisioned the lifestyles for 2030 under severe environmental constraints, the desired values changed. This indicates that the methodology of lifestyle design by backcasting was effectively applied by the employees of local municipalities and business professionals.

In fact, the environmental constraints that make possible spiritual affluence were widespread in prewar Japan. Most of the spiritual affluence discussed in Chapter 3 was spiritual affluence that could be experienced precisely because there were environmental constraints in prewar Japan. In today's society, because these constraints have been removed by technology, spiritual affluence has taken over that is defined by a condition *without constraints.* Mostly this is spiritual affluence defined by the satisfaction of personal desires. Thus, many of the new values that are conceived by backcasting actually existed in the lives of people living in prewar Japan, and these spiritual affluence defined only with constraints, not the satisfaction of personal desires both with and without constraints.

4.6 How to Advance Lifestyle Change

After designing a spiritually affluent future lifestyle using backcasting, the next step is to think about how to introduce that designed

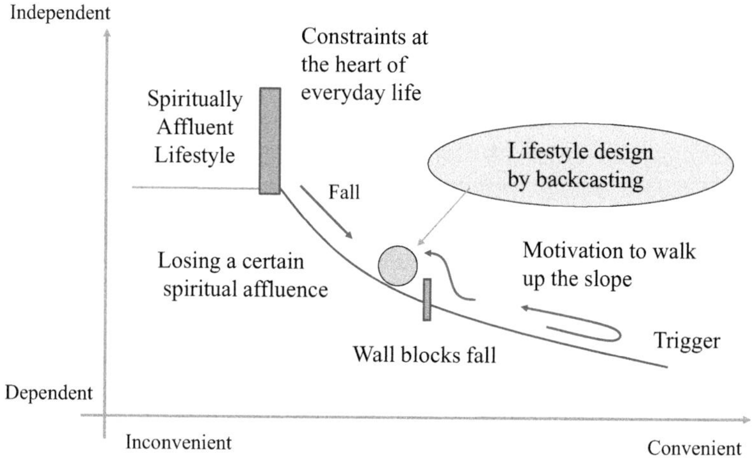

Figure 4.6 The slope of convenience.

lifestyle into society. As a system to make our designed lifestyles a reality, we have devised a "nature technology creation system," teaming up with the Nature Technology Research Consortium of the Monozukuri Nippon Conference to tackle research into the development of methodologies and identify issues for implementation and conduct the implementation itself. It's a system of monozukuri (crafting of items) that takes its inspiration from nature. As Fig. 4.6 shows, our lives have changed from an independent lifestyle to a dependent lifestyle, and while convenience has increased, we are in the process of losing certain types of spiritual affluence. If we call this trend the "slope of convenience [9]," then due to future environmental constraints we will likely again move toward less convenience, but by moving from dependent lifestyle to independent lifestyle, we can reduce our environmental impact and perhaps recover some type of spiritual affluence, which is currently being lost and is what many people still desire.

To do this, we have to stop those who are moving down the slope toward more convenience and give them an incentive to change, the motivation to go up the slope, and set constraints so that once they start to go up they do not turn around and slide down again. We should have them stop on their way up to notice the spiritual affluence under constraints that previously they had not been aware of. Life might become less convenient, but we must think about heading toward an independent lifestyle if we are to regain the spiritual affluence that is being lost. There are already people who have stopped riding in cars, who carry tea in their own bottle instead of wasting plastic disposable ones, and take the stairs instead of using escalators. These people are moving up the slope toward less convenience. When we analyze these examples, we see what factors are necessary. Deciding the target of constraints is especially difficult. But we are also made aware of, and made to question, the loss of these constraints in the modern world.

4.7 What Will Be the Motivation for Changing Lifestyles?

In this way, in order to gradually introduce new lifestyles into today's society, we must think about the gap between society's current values and the future values that will likely become dominant under severe environmental constraints. As to whether or not people would be receptive to these lifestyles, we can for example measure society's receptiveness via an online survey in which these lifestyles are described in words. Figure 4.7 shows measurements of social acceptability of 2030 lifestyles designed using backcasting in collaboration with Dentsu Grand Design Laboratory [10, 11] (carried out by Rakuten Research, September 6–8, 2010, 1000 participants, 20s–60s, male/female). Even among lifestyles designed by the same group, depending on the factors included in each lifestyle, certain lifestyles were desired by more than 70% of those surveyed, while others were desired by only around 20%. The reason for low rates of desire is simple. The spiritually affluent lifestyles were visualized under severe environmental constraints. For some, this kind of spiritual affluence seemed unrealistic under current weak environmental constraints. Other individuals embracing present-day values, the values near the bottom of the slope, near great convenience, could not relate to values that can only be appreciated at a point halfway up the slope. In other words,

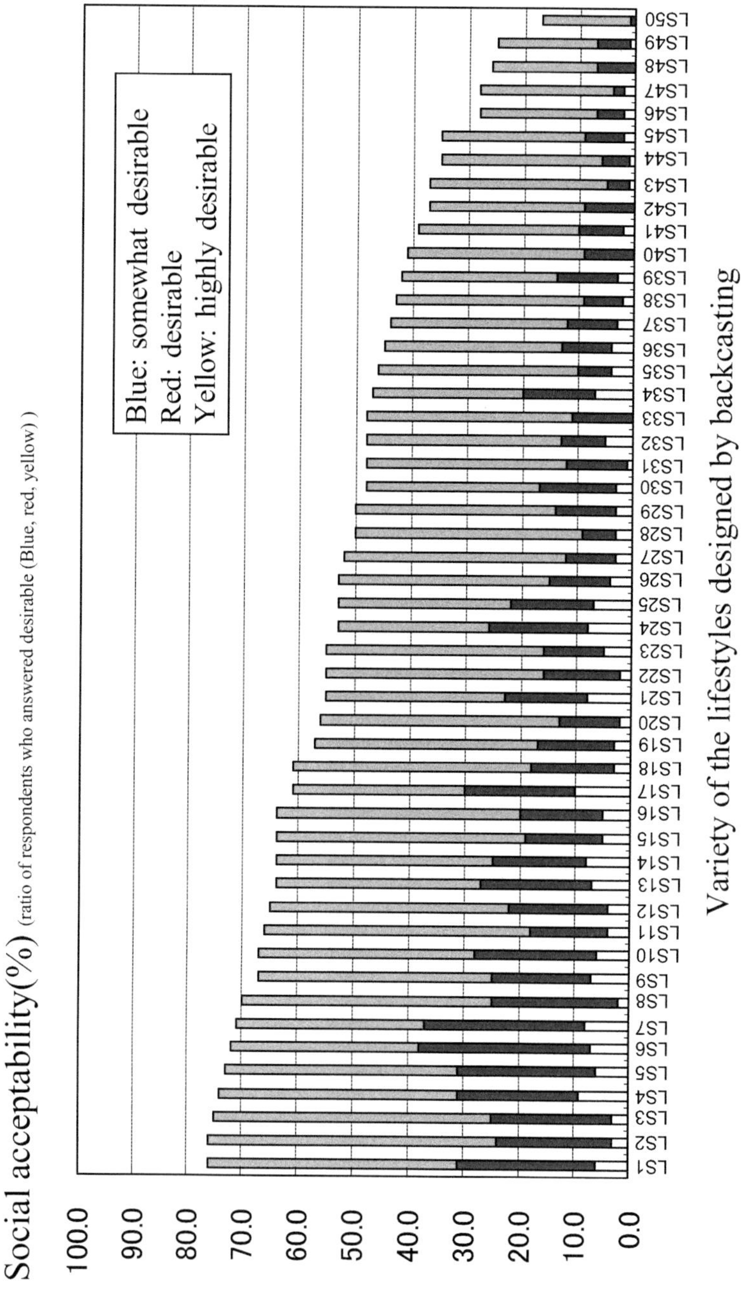

Figure 4.7 Social acceptability of the 2030 lifestyles.

even lifestyles that are not desired now, under weak environmental constraints, have the potential to become desired lifestyles in the future under severe environmental constraints. What we need to see is that in cases where we are introducing new lifestyles as a first step toward advancing lifestyle change, there is a high potential for failure if the social acceptability for that lifestyle is currently low. In order for us to introduce lifestyles that we have designed by backcasting, we must analyze the values included therein, i.e., confirm there are no obstacles to moving up the slope to less convenience so that the introduction of the lifestyle goes smoothly, and we must design a change in people's values with a strategy in mind.

To what degree will these 50 types of lifestyles be accepted in the society of today, and which elements will be relevant? Let's look at the measure of social acceptability and the underlying desires. This time, using the simplified 40 lifestyle evaluation factors that constitute the lifestyles, we quantitatively analyzed which types of factors played the most important role in social acceptability. These 40 factors are our former lifestyle evaluation factors and parts of the 70 lifestyle evaluation factors. These 50 types of lifestyles all visualized different scenarios, but we grouped those with similar elements using cluster analysis, and then analyzed the relationship between each cluster's constituent elements with respect to its social acceptability.

Next, we performed factor analysis on each cluster and ranked them by social acceptability (see table below). We named each of the factors that were extracted from the analysis. For example, we decided to name the first factor in Cluster 1 as "social unity," the second factor as "nature," the third factor as "personal growth," the fourth factor as "convenience," the fifth factor as "comfort," and the sixth factor as "nature/health."

The results of the cluster analysis were as follows: lifestyles with the highest social acceptability included the factors of "nature," "enjoyment," "personal growth," and "social unity," so it was clear that people had a latent desire for lifestyles with these characteristics. On the other hand, even though some factors that might be "inconvenient" were included, this did not make as strong an impact on ranking as anticipated. We therefore surmise that a certain degree of inconvenience is permissible.

Cluster #	Primary factor	Secondary factor	Tertiary factor	Quarternary factor	Quinary factor	Senary factor	Social acceptability (%)
Cluster 1	Social unity	Nature	Personal growth	Convenience	Comfort	Nature/objects	63.6
	27 | 30 | 36	11 | 9 | 12	18 | 17 | 16	4 | 2 | 3	33 | 35 | 8	21 | 12 | 35	
	3.85	4.18	4.01	3.30	3.93	4.20	
Cluster 2	Importance of nature/objects	Enjoyment	Convenience	Cleanliness/health	Personal growth	-	57.3
	9 | 10 | 11	33 | 31 | 32	4 | 2 | 6	20 | 37 | 23	18 | 17 | 16		
	3.92	3.69	3.34	3.59	3.56		
Cluster 3	Enjoyable/luxurious	Nature	Convenience	Self/affection	-	-	52.4
	33 | 38 | 40	11 | 9 | 12	4 | 2 | 3	18 | 19 | 17			
	3.49	3.79	3.10	3.62			
Cluster 4	Comfort together	Convenience	Importance of nature/objects	Affection together	Work	-	38.7
	33 | 35 | 31	4 | 2 | 3	11 | 12 | 10	19 | 18 | 30	5 | 27 | 29		
	3.38	2.94	3.08	3.36	3.40		
Cluster 5	Personal growth	Cleanliness/Health	Importance of nature/objects	Convenience	Enjoyment	-	32.8
	17 | 18 | 16	20 | 21 | 38	9 | 10 | 11	4 | 3 | 2	33 | 32 | 31		
	3.08	3.89	3.34	3.25	3.08		

The numbers beneath each factor correspond to the critical factors for individual lifestyles, with the three highest factor loadings listed from left to right.

Source: Hiroyuki Takido, Ryuzo Furukawa, Emile H. Ishida, Takuya Masuda; Proceedings of the 25th Annual Conference of the Japan Society for Science Policy and Research Management (2010).

Figure 4.8 Factor analysis of 2030 backcasted lifestyles.

Also, for lifestyles that showed low social acceptability, we tried adding these same factors to see whether this would increase social acceptability. And we found that we were indeed able to increase social acceptability by adding these factors. It became clear that people have an underlying desire to seek out nature, to be a part of society, to experience personal growth, and have fun.

However, it's not as if future lifestyles have all these factors. When making a choice of lifestyle, there are cases where a tradeoff has to be made: one of these factors is chosen and another must be discarded. If people make a mistake in their choices, then the environmental impact could conversely end up being even more serious. The "enjoyment" factor is an especially risky one. If the enjoyment is not of a kind that spreads and intensifies impact on the environment, then there is no problem. But seeking new enjoyment in a divergent and reckless way points us in a direction that always increases environmental impact, and this becomes a problem. If we seek "nature" we probably don't head in a direction where we are increasing environmental impact. But if we only seek personal growth, then in the end we may run the risk of increasing environmental impact.

In this way, if these latent desires can be included and balanced in a good way, there is no problem. However, if there is always some tradeoff, then there is the possibility of some increase in environmental impact. In the instant an ecological dilemma occurs, we must be aware of the possibility of a tradeoff among the factors consumers latently desire.

What is the relationship among these latent desires? Is there a chance that these latent desires are the cause of this ecological dilemma? In order to clarify what kind of tradeoff structure or dynamic exists, we examined the identified factors of latent desire in a conjoint analysis (see Fig. 4.9). And we discovered that as an overall trend, the greatest weight was placed on "convenience." Next came "nature" and "enjoyment" to an equal degree. And these were followed, far behind, by "personal growth," "cleanliness," and "social unity," in that order. This revealed the structure of the tradeoff.

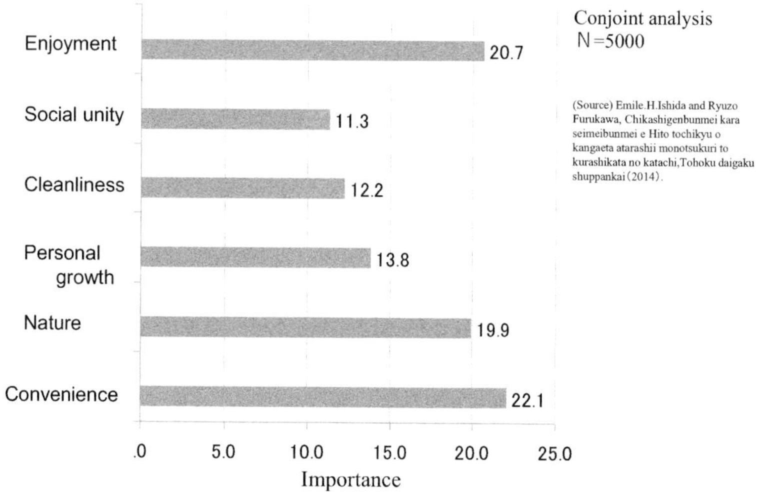

Figure 4.9 Comparison of the degree of importance of lifestyle elements ($n = 5000$).

How this tradeoff structure is activated is not clear, but it seems to be one of the reasons consumers do not gravitate toward the 2030 lifestyles that they latently desire. Consumers desire nature, but they do not take into consideration environmental constraints or the planetary carrying capacity, and rather seek out convenience and enjoyment that is incongruent with the world. No matter what one does, this ends up being the same thing as a vision of the future visualized through *fore*casting. Seeking nature, and at the same time seeking convenience *under* environmental constraints: this represents a path to the lifestyle required in 2030.

4.8 Implementing New Lifestyles in Regional Communities

Exactly where could revolutionary change in lifestyle happen? One thing I can obviously say from my experience interviewing many 90-year-old people throughout Japan is that Japan is aging rapidly and the birthrate is in sharp decline. There are villages whose very existence is in crisis. Also, regions facing the challenges of an aging society or lifestyle deterioration are in the global spotlight. In these areas, local leaders meet almost daily and think about the future. Local municipalities and businesses that are acutely aware of the crisis take a long-term view and have been among the first to express interest in lifestyle design. Many regions and businesses have begun to do something. In local areas, lifestyle change is already happening. Moreover, regions where there is still a connection to nature are considered to have the highest chance of achieving a shift toward a sustainable society because there is still a feeling in these regions of gratitude to nature for the gift of life—a spiritual affluence existence. Regions that have just started to move down the slope toward greater convenience have not completely lost important values that are likely to be important under environmental constraints. These regions are appropriate places to start implementing lifestyle change.

Lifestyle design projects in which I am involved have started in Toyooka City (Hyogo), Kitakami City (Iwate), Okinoerabu Island (Kagoshima), Shima City (Mie), and Akita City (Akita). In these regions, local residents take a long-term view and have a strong sense of urgency about the future. Nature is still abundant. Although Japan, with its long tradition of coexistence with nature, attained the status of a developed nation, people came to see that it had lost spiritually affluent traditional lifestyles along the way. But now Japan is a source of innovation, exploring implementations of spiritually affluent lifestyles within environmental constraints. It can share these innovations with the world.

The regions of Japan, through their own efforts, can become a wellspring of spiritual affluence. If regional lifestyles can evolve independently, then spiritually affluent lifestyles should one day be possible under even the severe environmental constraints that are likely to affect us in the future. It is precisely because lifestyles are dependent on the natural environment and natural resources that each region must, as an urgent priority, start to think about what it will pass down to future generations, and the actual identity of a spiritually affluent lifestyle. Such a major shift in values is likely to generate limitless business opportunities.

Acknowledgment

This work was supported by JSPS KAKENHI Grant number JP24120008 in Scientific Research on Innovative Areas: "Innovative Materials Engineering Based on Biological Diversity."

References

1. Robinson. J. Energy backcasting, *Energy Policy*, **10**, 337–344 (1982).
2. Furukawa, R., et al. Backcasting ni yoru lifestyle design shuho to innovation no kanosei, *Kobunshi ronbunshu*, **70**(7), 341–350 (2013).
3. Easterlin, R. A. Explaining happiness, *Proceedings of the National Academy of Sciences*, **100**(119), 11176–11183 (2003).
4. Vemuri, A. W., and Constanza, R. The role of human, social, built, and natural capital in explaining life satisfaction at the country level: Toward a National Well-Being Index (NWI), *Ecol. Econ.*, **58**, 119–133 (2006).
5. Sanui, J., and Kan, M. Repartory-grid hatten shuho niyoru jukankyohyoukakouzo no chushutsu: Ninchi shinrigaku ni motoduku juukankyohyouka ni kansuru kenkyu (1), Nihonnkenchikugakkai keikakukei ronnbun hokokushu, **367**, 15–22 (1986).
6. Kikakucho, K. Heisei 2 nen ban kokumin seikatsu hakusho, http://www5.cao.go.jp/seikatsu/whitepaper/h2/wp-pl90-000i1.html.
7. Kawakita, J. KJ ho: konton wo shite katarashimeru. Chuo koronsha, Tokyo (1986).
8. Ogawa, K., and Furukawa, R. Kokoro yutakana kurashi no katachi no kouzo bunseki-hyouka grid ho wo mochiite-, Kenkyu Gijutsukeikaku gakkai dai 29kai nenji gakujutsu taikai ippan koen Yoshi shu, 205–208 (2014).
9. Furukawa, R. Dainikai kurashikata wo minaosu-Naze ribenseitsuikyu niyori kokorono yutakasa ga ushinawarete shimaunoka, PEN (Public Engagement with Nano-based Emerging Technologies) Newsletter, August 2014, **5**(50), 22–26 (2014).

10. Ishida, E. H., et al. Kimiga otona ni narukoroni: kankyo mo hito mo yutakani suru kurashi no katachi,B&T Books nikkankogyo shinbunsha, pp. 43–66 (2010).

11. Ishida, E. H., and Furukawa, R. Chikashigenbunmei kara seimeibunmei e Hito tochikyu o kangaeta atarashii monotsukuri to kurashikata no katachi, Tohoku daigaku shuppankai (2014).

Chapter 5

Extracting Technologies from Future Lifestyles

Emile H. Ishida[a] **and Ryuzo Furukawa**[b]
[a]*Earth Village Research Lab. LLC, Okinoerabu Island, Kagoshima Z/C 891-9222, Japan*
[b]*Faculty of Environmental Studies, Tokyo City University, 3-3-1 Ushikubonishi, Tsuzuki-ku, Yokohama, Kanagawa 224-8551, Japan*
emile.h.ishida@gmail.com

An analysis of the social acceptability of lifestyles designed through backcasting clearly shows that, as a latent tendency, most Japanese seek nature and fun as much as they do convenience. The interviews with 90-year-olds, who presumably were able to find ways to enjoy life in spite of various constraints, led to the discovery of 44 elements crucial to the development of Japanese culture. Our research casts light on a structure for leading a life of spiritual affluence despite such constraints—a way of life that provides people with a sense of achievement and satisfaction by engaging their knowhow and skills to address minor inconveniences and constraints (obstacles), and thereby transition from dependence to a self-sustaining way of life. Unfortunately, the bulk of modern technology is designed to fuel dependence. The development of technologies and services based on conventional forecasting cannot be expected to support independence. That is why we have developed a new methodology for studying indicators.

Lifestyle and Nature: Integrating Nature Technology to Sustainable Lifestyles
Edited by Ryuzo Furukawa

ISBN 978-981-4774-62-8 (Hardcover), 978-1-315-14687-4 (eBook)
www.panstanford.com

5.1 Spiritually Affluent Lifestyles

The lifestyles envisioned for 2030 reflect a latent desire for values such as convenience, nature, pleasure, societal unity, and personal growth. These were all latent values that came to light in our analysis of lifestyles, designed through backcasting, with a high level of social acceptability. Meanwhile, the 44 elements revealed through the interviews with 90-year-olds had more to do with how they managed to lead a spiritually affluent lifestyle despite constraints: the essential ways of life upon which Japanese culture is based. Obviously there has to be a sense of continuity between the latent desires reflected in lifestyles designed through backcasting, which enable people to enjoy spiritual affluence under the constraints predicted for the future, and the factors that allowed our forebears to enjoy life despite constraints. Continuity is a must.

The 44 key phrases that surfaced through our interviews with 90-year-olds can be divided into the following categories: connection with nature; connection with the community and family; way of life; way of work; and association with life and death. These key phrases were a common thread in the evacuation centers set up in the wake of the Great East Japan Earthquake. Because of these key phrases or values, people were able to keep on smiling despite having experienced total loss. This experience reinforced my belief in the importance of keeping the values alive. Our heritage should provide the foundation for our future. Underpinning that future should be an awareness of an existence made possible by nature, a lifestyle that is in harmony with nature, a lifestyle that provides enjoyment through engaging with nature. However, no matter how affluent life was to people before the war, people today cannot be expected to embrace such a life wholeheartedly. There's no going back to the lifestyles of the past. This is for sure. Seen from a modern perspective, prewar lifestyles can never become the mainstream. Such lifestyles require time and effort, with no room to spare for novelty or luxury, which is why such a lifestyle would be difficult to accept. On the other hand, prewar lifestyles were highly valued for their abundant links with nature. People were fond of nature and lived a healthy life, enjoying strong ties with their family with whom they shared enjoyment. No food was wasted and children were well cared for. In other words, with the exception of convenience, prewar life was evaluated as having satisfied all the latent desires that surfaced in the analysis of lifestyles in 2030.

This suggests the existence of an underlying thread that connects the lifestyles designed through backcasting with the lifestyles revealed through the interviews with 90-year-olds. What if we were to juxtapose these two lifestyles? How would the level of spiritual affluence be measured?

People today would hardly consider prewar lifestyles to have mainstream potential. And they would sense no opportunity in them for novelty or luxury. This goes almost without saying. In order to appreciate prewar lifestyles, we would need to focus on the values embodied in such lifestyles. Prewar lifestyles lacked the conveniences of today. Everything took time and effort. But on the other hand, such experiences enabled people to come up with ingenious ways to have fun. They were able to engage with their environment. This is exactly the kind of latent desire for pleasure reflected in the analysis of lifestyles designed for 2030. Likewise, our analysis suggests that getting in touch with nature (having an awareness of an existence made possible by nature and engaging with nature) is not enough to satisfy the latent desire for nature. It is necessary to use knowhow and skills to make the most of nature in the face of various constraints.

By discussing such matters, we were able to envision a way of life that provided spiritual affluence (Fig. 5.1).

Recent studies on positive psychology suggest that 50% of one's happiness is determined by genetic factors, 10% by social status, fame, and income, and the remaining 40% by behavior modification, an illustration of which is shown in Fig. 5.1.

Needless to say, the way of life we envisioned that provides emotional fulfillment takes into account the severe environmental constraints predicted in 2030. Obviously, we are not talking about life-or-death situations. The premise is that the second tier of Maslow's Hierarchy of Needs has been cleared, revealing the factors that constitute a spiritually affluent lifestyle based on environmental constraints that satisfy a need for security and safety.

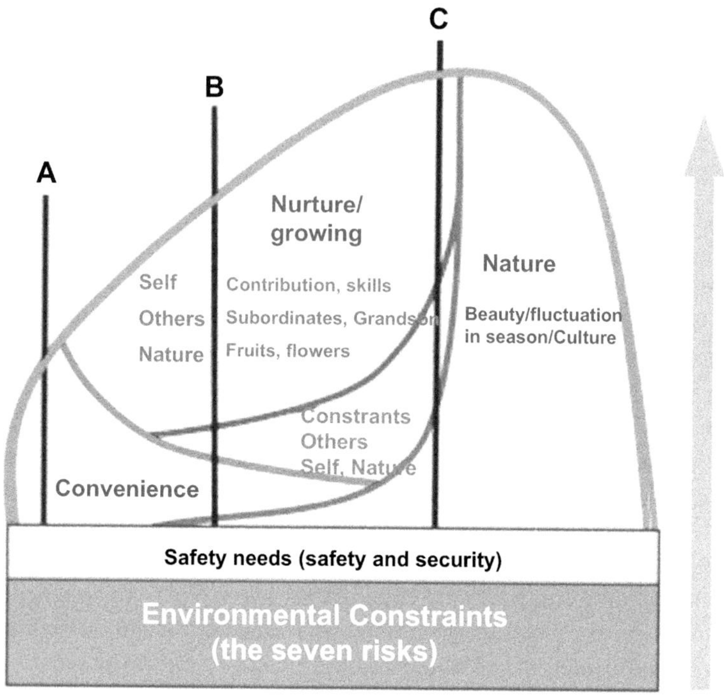

Figure 5.1 Structure of a spiritually affluent lifestyle.

Included in the factors are "convenience" and "nature." Prewar lifestyles and the lifestyles of 2030 share a desire for convenience and nature. In this context, nature refers to the cultural and aesthetic value of nature; it's the beauty of nature that provides peace of mind. The third factor is "growth." Personal growth and the nurturing of other people and nature is a source of joy, leading to satisfaction and a sense of accomplishment. An analysis of the lifestyles designed for 2030 shows a latent desire for pleasure, and clearly not the type of pleasure derived from computer games, films, or the internet. Analysis of prewar lifestyles shows how people were able to actively seek out ways to have fun. This falls into the category of "growth." To achieve "growth," one has to overcome restraint. Computer games are fun, but over time people come to feel a sense of futility. Why? Perhaps because computer games will not lead to personal growth. Your fingers may become nimbler, helping to prevent aging.

But there's little sense of achievement or satisfaction. How can the sense of achievement or satisfaction be heightened? This is the constraining factor. To achieve fulfillment from "growth," there needs to be a sense of overcoming constraints, both self-imposed and those imposed by others and by nature.

For example, think of what goes into cultivating a tree that bears delicious fruit. It's about cultivating nature. If it were simply a matter of watering the tree every day, it would be easy. But that's not how it goes. Some days it rains too much. Other times there will days of continuous drought. The temperature may be too hot or too cold for the tree. These are all constraints from nature. The amount or type of fertilizer needed will have to be adjusted according to the tree's growth or the type of soil. These are self-imposed constraints. The tree's branches may extend over your neighbor's property. Or, through consideration for your neighbors, you may choose not to use an organic fertilizer with a strong smell. These are all constraints imposed by others. By making the most of your skills and knowhow, you resolve these constraints or make effective use of them and grow volumes of delicious fruit. This is what provides you with a great sense of achievement and satisfaction.

By imposing self-constrictions and by setting a slightly higher hurdle for yourself, you seek to achieve personal growth, which ultimately will provide you with a sense of achievement and satisfaction. By achieving a consensus within a community split into multiple camps to create new rules or introduce new activities, you help to nurture the community, strengthen ties, and encourage your own personal growth. Successfully applying your knowhow and skills to overcome minor discomforts and inconveniences results in a greater sense of achievement, which is probably what lies at the heart of the subconscious desire for pleasure.

The vertical axis of Fig. 5.1 indicates the level of emotional fulfillment in life. B is higher than A and C is higher than B. For example, having an eco-friendly washing machine would fall in the realm of A. It has the same functions as a conventional washing machine but has less impact on the environment. It may be designed with more awareness of environmental constraints but in terms of functions, it's still a washing machine. It simply replaces one technology with another. You could say it's a product that focuses solely on improving convenience. As a result, it only has a short product life. People may be thrilled at first, but it's not long before the novelty wears off. It's the sort of product that is likely to cause an eco-dilemma. Unfortunately, a majority of eco-friendly air conditioners, refrigerators, and other products fall into this category.

What sort of technologies would be assigned to B? Assuming the technology will still be viable in a lifestyle with environmental constraints, let's compare a quartz watch with a hand-wound watch. Generally speaking, quartz watches are known for their high precision, whereas the appeal of a hand-wound watch is said to lie in the craftsmanship. In other words, different values are used to assess the product quality. Nevertheless, it should be possible to compare the two types of watches from the perspective of measuring time. How would the differences be reflected in Fig. 5.1? Under normal conditions, a quartz watch would not require any help from its user to keep time. It would continue to function as a highly precise timepiece. What's more, it's inexpensive. Obviously, it would fall in the realm of A, just like the eco-friendly washing machine.

How about the hand-wound watch, then? It would have to be wound every day or else it would stop. This would be a constraint. Therefore, it would fall in the realm of B. It's a minor constraint, but having to wind it up every day would make the user more aware of time, and somewhat concerned about whether the watch was running fast or slow. During this daily morning ritual, the user would unconsciously think of his or her schedule for that day. The constraints of having to wind the watch would eventually nurture an attachment to it. This would constitute a form of "growth." Unlike a quartz watch, it has to be wound up every day and is unable to keep precise time. It's usually more expensive, too. Why then do hand-wound watches continue to remain on the market? This must have something to do with the attachment people feel for them.

What about quartz watches, then? Is there no room for them in the realm of B? Yes, perhaps, in the case of a high-end model. The movement itself may cost only about 10,000 yen. But the case might be adorned with many diamonds, turning it into a luxury item costing millions of yen that would be unaffordable for most people. The introduction of an economic constraint might nurture a user's attachment to the watch. Personally, however, I would rather not defer to economic constraints. Besides, with the exception of antiques, most exorbitantly expensive products do not help reduce the burden on the environment.

The point I would like to make is that constraints are extremely important in bringing about spiritual affluence. But if the constraints were overwhelming and uncontrollable, they would lead to anguish. It's very important to strike the right balance. Take for instance, the 1000-day circumambulation undertaken by the monks of Enryakuji, a temple on Mount Hiei, which is considered to be the most austere of all ascetic religious challenges. If a monk succeeds in completing this seven-year challenge, which may well result in death, the monk will have achieved enlightenment and become a living Buddhist deity (God of Fire) or ajari. During this time he will have walked a distance equal to that of the circumference of the globe. He will also have to fast for nine days during which he can neither eat nor drink or sleep. He embraces overwhelming constraints and overcomes them to become a living Buddha. This would definitely be an extreme example of what would fall in the realm of C. Obviously, such constraints would cause nothing but pain to the vast majority of the population.

5.2 The Concept of the Missing Link

Let's look at Fig. 5.1 from the perspective of lifestyle. Lifestyles that fall in domain A value convenience and would therefore be defined as dependent lifestyles, whereas the lifestyles in domain B fall in the category of independent lifestyles because of the emphasis on personal growth and the nurturing of other people and nature. Numerous indications suggest that the way toward realizing a spiritually affluent lifestyle is in transitioning from a dependent to an independent lifestyle (internalization of lifestyles). Today, it's possible to pay for technologies and services that provide comfort and convenience. Obviously, this is less efficient than realizing comfort and convenience through one's own efforts. Increasing reliance on external sources leads to an exponential increase in the consumption of energy and resources. To illustrate this, let me cite an example that has already been mentioned. It only takes 308 Kcal to walk a distance of 10 km; bicycling, in which the person riding the bike serves as the engine, even less, just 118 Kcal, whereas a gasoline-powered vehicle would need around 9000 Kcal. People have come to believe that money can buy practically everything, even time, by making use of external sources. Even though it is obvious that this is far from what people truly desire, in reality, they are heading straight along that path. Why? Because of the wealth of technologies and services leading people to believe that constraints are all negative factors. Replacing constraints with convenience is the mainstream of modern business. This actually makes it increasingly difficult to achieve an emotionally satisfying way of life. Obviously, some businesses and services may be essential for confronting constraints that are impossible to overcome. But the fact is that the existence of constraints is crucial in realizing emotional fulfillment, which can only be achieved by transitioning from a dependent to an independent lifestyle. What we need now are services and technologies that support this transition.

5.2.1 Research on the Missing Link

Most people aspire to an independent way of life. But the majority of existing technologies and services lead to increased dependence. Cars that stop automatically without having to hit the brakes are no longer a rarity. The race is now on to develop a truly self-driving car. Home appliances are becoming fully automated. I'm opposed to this trend. But the problem is that every aspect of life has become automated, robbing people of options. The intentions of businesses and governments may be to improve quality of life, but that's not what is actually happening.

A totally dependent lifestyle is like being in need of care 24/7. It's like forcing a healthy person to stay in bed. At first, it might be nice to have everything done for you. You may feel elated at not having to lead a busy schedule, where you're always pressed for time. But what if you had to stay in bed day after day, without being allowed to do anything on your own? After a month or maybe just 10 days, it would become psychologically hard to bear. Unfortunately, most of the technologies and services in development today are headed in this direction, a way of life that requires full-time care. In Japan, since the mid-1980s, it's become clear that people are seeking spiritual rather than material wealth. Why then has technological development focused on fueling people's desire for material wealth? I believe the reason lies in the historical evolution of technology, a mindset that the value of technology lies in promoting material affluence, and an assumption that providing emotional or spiritual satisfaction is the domain of the service industry. Regarding the service industry, the popularization of services has led to an industry that plays an obsequious role, always at the beck and call of customers. In doing so, it has ceased to consider the essence of emotional fulfillment, the support of which should be its objective as a business. The same can be said of administrative services. For example, take Japan's welfare system. The system was developed to provide government support to help the needy get back on their feet as soon as possible. The number of welfare recipients showed little fluctuation throughout the 20th century, but began to soar in the 21st century. In 2000 the number of households on welfare was 750,000. In 2015 it reached 1.63 million, with a particularly large increase among those of peak working age—people in their 20s to 50s. This phenomenon has been analyzed by numerous research organizations, with the majority concluding that it is the result of a stratified society. Obviously, this is an important perspective. But another important way of looking at the phenomenon is the increase in people who are asserting their rights to be on welfare. Japan's welfare system was conceived to support those who are making an effort to become self-reliant. People were ashamed of being on welfare, as it meant they were unable to take part in society. Now, however, partly because welfare payments exceed the minimum wage, more and more people are asserting their right to be on welfare and are becoming less inclined to work, creating a vicious circle in which people continue to be on welfare for a long period of time.

Businesses are coming up with services that they believe will benefit their customers. Ironically, however, the harder they try, the more complaints they seem to receive. This suggests that corporate effort is geared towards services that are akin to strapping a healthy patient to the bed, resulting in frustration and complaints.

There's no doubt that society as a whole is seeking an independent way of life, a lifestyle in which the individual or community can overcome minor constraints or inconveniences through the application of skills and knowhow.

But the overall tendency is to associate an independent way of life with self-sustainment—a life in the countryside with no electricity, growing vegetables and rice. Such a life would have no need for technologies or services, nor room for industries other than primary industry. As a matter of fact, this is what led me to move to southern Japan in 2014. I wanted to further my research on lifestyles by living on Okinoerabu, one of the islands in the Amami Island chain. The first thing I did was to rent some land to grow vegetables for personal consumption, naturally without using agricultural chemicals. I bought eggplant and tomato seedlings, which I planted on my land. But in just three days, ladybirds had eaten the leaves and by the end of the week they'd finished off the stalks. For someone like me accustomed to life in the city, growing vegetables was too high a hurdle. In order to be able to renovate an old farmhouse, grow chemical-free vegetables, and rely on wood cut from the mountains for fuel, you need to undergo training. It's not a life for everyone.

The reason why so many people dream of becoming independent but are unable to take the plunge is because of the assumption that self-reliance demands self-sufficiency. The fact is that there is a wide gap between a self-reliant or independent lifestyle and a dependent lifestyle (Fig. 5.2). You don't have to grow your own vegetables from the word go. What's wrong with getting to know your neighbors and establishing a network of friends who will then provide you with freshly picked chemical-free vegetables from their own fields or gardens? Instead of attempting to renovate a house, why not start off by building a shelf? By applying the skills and knowhow you already have to overcome minor constraints, you'll acquire the experience and knowhow needed to achieve a higher level of independence. There's no need to rush.

In Chapter 1, I referred to a newspaper article titled "Mono hoshigaranu wakamono" ("The Young People Who Don't Want Things"). Do such people really exist? Probably not. Young people do not want things that further incite dependent lifestyles. What they do want are technologies and services that fill the gap between dependence and self-reliance. DIY, repairing and driving a vintage

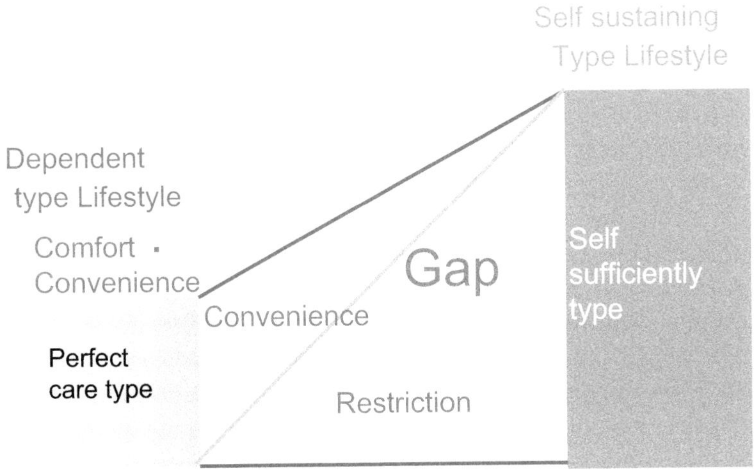

Figure 5.2 The missing link.

car, bartering at a flea market, home gardening, a growing interest in antiques, the revival of LP records. The interests of young people today are indicative of a desire to fill that gap. However, there are virtually no businesses or products designed to fill this gap. There exists a huge gap between a dependent and independent way of life. A huge link is missing. There's vast potential for new businesses and research studies. There's a wealth of opportunity in this gap. What we need to do now is to fill this gap.

5.3 Studying the Indications

How then should we elicit the development of businesses and policies that fill the gap? As has already been discussed, the key lies in backcasting. Envision a lifestyle that provides spiritual affluence despite severe constraints, and through backcasting ascertain the technologies and services necessary to sustain that lifestyle. Ten years ago, my colleagues and I opened a graduate course. Targeted mainly at working members of society, our aim was to nurture the necessary human resources needed for developing management strategies based on backcasting, taking into account the severe environmental constraints predicted for the future. We have provided similar consultation services for numerous corporations. Over the years we have discovered that people who have experienced success, and companies that have developed effective in-house manuals, find it difficult to accept the backcasting methodology.

For instance, let's consider the case of a young woman who came to me for advice after I'd finished giving a talk. She was out of breath as she had rushed to catch me before I left. She explained that after graduating from a university in Kyoto, she had gone to work for a company in Kyoto where she was currently employed. She'd met her boyfriend while attending university, but he was now working for a company in Tokyo, which meant they were conducting a long-distance relationship. After three years, they were both committed to their work, but she was starting to think seriously of marriage. Recently, however, they'd end up quarreling every time they met and were unable to calmly discuss their future together. She wanted my advice on what she should do.

Since her situation represents an ideal opportunity for studying how to solve a problem through backcasting, with her permission I have been using her case as an example in various seminars. What would you suggest she should do? Some of the many answers given so far are: "The quickest way to solve the problem would be to become pregnant." "They should both quit their jobs and go to work for a local company." "Obviously, the woman should quit her job and get the young man to marry her." "They should go and visit various places and make traveling a shared pastime." If this were an exam, such answers would obviously have received a failing mark.

As I've explained over and over again, backcasting starts from the premise that there are constraints. In this case, the constraint would obviously be "separated by a long distance." Any solution arising from backcasting would take this constraint into consideration, whereas with forecasting, the idea would be to eliminate the constraint. The first three of the four suggestions above are typical of forecasting. In this case, the constraint is obvious. But normally, it's not that easy to pinpoint the constraint with a simple question. What are the constraints that hamper the resolution of the imminent problem? How did the constraints come about? By asking ourselves these simple questions over and over again, we come to realize that many of the problems we thought existed are actually the result of a very simple constraint, a constraint that is causing multiple problems. The young woman's problem was ideal for training in backcasting, because the constraint was so obvious.

The fourth suggestion provided by seminar participants—making traveling a common hobby—merits points in that it recognizes the constraint. The problem is that it's forcing the young couple to take up traveling as a hobby.

Traveling is not a solution to overcoming the constraint. It's simply a divergence. A solution achieved through backcasting has to be the result of pinpointing the constraint by asking simple questions. In this case, it would therefore be necessary to acknowledge the constraint of being separated by a long distance and then consider a common objective that would bring them together.

For example, how about seeking a place where they could start a home? Tokyo and Kyoto would simply be places where they lived for the time being until they found a place where they could settle down for good—a place, for instance, surrounded by nature. They could then spend the few times they met once a month searching for a suitable place, traveling perhaps all over the country. Over time, they might start discussing how much they would need to relocate to their new home and how to come up with the money. If they both continued to work, they might be able to realize their dream quickly. If so, they might consider buying a place and using it as a second home to which they could each commute from their temporary residences. And indeed, after discussing such matters with the young woman, her face lit up with hope.

Unfortunately, more often than not, people who have continued to enjoy success in the conventional sense or are used to following a manual have a hard time changing their mindset. As a result, they tend to adhere to conventional methods of problem-solving that focus on eliminating constraints or diverging from the problem. Their track record of success (which has resulted in a rising burden on the environment) prevents them from seeing things in a different light.

What if it we were able to apply forecasting methods to provide an insight into the world as seen through backcasting? Forecasting,

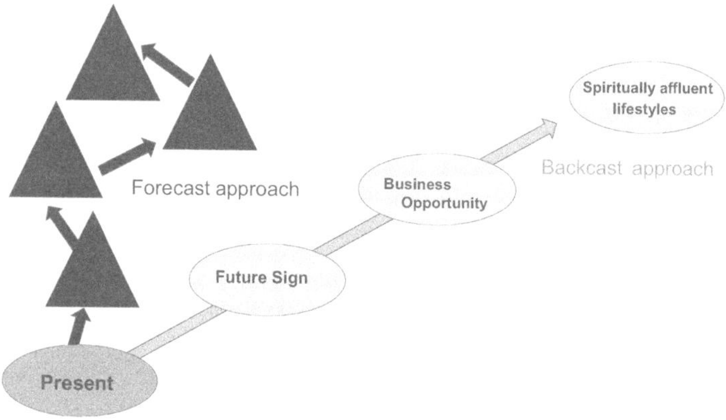

Figure 5.3 Studying the indications.

obviously, looks at the future from the present. There's no knowing which direction the future will take. But by taking social trends into consideration, wouldn't it be possible to control that direction? This is the premise for our research on indicators. All sorts of new trends are emerging throughout the world on a daily basis. What if we were to examine business trends to determine whether they were indicative of the technologies and services needed to transform and realize future lifestyles, or were mere fleeting trends? We could then use these indicators to forecast the future, which would allow us to offer seeds for new businesses. On the basis of this hypothesis, we began our research on indicators (Fig. 5.3).

Details of what constitutes an indicator will be described in the next section. But for now, suffice to say that society is clearly starting to reconsider its lifestyle of mass production and mass consumption, and to reexamine individual values. As mentioned before, in Japan in the mid-1980s, the number of people seeking emotional satisfaction surpassed the number of people seeking material affluence. The divide is expanding, with those seeking spiritual affluence leading by over 30 points. More and more young people are saying cycling is more stylish than driving. Relaxing in a natural environment has become a popular way of spending the weekend. There is no negative connotation to bartering. Vegetable gardens and gardening

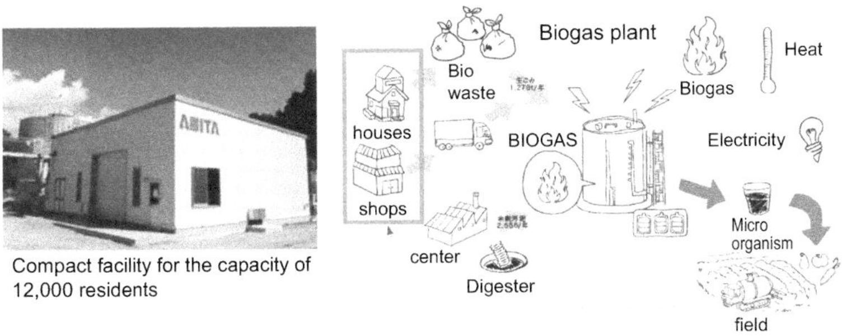

Figure 5.4 Minamisanriku (Tohoku district) bio gas facility by Amita Corp.

as a whole have become synonyms for stylish living, as with DIY, and repairing and maintaining products for longtime usage. An increasing number of people are making decisions based on personal rather than monetary values.

Take for example, the biogas facility, BIO in Minamisanriku, Miyagi Prefecture, run by AMITA Holdings Co, Ltd. The town of Minamisanriku was devastated by the Great East Japan Earthquake of March 2011. Most of its major public facilities were destroyed, including its sewage treatment facilities. Rather than rebuilding a conventional waste-processing facility, local authorities in collaboration with AMITA came up with the idea of opening the biogas facility illustrated (Fig. 5.4). Organic waste collected from households and businesses undergoes fermentation to produce biogas, which is converted into heat energy to generate electricity, while liquid waste is used to produce fertilizer. This in itself is not unusual; it's what takes place at a standard biogas facility. The big difference is in that BIO does not perform any separation of waste. By doing away with waste-sorting facilities, the town was able to minimize initial start-up and personnel costs. So where does the sorting take place? At individual households and businesses. It's their responsibility to follow the guidelines for separating waste for collection. This system is what makes it possible to operate BIO. At present, 98–99% of the waste is properly sorted. Many of the town residents drop by to see the facilities. This is a typical example of how the application of knowhow and skills to overcome minor inconveniences can lead

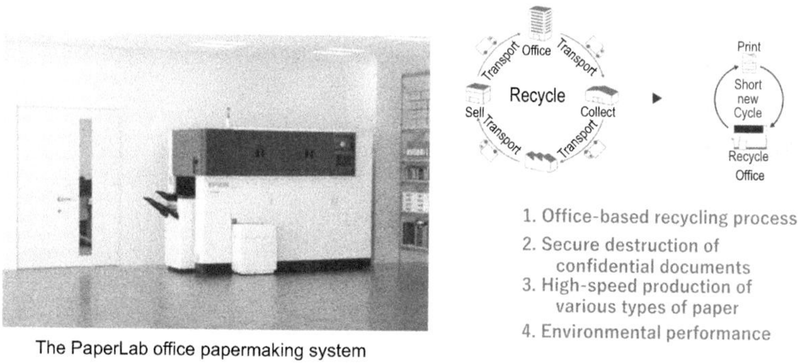

Figure 5.5 Epson develops the world's first office papermaking system that turns waste paper into new paper.

to spiritual affluence and attachment. It's an indication of how to fill the "gap."

Seiko Epson Corporation's papermaking system "PaperLab" was showcased at the 2015 Eco-Products Fair (Fig. 5.5). Full-scale marketing is scheduled for 2017. The small, compact papermaking system can destroy documents and develop new sheets of paper in three minutes just by pressing the Start button. The system can output about 14 A4 sheets per minute and 6,720 sheets in an eight-hour day. It can produce different types of paper of various thicknesses, from A4- and A3-size office paper to paper for business cards. It can even produce color paper and scented paper. PaperLab features a mechanism that breaks down waste paper into long, thin cotton-like fibers without using water. This process completely destroys confidential documents. A variety of different binders can be added to the fiberized material to increase the binding strength or whiteness of the paper or to add color, fragrance, flame resistance, or other properties needed for a given application. The density, thickness, and size of paper can be controlled to form A4- and A3-size office paper, paper for business cards and various other applications. Several years ago, the idea of recycling paper in-house was unheard of. But now, the concept of self-sufficiency has started to spread through businesses. This is a clear indication of what the future holds.

Figure 5.6 La Collina in Ohmihachiman, Shiga Prefecture.

Meanwhile La Collina in the city of Omihachiman in Shiga Prefecture is the name of a confectionery theme park operated by Taneya, one of Japan's leading manufacturers of traditional Japanese sweets. The shops, restaurants, and production facilities are scattered around a paddy field. The rooftops are covered in a lush green. The park reflects the manufacturer's awareness of being nurtured by nature and conveys a powerful message of creating sweets that make the most of the bounties of nature (Fig. 5.6).

We're also witnessing a sharp rise in lifestyle stores that handle everything from clothing and housewares to food needed for a curated lifestyle. Nationwide retail chains expanding from cities to rural areas are in decline and are being replaced by stores that offer curated lifestyles. LP records are enjoying a revival, with stores specializing in LPs opening throughout the country. Antique stores are drawing young people who are looking for items that appeal to them on a personal level. What's different about these stores is that rather than dealing in vintage products, the emphasis is more on providing people with quality products that will last a long time with proper maintenance. Such indicators have not yet become the mainstream. But they have clearly gained social acceptability and are milestones on the path to the future.

5.4 Studying Indicators

In May 2016, we set up a project to study indicators. Our invitation to businesses to take part resulted in over 70 applications in just half a day. This provided us with a glimpse of the difficulties faced by many companies in developing next stage business strategies. After meeting with the applicants, we invited 29 companies to take part in what we envisioned as a five-month project. We defined an indicator as a trend that satisfied two conditions. First, it had to be a trend resulting from environmental constraints; second, it should provide spiritual affluence and embrace several of the 44 lifestyle elements gleaned from the interviews with 90-year-olds that formed the basis of traditional Japanese lifestyles (Table 5.1). We differentiated an indicator from a temporary trend by determining whether or not it satisfied these two factors. Based on this definition, we created an Indicator Study Sheet, which project members used to analyze various indicators and state their findings. The sheet comprised four sections: 1. What is the indication? 2. Why can it be considered an indicator? (What environmental constraints inspired it? Which of the 44 lifestyle elements does it conform to?) 3. Based on this indicator, what sort of lifestyle can be expected in the event of further constraints? (Describe in 150–200 words.) 4. What sort of new businesses would be needed to realize this lifestyle?

For example, if we were to focus on the DIY trend, the completed sheet would probably look like this:

(1) What is the indication?
Visitors to annual DIY shows are steadily increasing. There's also been an increase in DIY workshops.

(2) Why can it be considered an indicator? (What were the environmental constraints that inspired it? Which of the 44 lifestyle elements does it conform to?)
Growing apprehension about the depletion of energy and other resources resulting from mass production and mass transportation inspired the trend. With regard to the 44 lifestyle elements, it embraces: making everything by hand; mending and careful use; building on ingenuity. (If working together with friends, then:

Table 5.1 44 lifestyle values which make up the Japanese culture

1. Living in tune with nature	23. Willingness to share
2. Nature knowhow	24. Fun times together
3. Food from mountains, river, ocean	25. Giving guests a good welcome
4. Getting your own food	26. A place to run into people
5. Expending time and effort on preserved foods	27. The fun of festivals and markets
6. Simple everyday meals	28. Upholding customs
7. Special days, special food	29. Life and death up close
8. Playing to your heart's content in the hills	30. Lots of people living together
9. Making good use of water	31. Feelings for family
10. Fuel from local mountains and woods	32. Roles for everyone
11. A hearth at the heart of the home	33. Kids working too
12. Gratitude for nature	34. Learning through living together
13. Trees at home supporting everyday life	35. One life, many livelihoods
14. A home that reflects the way you live	36. Trading value without money
15. A one-year stock	37. Towns linked with villages
16. Making everything by hand	38. Small shops, big bustle
17. Mending and careful use	39. Selling on foot and by weight
18. Using things until they're useless	40. Walking everywhere
19. Building on ingenuity	41. Little luxuries
20. Creatures all around	42. Interesting stories
21. Songs at the heart of everyday life	43. Just right
22. Ways to help each other	44. Feeling glad to be alive

nature knowledge, ways to help each other, and fun times together would also apply).

Since this trend was inspired by concern about the global environment and reflects some of the 44 elements, it is an indicator.

(3) Based on this indicator, what sort of lifestyle can be expected in the event of further constraints? (Describe in 150–200 words.)

It has become standard practice to build detached houses and apartments that last over 100 years. The initial investment is relatively high, but from then on, the structure will only require maintenance and this will result in considerably less impact on the environment. Today, the entire family is helping to build a wall. Since the youngest child has entered elementary school, the family has decided to build a children's room. The father and eldest brother have attended various DIY schools and can make practically everything. They have the respect and admiration of the entire family. After dismantling the walls, they'll lay the groundwork for the new room, and then everyone will help finish it. They've decided to cover the walls with clay from the mountains. They knead the clay and layer it on by hand, covering the wall with handprints as a memory of the day. They're all happy with the effect. Next month, they're going to try remodeling the generator so that they can make more use of natural energy resources.

What sort of new businesses would be needed to realize this lifestyle?

DIY schools run by housing manufacturers to service the houses they sell that last over 100 years (connection with customers for several generations), and sales of DIY supplies.
Various DIY schools specializing in woodwork, metal processing, as well as the fitting of electrical appliances and accessories.
Schools that teach how to appraise various antiques.
Dispatch of DIY senior helpers.
Tool shop for children.
Various parts stores.

If each of our corporate project members were to discover five indicators a month, we'd be able to gather around 150 indicators per month, allowing us to create a map of new businesses. That was our plan. Unfortunately, however, we encountered several obstacles.

As we proceeded, we discovered that while the majority of our members were able to discover indicators, they were unable to envision how the indicator would develop into a lifestyle that would offer emotional fulfillment in the face of severe environmental restraints. As a result, we were unable to come up with ideas for new businesses for future lifestyles based on the indicators.

An indicator, it should be recalled, was a new trend inspired by a desire to restore some of the 44 lifestyle elements that were lost in making the transition toward a dependent lifestyle. In order to come up with ideas for new businesses, the project members had to be able to envision a lifestyle that would restore the lost elements and provide emotional fulfillment under even more severe constraints. That proved to be virtually impossible. In other words, despite having indicators as cues, without understanding how fulfillment was a by-product of living with constraints, people were unable to come up with new business ideas. In other words, this underlined the importance of adopting the perspective of backcasting.

In the end, we spent five months completing the Study Sheets on the indicators that our members had discovered. This offered us training and an appreciation of how difficult it was to generalize a way of thinking based on backcasting.

5.5 Structure of Indicators

On the basis of the indicators discovered by the Indicator Study Group and my own research group, we came up with 145 business indicators that we then set about analyzing to determine how they were structured.

We identified roughly seven environmental constraints that inspired the indicators. Although biodiversity seemed to have a slightly larger influence than the others, the indicators were distributed almost equally among the seven environmental constraints. We then divided the indicators into 10 categories for analysis (Fig. 5.7).

For instance, 13 of the 145 indicators could be placed under the "food" heading. The extent to which the food indicators complemented the 44 essential lifestyle elements is shown in Fig. 5.8.

There was a considerable discrepancy in the extent to which the food indicators complemented each of the 44 elements. For example, elements that were enhanced by over 40% were: *3. Food from mountains, river, ocean*; *4. Getting your own food*; *16. Making everything by hand*; *18. Using things until they're useless*; *19. Building on ingenuity*; *22. Ways to help each other*; *25. Giving guests a good welcome*; and *42. Interesting stories*.

The other 9 categories besides food displayed similar tendencies. Overall results for the 145 indicators are shown in Fig. 5.9.

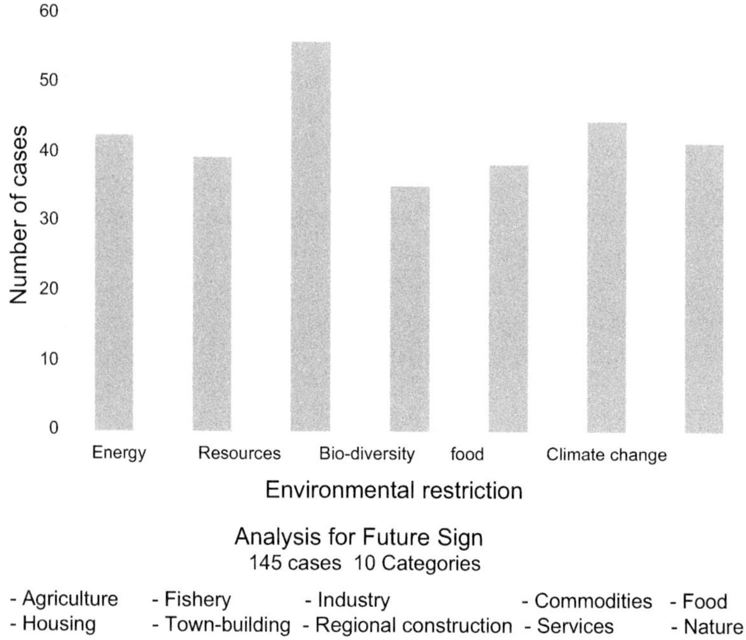

Figure 5.7 Environmental constraints for 145 future signs are dispersing approximate uniformly, even though a little more in bio-diversity and were categorized into 10.

After collating the results for each category, we were able to determine which of the elements were complemented by over 50% and 30% as shown in Fig. 5.10. Since we are still in the early stages of research, it is difficult to correctly evaluate the significance of the 50% and 30% levels. But it's safe to say that lifestyle elements that are enhanced by over 50% or 30% in each category are important values for the indicators.

For instance, *19. Building on ingenuity*, is an element that comes up in all 10 categories. Although *16. Making everything by hand* is an element enhanced by 50% or more in just two categories, it's enhanced by more than 30% in seven categories. The elements that were compensated for by over 50% or 30% in more than half of the 10 categories were: *1. Living in tune with nature*; *9. Making good use of water*; *10. Fuel from local mountains and woods*; *16. Making*

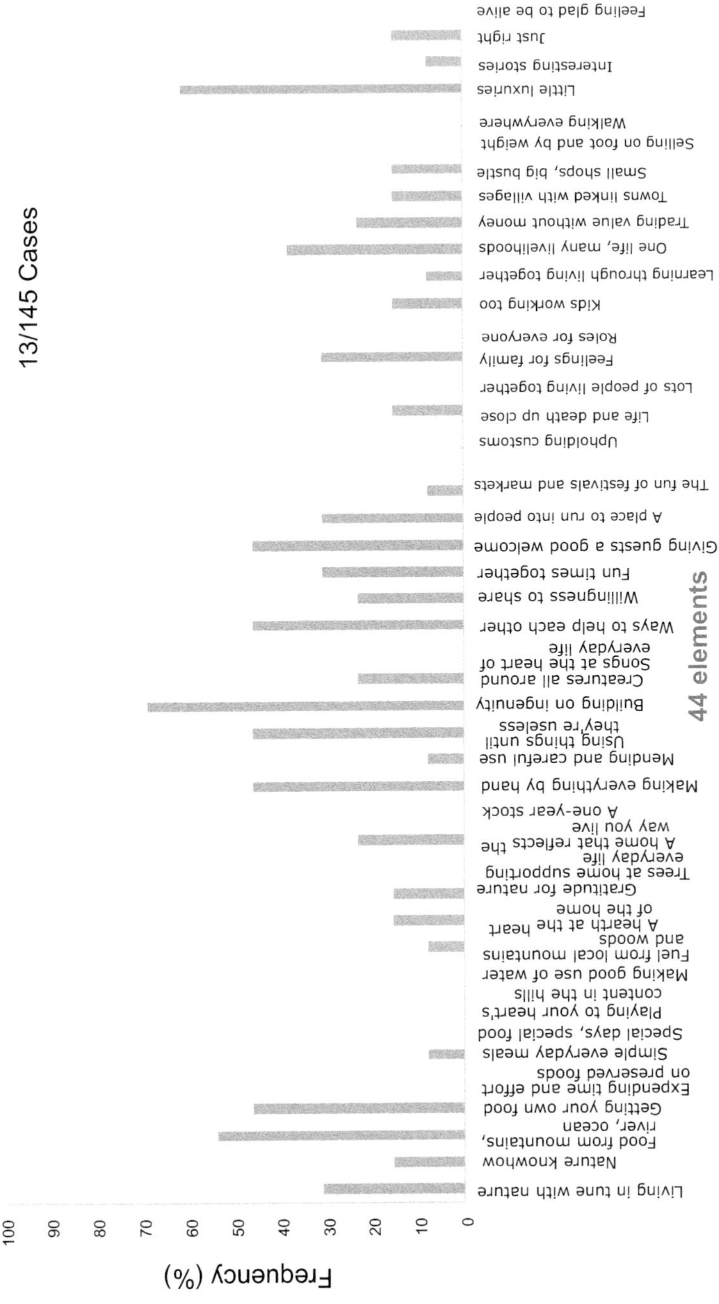

Figure 5.8 Example of analysis for food on the 44 lifestyle values.

Figure 5.9 Analytical results for 145 cases on the 44 lifestyle values.

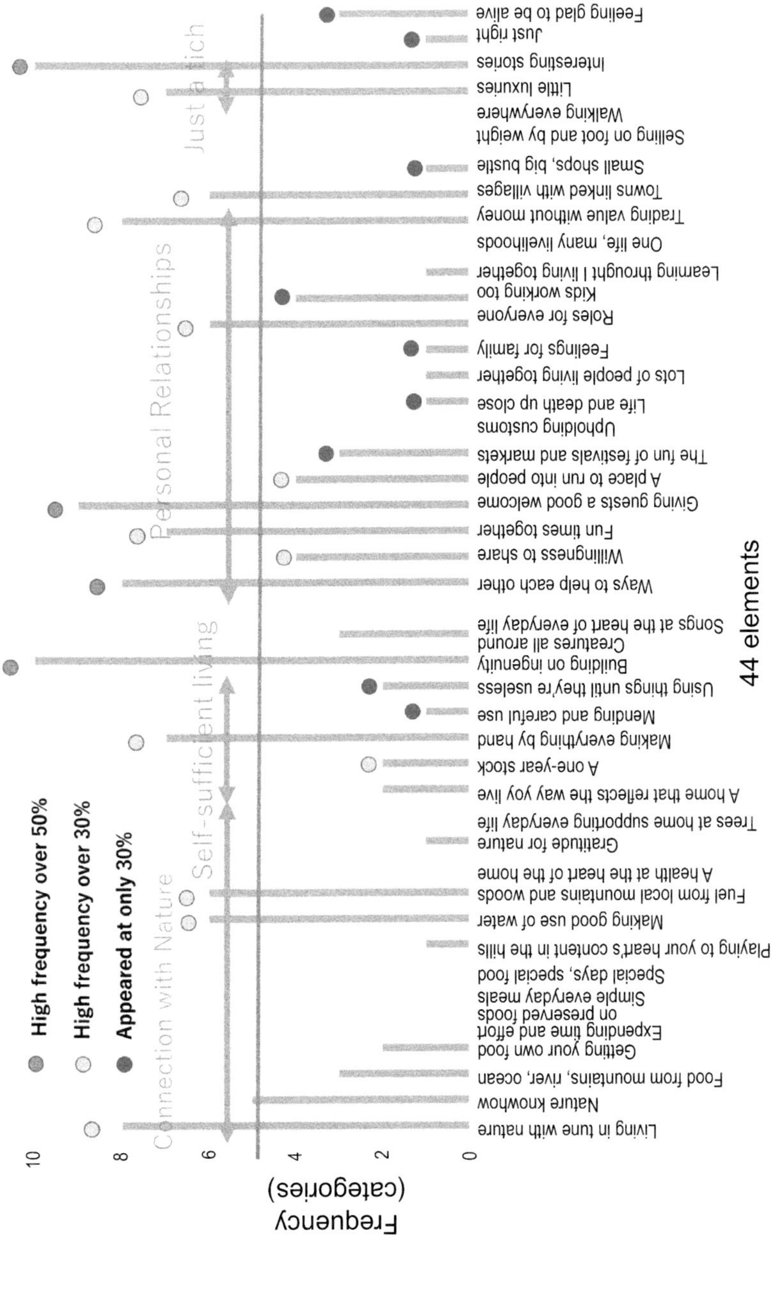

Figure 5.10 Distribution of the lifestyle values having the commonality against 10 categories.

everything by hand; *19. Building on ingenuity*; *22. Ways to help each other*; *24. Fun times together*; *25. Giving guests a good welcome*; *32. Roles for everyone*; *36. Trading value without money*; *37. Towns linked with villages*; *41. Little luxuries*; and *42. Interesting stories*. Therefore, suffice it to say that these can be considered to be important lifestyle values across various fields.

The expansion of human activity is at the heart of global environmental problems. We need to stem the spread of human activity and reduce it. But in order to lead spiritually affluent lifestyles without having to be made to feel that one has to struggle and persevere, we must find ways to fill the gap between a dependent and independent lifestyle. Our research is gradually casting light on the lifestyle elements that are important in this process (Table 5.2).

Table 5.2 Expansion of human activities will be stopped or shrunk without patience by controlling some of the 44 values (Filling the Gap)

○	1. Living in tune with nature		20. Creatures all around
●	2. Nature knowhow		21. Songs at the heart of everyday life
	3. Food from mountains, river, ocean	○	22. Ways to help each other
	4. Getting your own food		23. Willingness to share
	5. Expending time and effort on preserved foods	●	24. Fun times together
	6. Simple everyday meals	○	25. Giving guests a good welcome
	7. Special days, special food		26. A place to run into people
	8. Playing to your heart's content in the hills		27. The fun of festivals and markets
●	9. Making good use of water		28. Upholding customs
●	10. Fuel from local mountains and woods		29. Life and death up close
	11. A hearth at the heart of the home		30. Lots of people living together
	12. Gratitude for nature		31. Feelings for family
	13. Trees at home supporting everyday life	●	32. Roles for everyone
	14. A home that reflects the way you live		33. Kids working too
	15. A one-year stock		34. Learning through living together
●	16. Making everything by hand		35. One life, many livelihoods
	17. Mending and careful use	●	36. Trading value without money
	18. Using things until they're useless	●	37. Towns linked with villages
○	19. Building on ingenuity		38. Small shops, big bustle
			39. Selling on foot and by weight
			40. Walking everywhere
		●	41. Little luxuries
		○	42. Interesting stories
			43. Just right
			44. Feeling glad to be alive

● Over 5 of 10 categories <higher commonality>

○ Over 8 of 10 categories <highest commonality>

Using these lifestyle values as guidelines will prove effective in developing new businesses. As we make the transition from a dependent to an independent way of life, we will increasingly see these lifestyle values being realized.

How can we apply such lifestyle values in developing ways to fill the gap? We still need to give thought to this, but we believe the increase in indicators will definitely help inspire ideas for new businesses.

Acknowledgment

This work was supported by JSPS KAKENHI Grant number JP24120008 in Scientific Research on Innovative Areas: "Innovative Materials Engineering Based on Biological Diversity."

References

Alexander, S., and Ussher, S. The voluntary simplicity movement. *Journal of Consumer Culture*, **12**(1), 66–86 (2012).

Chittister, J. *Happiness*, Darton, Longman and Todd, London (2011).

Christensen, C. M. *How Will You Measure Your Life?* HarperCollins, London (2012).

Dreborg, K. H. Essence of backcasting. *Futures*, **28**(9), 813–828 (1996).

Elgin, D. *Voluntary Simplicity: Toward a Way of Life That Is Outwardly Simple, Inwardly Rich*. William Morrow and Company, New York (1981).

Gilbert, D. *Stumbling on Happiness*, Harper Perennial, London (2006).

Harleman, A. *Happiness*, University of Iowa Press, Iowa City (1994).

Hilty, C. (trans. Greenwood-Peabody, F.), *Happiness: Essays on the Meaning of Life*, Macmillan, New York (1903).

Hiroi, Y. *Posuto shihonnshugi* (Post-capitalism), Iwanami-shotenn, Japan (2015).

Ishida, E. H., and Furukawa, R. *Nature Technology: Creating a Fresh Approach to Technology and Lifestyle*, Springer, Japan (2013).

Leonard-Barton, D. Voluntary simplicity lifestyle and energy conservation. *Journal of Consumer Research*, **8**(3), 243–252 (1981).

Maeno, T., *Shiawase no mekanizumu* (Mechanism for the Fulfilling Life), Koudannsha, Japan (2013).

Messenger, L. *Happiness Is . . .*, Messenger Publishing, Sydney (2004).

Miyamoto, Y. *Miyamoto Yuriko Zenshuu Dai 21 Kan* (Complete Works of Yuriko Miyamoto, Vol. 21), Nikkei Business Publishing, Tokyo (2012).

The SPREAD Sustainable Lifestyles 2050 Project (2011–2012). http://www.sustainable-lifestyles.eu/

Vergragt, P. J., and Jansen, L. Sustainable technological development: the making of a Dutch long-term oriented technology programme. *Project Appraisal*, **8**(3), 134–140 (1993).

Yamada, H. (ed.). *Kaburagi Kiyokata Zuihitsushuu* (Recollections of Kiyokata Kaburagi), Iwanami Shoten, Tokyo (1987).

Yuki, T. *Tohoku wo Aruku* (Walking Tohoku), Shinjuku Shobo, Tokyo (2008).

Chapter 6

Matching Lifestyle to Technology

Yuko Suto[a] and Ryuzo Furukawa[b]

[a] *School of Engineering, Tohoku University, 6-6-04 Aramaki-aza-Aoba, Aoba, Sendai, Miyagi 980-8579, Japan*

[b] *Faculty of Environmental Studies, Tokyo City University, 3-3-1 Ushikubonishi, Tsuzuki-ku, Yokohama, Kanagawa 224-8551, Japan*

yuko@tohoku.ac.jp

To realize new lifestyles that don't yet exist we need to create new products, systems, and services. All of those will require new types of technology. But what kind? Lifestyle design focuses on everyday life, moments from an ordinary day. Naturally, it doesn't offer a concrete insight into potential technologies. A single "slice of life" image is surrounded by countless other such slices, each containing a myriad of necessary technologies.

Teasing out the technology that could shape such lifestyles first requires us to expand and carefully explore those lifestyles to provide a more concrete shape. This chapter will explain how to extract technology ideas by expanding and examining lifestyles, using a conceptualization method and/or an action decomposition approach based on ontology engineering.

This chapter will look at the system frameworks created for nature technology in Chapter 2. It will examine how we can extract the necessary technology from a lifestyle, and the process of learning those seeds from nature. Lifestyles seen through backcasting

Lifestyle and Nature: Integrating Nature Technology to Sustainable Lifestyles
Edited by Ryuzo Furukawa

ISBN 978-981-4774-62-8 (Hardcover), 978-1-315-14687-4 (eBook)
www.panstanford.com

(a concept explained in the previous chapter) are new ideas that do not yet exist. They require new technologies to be realized. In other words, we must extract the technological elements of those lifestyles. We must also discover whether hints for such technologies can be found in nature. We'll use two lifestyles created by backcasting as an example. First we will extract the necessary technology, then try to match it with technology "seeds" found in nature.

6.1 Use of the Simplest Method

For our backcasting lifestyle examples, let's turn to the 50 lifestyles imagined by 10 members of the 2012 NatureTech Research Consortium. We will use the scenarios that received greatest acceptability in a survey of a thousand men and women of all ages. Let's start with the following lifestyle idea:

Wooden Utility Poles as a Hub for All-Purpose, Small Generators (Utility Pole Lifestyle)

Utility poles are often unpopular and criticized as unattractive. However, their potential importance as an infrastructure for small generators has been noticed. Windmills or solar panels are attached to them to generate electricity. They can also use rain to create hydropower. The electricity travels along wires to provide individual homes with power. Rechargeable batteries attached to poles also store such power, before being used for mobile phones and game consoles. Large utility poles are capable of powering electric cars, helping to promote this type of vehicle.

This lifestyle achieved 71% social acceptability [1]. This is calculated by the ratio of people who said they would like to change their lifestyle to the one depicted. It's imagined to take place under severe environmental constraints of the kind predicted to come about by 2030. Even when appraised by current values, this lifestyle is an attractive prospect.

So, what technology would be needed to achieve this? Rereading the lifestyle description, we apparently need wooden utility poles with generators attached to them. We also require small wind-, solar-, and rain-powered generators on these poles. The produced electricity will not just be sent along wires; it will also be used to charge batteries which can be directly removed from the poles. That will require small-scale charging equipment. And finally, we need large chargers for electric cars.

So we've extracted the necessary technologies. Now, can we match them to technology "seeds" found in nature? We could turn to the support systems that will be examined in the next chapter, but it's simplest to start with existing databases of natural technologies, so this chapter will focus there. The NatureTech Research Consortium and Biomimicry Institute are currently building databases. However, in this chapter we'll use the Amazing Natural Showroom (http://nature-sr.com/) created by NatureTech. If you're based outside Japan, use an international database such as Ask Nature.

The Amazing Natural Showroom site currently has nearly 200 technologies from nature. This is the highest number in Japan. Each article focuses on four areas: "Why Is It Amazing?" "How Can We Use It?" "What Research Is Being Done?" and "'How Can We Develop It?" The database divides these natural technologies into 6 functions and 16 uses. You can read up on the functions and uses that seem related to the specific technologies that interest you. You can also search every category by using on keywords. If you're doing a keyword search, articles using that word will also be returned, which allows you to read up on related issues. We will do a broad keyword search and look for articles that seem helpful.

Next we find appropriate keywords for the technologies we extracted and do a search on them. Let's begin with "wind energy," "solar energy," and "rain energy." The articles returned are all focused on using wind, solar, and rain power. The term "generation" was not included in the search because natural mechanisms rarely "generate" electricity directly. It will also overlap with other articles we've already found. When searching for ways to charge and store electricity, we searched using the words "charge" and "store," but *do* not search using the word "electricity." Here are the results (Table 6.1):

Table 6.1 Showing the match between necessary technologies for "the Utility Pole Lifestyle" and technology "seeds" found in nature

Keyword search	Articles found	Actually related articles	Titles of related articles
Wind power	7	2	Humpback whale fins are designed to conserve energy Dragonflies suggest new wing shapes
Sunlight	10	6	How photosynthesis delivers electrons and creates energy Butterflies stay warm by collecting light through their scales Carbon nanotubes make artificial photosynthesis possible?! Plants that create clean energy Rainforest trees with roots above ground Light moves the maranta plant
Rain power	3	3	Rainforest trees with roots above ground Seed dispersal methods of subtropical rainforest herbs Fungi that jump to other leaves on rainy days
Charging	3	1	How desert lizards control temperature and rehydrate
Stored	11	1	Plants that create clean energy

Now let's look at another lifestyle designed by backcasting.

Domestic Electricity from Trash-Based Biofuel (Microbe Lifestyle)

All packaging and material used to wrap food and other products is now made from plant-based products. Each home keeps bacteria that decompose this packaging and transform it into liquid fuel. The invention of a device that burns this fuel to generate electricity means every home can meet its own energy needs. Packaging and paper is no longer discarded. All trash is taken home and fed to bacteria. Children and adults are now used to picking up trash on the street and taking it home. This has resulted in clean streets. There are different varieties of bacteria, some to deal with paper, others with kitchen waste, while still others create biofuel. Each variety has been selectively bred.

Social acceptability for this lifestyle is high at 63% [1]. As before, even when viewed with today's values, this lifestyle is attractive. The scenario is predicated on plant-based packaging which can be decomposed. In addition, it requires bacteria that can be kept at home and that decompose plant-based packaging and kitchen waste, and turn it into biofuel. The resultant fuel is turned into energy in the home, so we also need a home generator that uses biofuel.

Next we'll use the same database to search for possible technology "seeds" for these lifestyle technologies. Since "trash" is unlikely to be part of a natural mechanism, we used the keywords "plant matter" and "organic matter" instead. Since fuel would have the same problem, this was also changed to "organic matter." The words "decompose" and "generate" were also used. "Transform" was another keyword, since it sums up the main thrust of the lifestyle technologies. The results were as follows (Table 6.2):

Table 6.2 The match between necessary technologies for "the Microbe Lifestyle" and technology "seeds" found in nature

Keyword search	Articles found	Actually related articles	Titles of related articles
Decomposition	15	2	Gorgeous wild silk A 21st-century savior? Termites that love wood
Generate	8	1	A 21st-century savior? Termites that love wood
Transform	11	0	
Plant matter	4	2	Light, strong bamboo Okra's total control of water
Organic matter	7	2	Microalgae that make oil Biofilm: an extraordinary aggregate of tiny creatures

The next step is deciding which technology "seeds" discovered in the search can be of practical use. Since that takes us into the technology development stage, we won't discuss it in this chapter. The first step has been achieved: we've found some seeds.

6.2 Why Expand and Dig into Each Lifestyle?

We've just looked at how to extract three or four technologies and their keywords from two lifestyle explanations. But it would be a waste to stop there. Why? Because there's many more technologies waiting to be discovered.

As the word makes clear, a "lifestyle" is a way of living, and it paints a scene of everyday life. When designing a lifestyle, its creator has an image in their mind's eye, whether or not they're aware of every detail. What kind of environmental constraints and social issues are they trying to solve? Which spiritual affluence are they trying to get? These concerns will be part of the scene that they create. When turning that picture into a short paragraph, they become very specific. Location: Are they at home or outside? Are they at work? The local neighborhood? On vacation? Time: Is it morning, midday, or night? Is it a weekday or a weekend? People: Is this the lifestyle of a married couple? Children or the elderly? Are they local residents or visitors? All of these choices must be made for the lifestyle to be described in a few sentences. Without clarity, the overall design gets lost.

However, when it comes to extracting technology, that specific, concentrated lifestyle has to be unpacked to cover every location, every time, and every possible person. We also need to make the actual flow of events more specific for each of those locations, times, and people. In other words, we use that single "scene from life" that we already described to reveal details about the many similar scenes of life that surround it. That gives us a broader and more specific set of technologies that is as close as possible to the original designer's image. This requires research into methods of conceptualization and action decomposition tree.

6.3 Expanding Lifestyles: Method of Conceptualization

Concepts occupy a space between the concrete and the abstract. One way to expand a lifestyle is to examine how such concepts evolve (Fig. 6.1). As mentioned earlier, a lifestyle design comes from

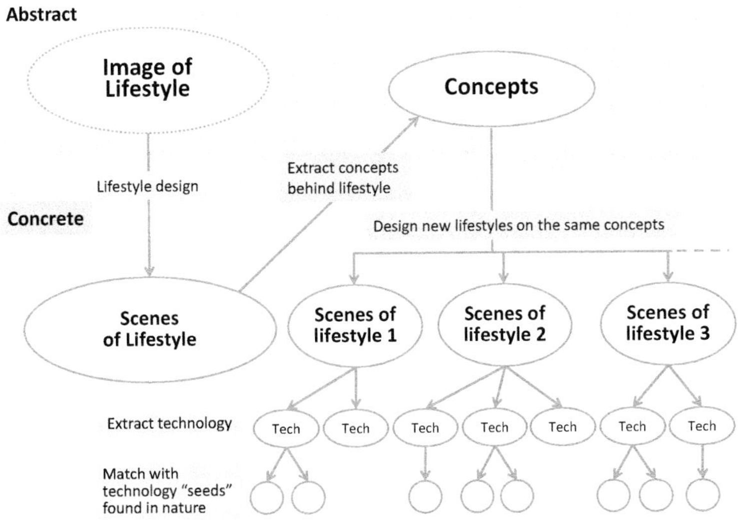

Figure 6.1 Outline of conceptualization method [1].

a mental image of an abstract world. A few short sentences make it concrete, defining location, time, and people. We enter a more specific world. By laying out concrete ideas, the lifestyle becomes sharper and clearer, but it does not reveal the entire lifestyle that was in the designer's mind. So we need to return to that abstract world and discover the core concept behind it. Then we make our way back to the specific and write out several lifestyles based on the same concept. This method of conceptualization helps to expand the original lifestyle.

A key precept when extracting a concept is that it must not be too concrete or restrictive. Take a lifestyle where people share the batteries they've charged in the home with their neighbors. The concept for this should be "sharing everyday items." If this is reduced to "sharing batteries," the designer will be limited. It's much harder to create similar lifestyles with the same concept.

We'll take a closer look at these methods using the two lifestyles introduced in Section 6.1. Conceptualization was carried out by 10 members of the NatureTech Research Consortium, including the two persons who created the example lifestyles.

6.3.1 Finding the Key Concept behind a Whole Lifestyle (Going Back to the Abstract)

First each person finds a key phrase that could serve as a concept for the lifestyle. They bring those ideas to a group discussion, collating similar concepts before deciding on a final main concept. Group discussion helps everyone share an understanding of the environmental restrictions and social issues the lifestyle hopes to overcome. It also allows discussion of the spiritual affluence, overall goal, and the nuances surrounding the lifestyle.

6.3.1.1 Conceptualization of the Utility Pole Lifestyle

As the chart (Fig. 6.2) shows, several key phrases were brought to the discussion: "making, storing, charging electricity anywhere," "people gather at the utility poles," and "systems for generating and storing electricity throughout town." These were discussed in a group to produce more key phrases, creating a total of 32. Some of the concepts people saw between the lines of the lifestyle description overlapped, while others did not. For instance, one person saw the shared energy and the potential to generate it anywhere, but did not foresee that people would gather at utility poles. This is why group discussion is important; diverse visions of the lifestyle can be integrated and shared.

The 32 key phrases were discussed again in order to collate similar ideas. The core idea lay in power generation, storage, and charging "anywhere, anytime." This concept would be "multi-functional," "shared," and "accessible/easy to use." Accessed "on demand" by "your choice," it would be part of "the landscape." The 11 keywords in the chart below emerged from the original set of 32.

Finally, these keywords were collated to produce the following two key concepts:

"Power stations available anywhere, at any time: utility poles are fun places where people gather"

(This emphasizes "anywhere, anytime," "fun," and "gather")

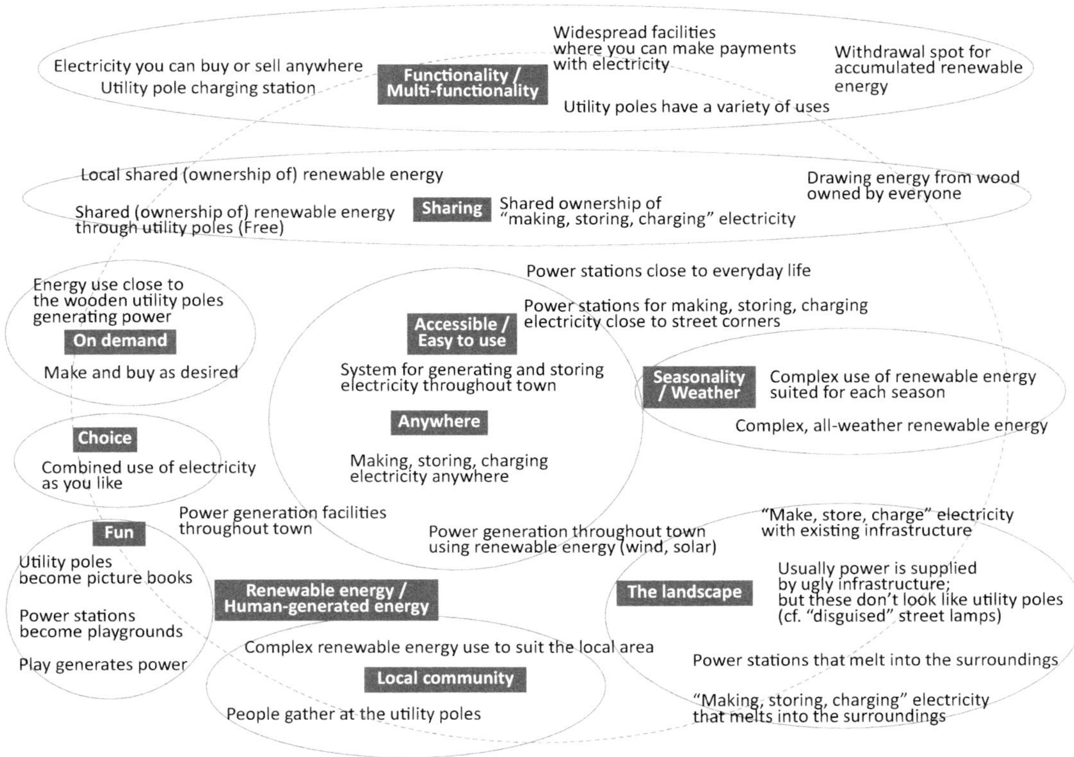

Figure 6.2 Collated keywords for the Utility Pole Lifestyle from the group discussion.

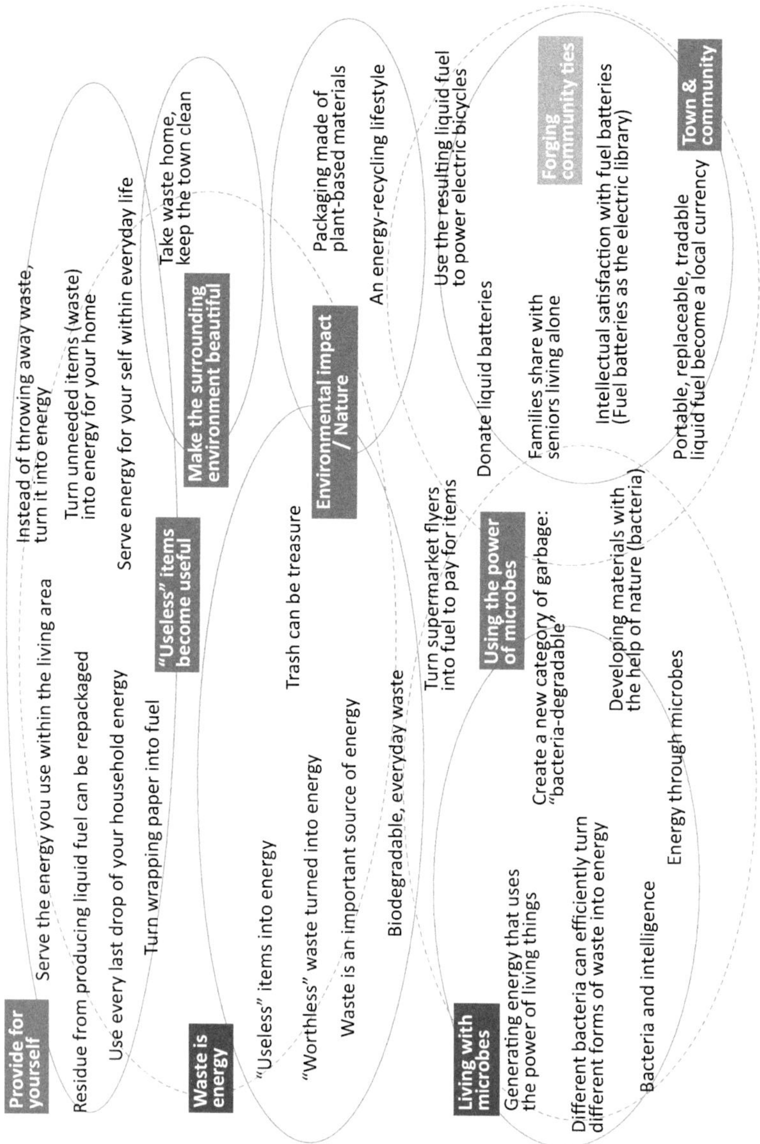

Figure 6.3 Collated keywords for the Microbe Lifestyle form the group discussion.

"Power stations available anywhere, at any time: utility poles are places to help and communicate with others"

(This emphasizes "anywhere, anytime," "community," and "gather")

6.3.1.2 Conceptualization of the Microbe Lifestyle

Phrases brought to the discussion included "the energy you use serves your lifestyle," "taking waste home keeps the town clean," and "waste into energy." In the same way as before, a group discussion created more key phrases. A total of 28 phrases emerged, which can be seen below (Fig. 6.3).

Similar ideas were collated into nine keywords: the emergence of a material with low "environmental impact," "living with microbes," and "using the power of microbes" mean that "waste is energy." "Useless items become useful," and people can "provide for themselves."

Finally these keywords were collated to two key concepts:

"Previously useless waste is now a household resource"

(This emphasizes "waste is energy," "useless items become useful," and "provide for yourself")

"Keeping microbes at home to keep the streets clean"

(This emphasizes "living with microbes," "using the power of microbes," "make the environment beautiful," "environmental impact and nature," and "town and community")

6.3.2 Create New Lifestyles with the Same Concept (Returning to the Concrete)

Each of the two lifestyles we've examined have been assigned two key concepts each. Now each person draws up other lifestyles that fall under those key concepts. Although about 10 new lifestyles were created for each concept we'll focus on two examples for each concept.

"Power stations available anywhere, at any time: utility poles are fun places where people gather"

Human Power Generation at Utility Poles

Utility poles are hubs for electrical power. Many people use these poles in various ways. Stores and services offer lower prices to customers who've generated power at a nearby utility pole and sent it to that establishment. People heading to bars for a drink after work first stop at a nearby pole generator to do some exercise. The power they generate lowers their bill at the end of the night. Moms eating out with their kids let the kids work off some energy playing at a utility pole near the restaurant. Since the pole can be seen through the window the kids can play as long as they want. Meanwhile parents can enjoy a long chat without the kids complaining. People going to a hairdresser will head to a pole to cycle for an hour before their reservation. They'll generate enough power to run the hairdryer at the salon. Shops and services get this power at a much cheaper rate than from the electric company, so they welcome the extra generation.

Children's Play Generates Power

Swings, seesaws, slides, and other playground equipment generate power when played with by children. The electricity flows to the park utility pole and is collected in batteries that can be taken home and used for a grandmother's low-frequency massager, or in a child's game console. At the local community center are running and cycling machines where kids can generate electricity and stay healthy through games. Game scores are recorded, with local and national competitions held to encourage users. Excess electricity is stored in batteries that are kept at the community center. The center doubles as an emergency evacuation hub.

"Power stations available anywhere, at any time: utility poles are places to help and communicate with others"

Utility Poles Support Digital Nomads

Since utility poles can be found everywhere, and use a variety of natural energies to power their small generators, people can now travel much further in electric cars with no concerns about powering their ride. This leads to a rise in telecommuting with people using smartphones and computers to work in parks and other places where they feel connected to nature. Eventually, people who get along will gather at the same utility pole. Strangers will strike up conversations. People exchange information on utility poles that provide a full charge on windy days, or which have extra power available at the weekend. Natural power generation is now shared between neighborhoods.

Gifting Electricity to Schools

Utility poles now have chargeable batteries and are surrounded by equipment that permits human power generation. Generating electricity earns a person points which can be donated to local elementary and junior high schools. The schools can certainly use the points against their electricity bill or to buy gym equipment or musical instruments. Parents who are too busy to attend school events will make up for it by donating points. They might visit utility poles near their office in between meetings, or while on business trips. Elderly residents also donate points to the school and in return are invited to school concerts and sports events where they can eat with the young students. Once they know each other the children and elderly residents begin to chat near utility poles as they exercise to generate electricity.

"Previously useless waste is now a household resource"

Fuel Cartridges Using Domestic Waste

Nearly all combustible rubbish is now used for household fuel since the spread of small-scale solid fuel manufacturing devices. Trash is thoroughly dried then put inside these machines to produce cartridges of solid fuel. Clean and portable, they provide safe energy for heating, cooking, and other uses. Any waste you produce can be consumed by your household as fuel. While those who don't produce much waste may be concerned they'd run out of fuel, cartridges last a long time, unlike electricity. Tidying the garden will produce lots of branches and leaves, organizing a room will result in unwanted items. Stock up on cartridges at these moments to keep your supply of fuel steady.

Children Recycle at Home

Most families have two working parents, so children play an active role in household chores. They might make paper-based necessities such as tissue paper and toilet paper. How do they do that? Well, manufacturers sell "Toy Creation Sets," which contain simple recycling kits. Materials are placed inside so kids can simply and safely create necessities. The materials will be waste paper produced by the home. Parents are delighted at the lack of waste paper, and at the reuse of resources. Children not only help out but enjoy adding different colors and patterns. The work promotes creativity and a feel for beauty—another reason the kits are popular. A variety of recycling kits are developed and sold under the banner "Little Helping Hands: Eco Series."

"Keeping microbes at home to keep the streets clean"

Popular Bacteria Kits Help Children Learn and Play

Children get up early on Saturday mornings to collect trash. Home-study courses offer free bacterial growth kits for biodegrading waste. The bacteria have to be fed on the various scraps of waste found on the street. Kids compete over who can collect waste more quickly, and take it home to feed their bacteria. The kits help them learn that certain types of bacteria work more efficiently with certain types of waste. Collecting and sorting waste becomes a popular hobby. Parents were initially unhappy that kids were bringing home trash, but chic bags for transporting waste, less trash on the streets, and the safety of the bacteria sets mean they're now happy with the situation. The streets are cleaner and kids learn through play as they raise their bacteria.

Everyone Raises Multipurpose Bacteria

Projects using bacteria as a way to promote picking up trash are everywhere. The trash is decomposed and transformed into a resource, and into energy. This in turn has led to a lot of work in selective breeding of bacteria. The most common are bacteria that decompose waste and turn it into fuel. Researchers are finding ways to improve fuel power and to break down lots of trash at once. Schools hold exhibitions and contests of bacteria raised at home. There are competitions over decomposition volume, and bacteria have become very familiar to most people. Children who became used to bacteria very young maintain their interest, and more kids dream of becoming researchers. New varieties of bacteria remove smells, or function as lights through luminescence. Japan becomes a global leader in bacterial research.

6.3.3 Matching Redesigned Lifestyles and Technology

Now it's time to extract technologies from the newly designed lifestyles we've just created. As can be seen below (Tables 6.3 and 6.4) some of these technologies will overlap, but as with the original lifestyle examined in Section 6.1, about five technologies can be found within each paragraph. The result of conceptualization is to double the number of new lifestyles. With the Utility Pole Lifestyle idea, the original lifestyle before conceptualization gave us four technologies. After conceptualization, another four lifestyles created 14 technologies. With the Microbe Lifestyle concept, we went from three technologies to 11 after designing four additional

Table 6.3 The match between necessary technologies for the post-conceptualization Utility Pole Lifestyle and technology "seeds" found in nature

Keyword search	Articles found	Actually related articles	Titles of related articles
Human Power Generation at Utility Poles			
Rotation	14	3	New fan shapes: learning from natural spirals in eddies Bacteria flagella are the world's smallest and strongest motors New wave-power generation from giant kelp
Pressure	15	2	A Venus flytrap snaps its mouth closed No stomach for pressure? The *Mimosa pudica*
Identification	1	0	

(*Continued*)

Table 6.3 (*Continued*)

Keyword search	Articles found	Actually related articles	Titles of related articles
Children's Play Generates Power			
Rotation	14	3	New fan shapes: learning from natural spirals in eddies Bacteria flagella are the world's smallest and strongest motors New wave-power generation from giant kelp
Charging	3	1	How desert lizards control temperature and rehydrate
Measuring	2	0	
Identification	1	0	
Stored	11	1	Plants that create clean energy
Utility Poles Support Digital Nomads			
Charging	3	1	How desert lizards control temperature and rehydrate
Information	19	5	Weevils reveal the potential for light computer chips Venus's flower basket creates delicate masterpieces in glass Rainforest trees with roots above ground Buzzer midge young survive without air The smoke sensors of jewel beetles
Gifting Electricity to Schools			
Identification	1	0	
Choice	3	1	Bacteria flagella are the world's smallest and strongest motors

lifestyles after conceptualization. We'll use the same method of keywords, as explained in Section 6.1, to look for seeds for these technologies in nature. The results are also shown in Tables 6.3 and 6.4.

Table 6.4 The match between necessary technologies for the post-conceptualization Microbe Lifestyle and technology "seeds" found in nature

Keyword search	Articles found	Actually related articles	Titles of related articles
Fuel Cartridges Using Domestic Waste			
Transport	4	1	Feather ticks use a hooklike limb for stabilization
Warmth	13	2	Warm but smell: a skunk cabbage home Butterflies stay warm by collecting light through their scales
Children Recycle at Home			
Recycling	4	1	Recycling their own selves: secrets of the crayfish molt
Paper	8	0	
Popular Bacteria Kits Help Children Learn and Play			
Biodegradation	1	1	Gorgeous wild silk
Breeding	11	3	Warm, resilient down Microalgae that make oil The interferon factory of silk worms
Cleanliness	2	2	Wax rolls aphid nectar Clean, healthy algae rom furanone
Everyone Raises Multipurpose Bacteria			
Odor	5	1	The comfortable nests of termites
Decomposition	15	3	Antimicrobial peptide melittin in bee venom Biofilm: an extraordinary aggregate of tiny creatures The comfortable nests of termites
Luminescence	2	1	*Papilio nireus* decked out in the latest LEDs

By conceptualizing the original lifestyles we reveal new lifestyles that share the same core concepts. It provides us with a clearer, broader vision of different aspects of everyday life. However, an increase in scenes of life does not always lead to an increase in natural technology seeds.

6.3.4 Broadening Lifestyles Even Further (Focusing on Key Technologies and Illustrations)

As explained above we've broadened our understanding of lifestyles by conceptualizing them. This led us to the technology seeds that lie in nature. Although we used the Amazing Natural Showroom as an example, engineers and researchers can use a variety of databases and information sources. The NatureTech Research Consortium focused on the Microbe Lifestyle in particular. Members accessed other information sources and discovered other technology seeds which helped to expand their lifestyle ideas.

One such technology seed was the turkey tail mushroom. This fungus is known for decomposing waste and transforming it into bioethanol. The mushroom was quickly placed as a key technology for the Microbe Lifestyle. A key phrase for the lifestyle's conceptualization was "provide for yourself." Naturally, this led to a household providing for itself or a neighborhood providing for itself. Key phrases from the Utility Pole Lifestyle conceptualization were "people gather" and "communication." These elements were added to this new idea: local residents bring in their household waste and takehome bioethanol made by fungi. While they wait they grab a coffee and chat to friends. The hub for this activity would be an "Upcycle Station."

The group also imagined the scenes of everyday life that would surround this idea. Waste is usually brought in by mothers in the morning after seeing off the kids at the school bus stop. Once school is over, kids arrive and generate power while playing on the grounds. Local residents walking their pets will stop for a cup of tea and their pets will generate power through play. In the evening people coming home from work will rack up points by generating power at a sports bar, then cashing in those points for a beer. On weekend mornings a local grandpa who's an experienced farmer goes to an allotment next to the center. He tells kids what to do with the residue left over after the fungi have made bioethanol. All these ideas are shown in the illustration (Fig. 6.4). The written lifestyle explanation is below:

Figure 6.4 Illustration of an upcycle station.

Upcycle Stations

"Good morning!" People arrive holding household waste. At the upcycle station, this much-hated waste is turned into energy and fertilizer. The stream of people led to a park and café being opened so they can relax. The fertilizer is used in the very popular vegetable patch. The park and aerobics equipment are linked to a power generator and the electricity is stored in the station's pre-existing system. The energy generated earns the exerciser points so more people are coming just for that, instead of only when they have waste.

By extracting technologies and matching potential seeds from nature, an illustration of key technologies helps flesh out the original lifestyle. The group can extract new technologies from this idea and run more matches. However, this larger expansion is only possible because all the members worked together on the whole process: conceptualizing the original lifestyle, creating new, similar lifestyles, and matching technology seeds.

6.4 Digging Deep into Lifestyles: Method of Action Decomposition Tree That Introduces Ontology Engineering

Another way to expand and dig into a lifestyle is the method of action decomposition tree (Fig. 6.5). This introduces ontology engineering and creates a structure for lifestyles. Ontology engineering is very effective at providing a shared vocabulary, as well as a medium, or structure, for sharing and understanding knowledge. A function decomposition tree can be used to understand production facilities or functionality; it provides a systematic view of the functions of an artifact [2, 3]. With human behavior there are psychological elements that are separate from such functions of artifacts. However, lifestyles that include such intangible aspects can also be examined through ontology engineering, and an action decomposition tree can be made. This will give us a systematic structure for that lifestyle.

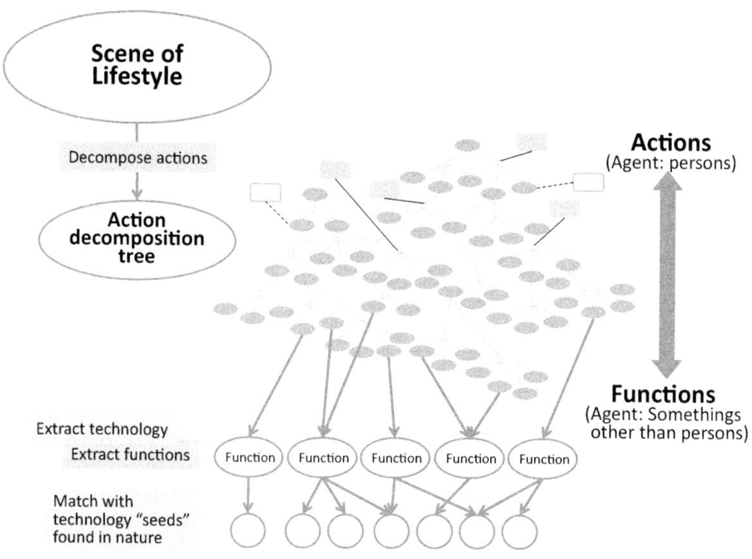

Figure 6.5 Outline of action decomposition approach.

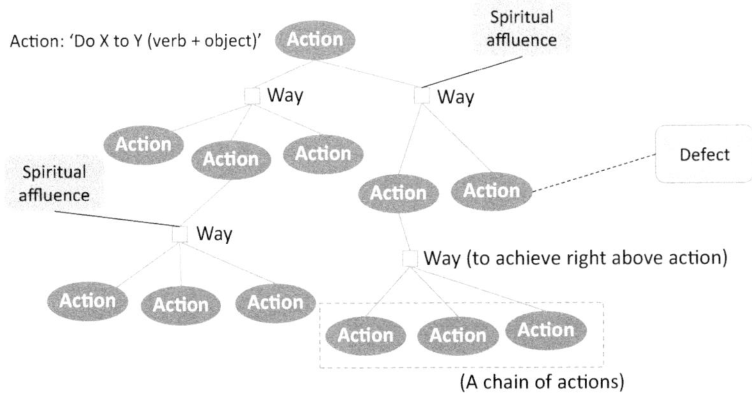

Figure 6.6 Structure of action decomposition tree.

6.4.1 Creating an Action Decomposition Tree

An action decomposition tree expands on a function decomposition tree of artifact in order to express aspects of a lifestyle. As the chart below (Fig. 6.6) shows, it's a tree structure made up of actions and the ways and sub-actions needed to achieve those actions. Each action is expressed with a simple phrase: "do X to Y (verb + object)." The action is "what to achieve" (goal) and the way is "how to achieve that goal." When making such a tree, each action is expressed using three types of action node. One is used for simple actions. (This is called an "action" node and will be in blue). Another is an action node which causes discomfort or problems for society, the environment, or residents. (This will be called the "defect" node and will be yellow.) The third node is for actions which should lead to spiritual affluence for residents. (This is the "spiritual affluence" node, in green.) This third, spiritual affluence node is a special node added in order to address the aspects of an action decomposition tree for lifestyle.

The action, defect, and spiritual affluence nodes, along with way nodes, help us structure the actions that will make up a lifestyle. It also creates a visual reference for the environmental constraints, social issues, and spiritual affluence that give rise to the lifestyle. The top of the decomposition tree starts with broad, conceptual actions. As we travel down, it should show more specific, detailed actions.

The more we decompose our actions, the deeper we can dig into the necessary actions for each lifestyle.

When it comes to the three types of action nodes, the agent acting in each case is a person. If a specific person such as an elderly person or child is required, then that should be attached to the node. As action decomposes continue, the agent will shift from humans to something other than humans, such as a mechanical device or a container. This will reveal a function decomposition tree. If the agent is not a person, this should also be attached to the node.

In order to begin the decomposition, take another look at the lifestyle text and extract actions in the form "do X to Y." Then add the actions needed to bridge the gap until the next action. In other words, reveal the other, hidden actions that are needed for the described actions to take place. Naturally, part of the process will involve changing some of the initial actions you found in the text, or readjusting other actions to make them simpler. Since the work involves adding new ways in order to achieve certain actions, this will also help expand the original lifestyle.

Let's use the two lifestyles introduced in Section 6.1 and create action decomposition trees for both. Both trees were created by four members of the Graduate School of Environmental Studies at Tohoku University.

6.4.1.1 Making an action decomposition tree for the Utility Pole Lifestyle

Actions we can find in the lifestyle text include "put windmills on utility poles," "put solar panels on utility poles," "generate wind power," "generate solar power," "generate rain power," "put chargeable batteries on utility poles," "store electricity," "send electricity to local homes," "use power for electronic devices," and "use power to charge electric cars." This is what our action decomposition tree will be based on. The resulting tree in its entirety is shown in Fig. 6.7.

This lifestyle involves a variety of ways to generate power: wind, sunlight, and rainwater. It also calls for a variety of different generator types and use cases. However, the required flow of generating power from natural energy and using it is the same.

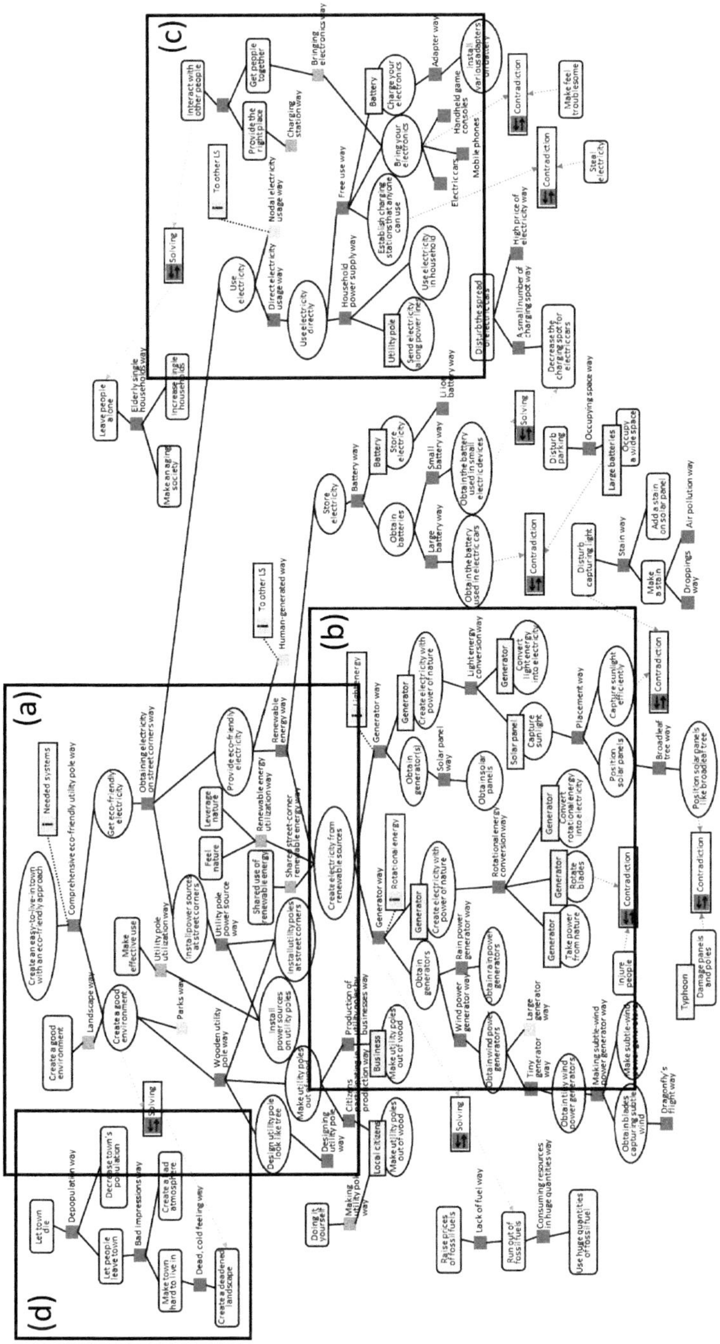

Figure 6.7 An action decomposition tree for the Utility Pole Lifestyle.

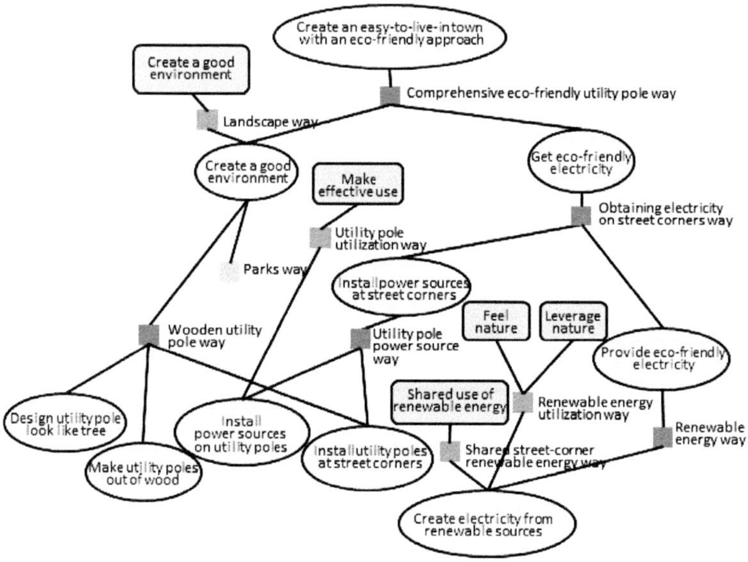

Figure 6.7 (a) A part of the action decomposition tree for the Utility Pole Lifestyle.

So the higher-placed actions should be "install power sources in utility poles," "create electricity from renewable sources," and "use electricity" (Fig. 6.7a). Remember, we also inferred that this lifestyle calls for human power generation as well as power generation from natural resources. Another top action is "provide eco-friendly power." This can be achieved by "renewable energy way" and "human-generated energy way." Also note that power sources are not only located at utility poles. We can infer that they might be placed in other locations, and therefore add the action "install power sources at street corners." Add a way node for "utility pole power source way." The lifestyle also specifies wooden utility poles because it is interested in more than just a convenient supply of electricity; it also wants to preserve the landscape. Another high-placed action should be "create an easy-to-live-in town with an eco-friendly approach." We have considerably expanded on the original lifestyle by examining the granularity of each action to see if it can be broadened, then examining each action to see if different ways could achieve the same outcome.

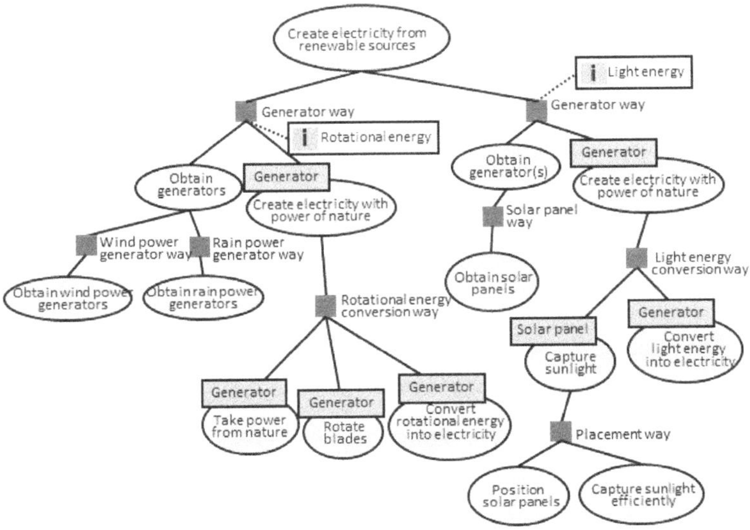

Figure 6.7 (b) A part of the action decomposition tree for the Utility Pole Lifestyle.

Meanwhile, actions in the lower sections can be decomposed further. Take solar power generation. Under "obtain generator" is "obtain solar panels." Below "create electricity with the power of nature" is "capture sunlight" and "convert light energy into electricity" (Fig. 6.7b). To achieve "capture sunlight" we need two further actions, "position solar panels" and "capture sunlight efficiently." When it comes to using power, we have two end-use cases. One where electricity is sent to homes and the other where it's used at the pole. This is split into two sections. When power is used on-site, it's decomposed into a series of actions: "free-use way," "establish charging stations that anyone can use," "bring your electronics," and "charge your electronics" (Fig. 6.7c).

In this way simple actions and ways reveal a visual guide to the necessary steps and structure to create this lifestyle. However, a lifestyle also solves environmental constraints and social issues and promotes spiritual affluence through action. This also needs to be added to the tree.

Let's consider what kind of environmental constraints and social issues this lifestyle is addressing. It describes using renewable

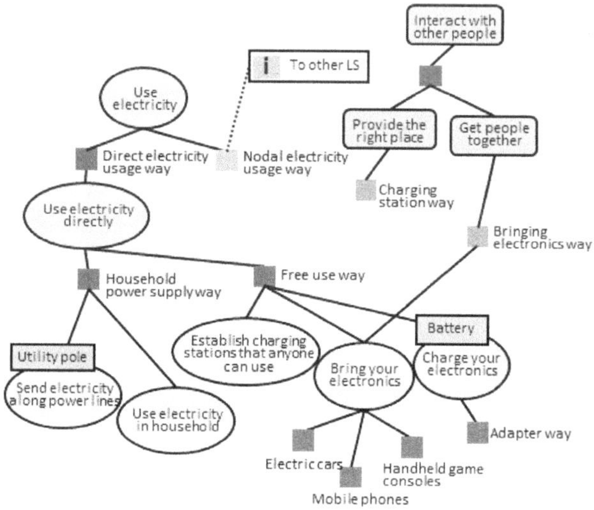

Figure 6.7 (c) A part of the action decomposition tree for the Utility Pole Lifestyle.

energy, including wind, solar, and rain power. This addresses the future shortfall of fossil fuels. So we add a defect node called "run out of fossil fuels" and link it to the solution, "create electricity from renewable sources." In order to achieve "run out of fossil fuels," we need to add an action: "use huge quantities of fossil fuels." If "run out of fossil fuels" takes place, then an obvious result is "raise prices of fossil fuels." The use of wooden utility poles implies a desire to build a more natural vista, instead of one of concrete and metal. Let's add "create a deadened landscape" as a defect node and link it to the solution, "create a beautiful landscape." "Create a deadened landscape" leads to "make town hard to live in" and "create a bad atmosphere," which lead to "let people leave town." That in turn takes us to "decrease town's population" and "let town die." Using defect nodes in this way creates a visual reminder of the environmental and social issues on which the lifestyle is predicated.

So what about the spiritual affluence the lifestyle should promote? Since power is generated by wind and sunlight, we achieve "leverage nature" and "feel nature." These are achieved through the action of "create electricity from renewable sources," so let's

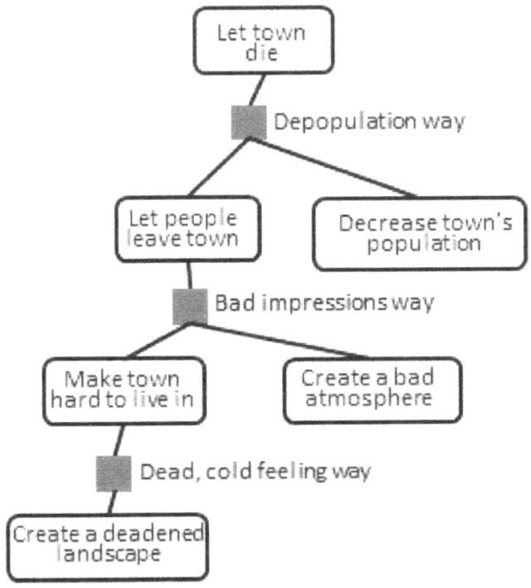

Figure 6.7 (d) A part of the action decomposition tree for the Utility Pole Lifestyle.

link those together (Fig. 6.7a). The efficient use of infrastructure through utility poles means "install power sources on utility poles" leads to the spiritual affluence node of "make effective use." The chain of actions "establish charging stations anyone can use," "bring your electronics," and "charge your electronics" links to the spiritual affluence node of "interact with other people" (Fig. 6.7c). Introducing spiritual affluence nodes reveals aspects of affluence that are hidden in the original text, and gives a visual guide on the actions that might bring that affluence.

6.4.1.2 Making an action decomposition tree for the Microbe Lifestyle

The lifestyle text reveals the following actions: "make packaging from plants," "decompose packaging and convert it into liquid fuel," "keep bacteria at home," "combust liquid fuel to make electricity," "create all necessary electricity at home," "pick up rubbish," "give waste to bacteria," and "breed improvements to bacteria." The resulting tree in its entirety is shown in Fig. 6.8.

Although "create all necessary electricity at home" should be placed highly as an overall concept, it's also possible that each home

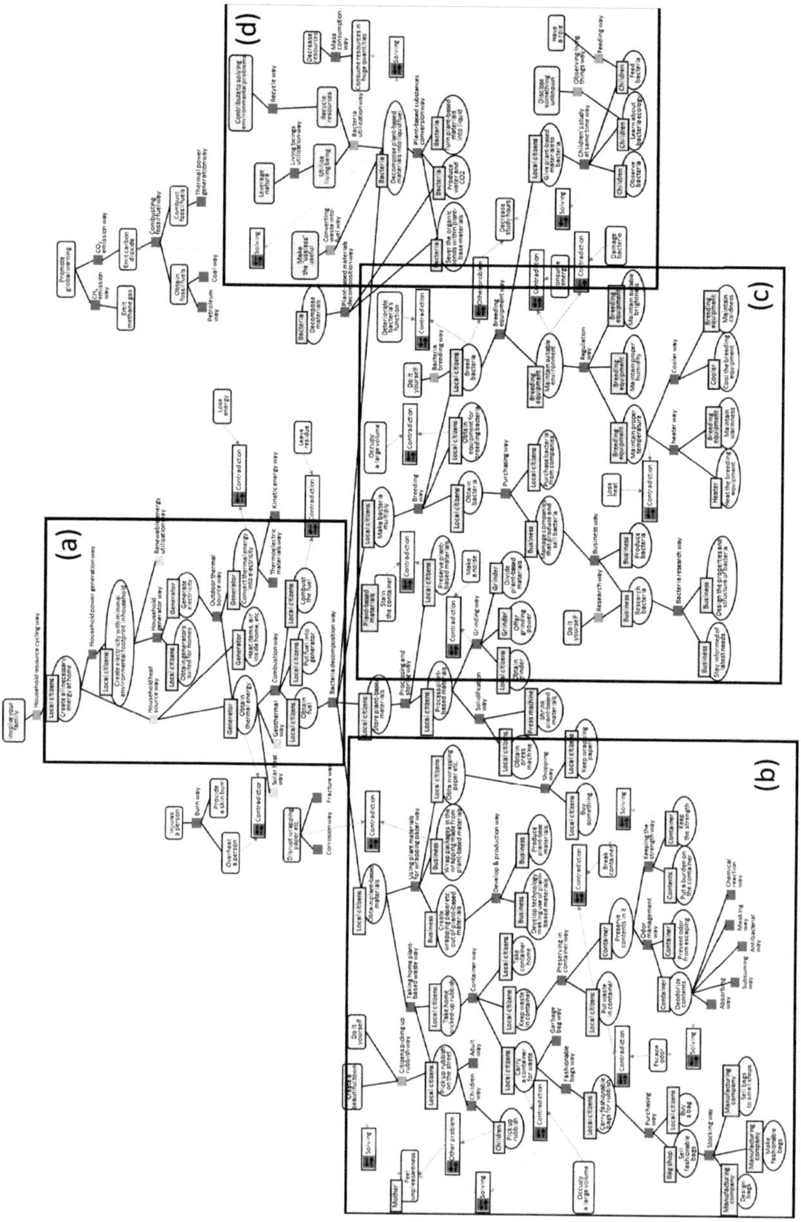

Figure 6.8 An action decomposition tree for the Microbe Lifestyle.

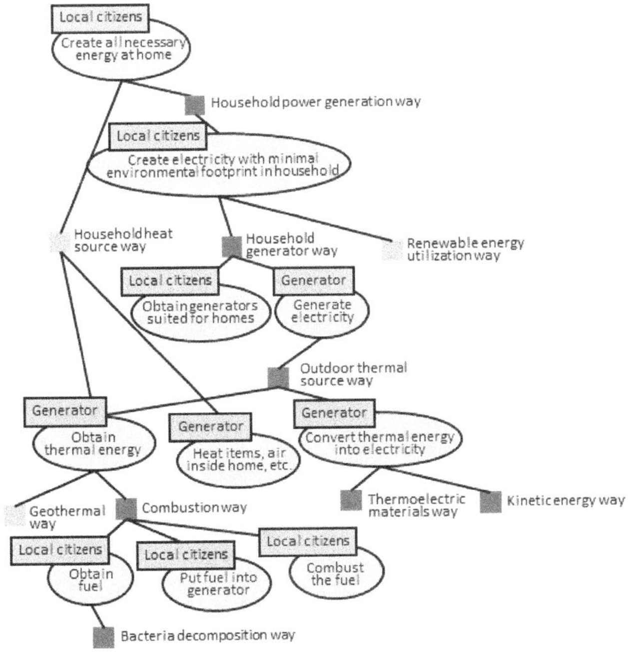

Figure 6.8 (a) A part of the action decomposition tree for the Microbe Lifestyle.

will generate other forms of energy. So "create all necessary energy at home" is added even higher (Fig. 6.8a). There's a gap for the process of burning and generating fuel between "combust liquid fuel to make electricity" and "create all necessary electricity at home." We can decompose that process and add it to the tree. "Combust liquid fuel to make electricity" can be decomposed into "obtain fuel," "put fuel into generator," and "combust the fuel." This leads to "obtain thermal energy."

The action "obtain fuel" can be decomposed further into a chain of actions as the "bacteria decomposition way": "obtain plant-based materials," "store plant-based materials," "make bacteria multiply," and "decompose plant-based materials into liquid fuel." To achieve the action "obtain plant-based materials," we have the "taking home plant-based waste way." This can be divided into "pick up rubbish on the street" and "take home picked-up rubbish."

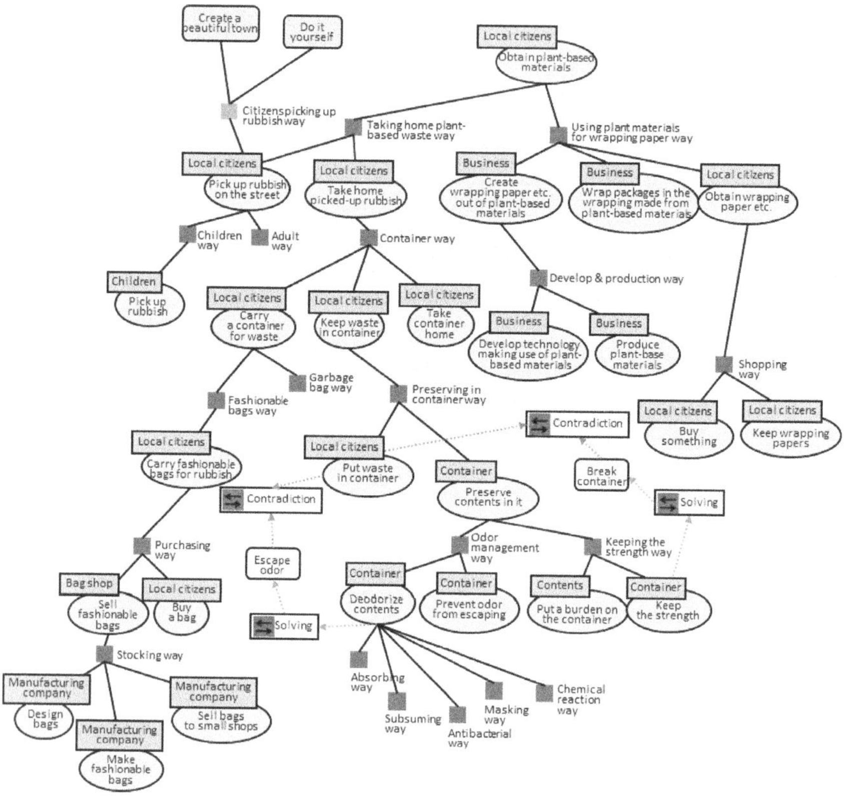

Figure 6.8 (b) A part of the action decomposition tree for the Microbe Lifestyle.

"Take home picked-up rubbish" can be further decomposed into "carry a container for waste," "keep waste in container," and "take container home" (Fig. 6.8b). The "breeding way" for "make bacteria multiply" includes "obtain bacteria," "obtain equipment for breeding bacteria," and "breed bacteria" (Fig. 6.8c). This final action leads to "maintain suitable environment" and "give plant-based materials to bacteria." "Maintain suitable environment" includes "maintain proper temperature," "maintain proper humidity," and "maintain suitable brightness." "Give plant-based materials to bacteria" involves "observe bacteria," "learn about bacteria ecology," and "feed bacteria" (Fig. 6.8d).

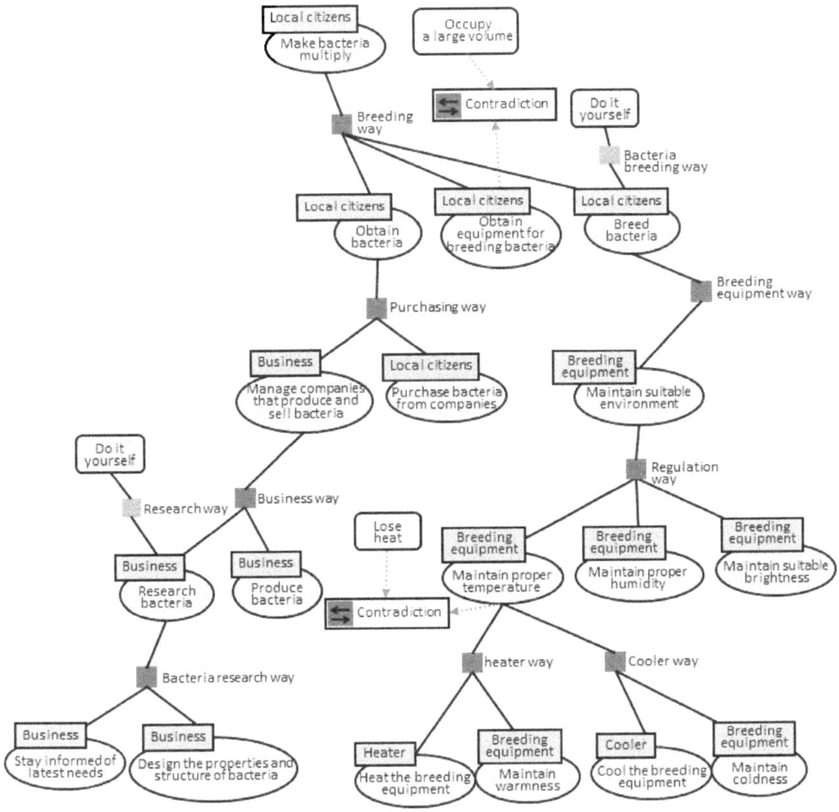

Figure 6.8 (c) A part of the action decomposition tree for the Microbe Lifestyle.

Now let's add environmental constraints, social issues, and spiritual affluence to the tree. This lifestyle describes recycling waste into liquid fuel to generate electricity. This tackles the problem of "consume resources in huge quantities" and provides affluence by "contributing to solving environmental problems" (Fig. 6.8d). By raising bacteria and generating power at home, people enjoy the affluence that comes from "doing it yourself" and "having a role" (Fig. 6.8c).

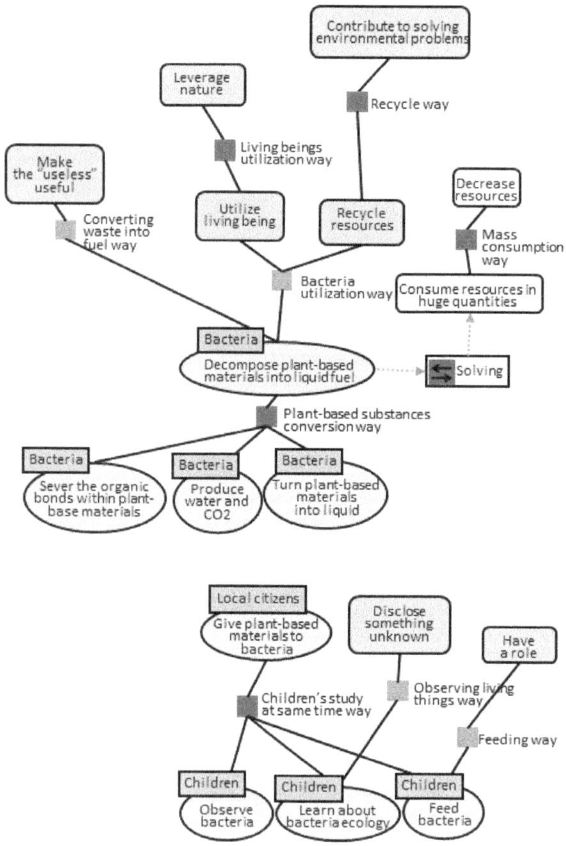

Figure 6.8 (d) A part of the action decomposition tree for the Microbe Lifestyle.

6.4.2 Technology Matching on the Functions of the Action Decomposition Tree

First extract the functions from the action decomposition tree; functions are those actions in which something other than people is acting. In the Utility Pole Lifestyle, we can find nine such functions, and 16 in the Microbe Lifestyle. Use the "do X to Y" phrase to find a search string and use the "Amazing Natural Showroom" of Section 6.1 to do a search.

Table 6.5 The match between the functions of the Utility Pole Lifestyle action decomposition tree and technology "seeds" found in nature

Keyword search	Articles found	Actually related articles	Titles of related articles
Take (mechanical)	18	2	The water absorption secret of kombu and wakame kelp The limber pine unbroken by strong winds
Rotation	14	3	New fan shapes: learning from natural spirals in eddies Bacteria flagella are the world's smallest and strongest motors New wave-power generation from giant kelp
Take (light)	18	2	How photosynthesis delivers electrons and creates energy Light-sensing material shared from archaeans to humans? Insect compound eyes see a broad field
Store	11	1	Plants that create clean energy
Send	16	0	

The nine functions found in the Utility Pole Lifestyle are shown below and the search result using those words is shown in Table 6.5.

- "Create electricity with the power of nature"
- "Take power from nature"
- "Rotate blades"
- "Convert rotational energy into electricity"
- "Capture sunlight"
- "Convert light energy into electricity"
- "Store electricity"
- "Send electricity along power lines"
- "Charge your electronics"

The 16 functions found in the Microbe Lifestyle are shown below and the search result using those words is shown in Table 6.6.

- "Shrink plant-based materials"
- "Combust the fuel"
- "Decompose plant-based materials into liquid fuel"
- "Obtain thermal energy"
- "Preserve contents in it"
- "Prevent odor escaping"
- "Maintain proper temperature"
- "Maintain proper humidity"
- "Maintain suitable brightness"
- "Heat the breeding equipment"
- "Cool the breeding equipment"
- "Divide plant-based materials"
- "Convert thermal energy into electricity"
- "Maintain warmness"
- "Maintain coldness"
- "Offer grinding power"

Table 6.6 The match between the functions of the Microbe Lifestyle action decomposition tree and technology "seeds" found in nature

Keyword search	Articles found	Actually related articles	Titles of related articles
Shrink	5	1	The Miura folds of leaf buds and insect wings
Combustion	1	1	The smoke sensors of jewel beetles
Decomposition	15	3	Antimicrobial peptide melittin in bee venom Biofilm: an extraordinary aggregate of tiny creatures The comfortable nests of termites
Thermal energy	1	0	
Maintain	7	3	Light, strong bamboo How the sea-hating *Littorina brevicula* prevents dryness A tree that excels at stress relief
Odor	5	1	The comfortable nests of termites

(*Continued*)

Table 6.6 (*Continued*)

Keyword search	Articles found	Actually related articles	Titles of related articles
Keep temperature	4	4	Wrapped in air, the sheep is cool in summer and warm in winter Frizzy but not stuffy: sheep's wool Warm, resilient down Froghoppers stay comfortable in foam
Keep humidity	3	3	The water absorption secret of kombu and wakame kelp Gorgeous wild silk Okra's total control of water
Brightness	3	3	The shadows of a black rose Moth eyes don't reflect light The versatility of penguin's eyes
Heat	13	5	Warm but smell: a skunk cabbage home Butterflies stay warm by collecting light through their scales Wrapped in air, the sheep is cool in summer and warm in winter Warm, resilient down Blood circulation of bumblebee
Cool	13	5	New fan shapes: learning from natural spirals in eddies Big noses help elephant seals capture water How desert lizards control temperature and rehydrate The comfortable nests of termites: passive cleaning The thorny dragon has a built-in mechanism for automatic water supply
Divide	1	0	
Separate	13	0	
Transform	11	0	
Grind	0	0	

Action decomposition reveals functions that remained hidden when we extracted technology from the lifestyle text in Section 6.1. This leads to an increase in the number of technology seeds.

Decomposing actions allows us to dig deeper into the structure of a lifestyle. As we move down the action decomposition tree, the performer of each action tends to become an object, a system, or a service. In other words, the action becomes a function. In terms of the actions required to achieve that lifestyle, it is always possible to climb upwards from the actions in the text. Continue to climb until you almost reach conceptualization. Moving upwards helps to reveal other ways by which something may be achieved. That will lead to further, downward actions. This is similar to designing new lifestyles around the same concept in the conceptualization method. It also expands the original lifestyle idea.

Quite apart from the expansion of a lifestyle, decomposing actions has the extra merit of using simple vocabulary with limited numbers. It is an extremely useful system for reaching a common understanding between people of different educational backgrounds and cultures. For more on its role as an information system, see Chapter 12.

6.5 The Limitless Business Seeds That Lie in Lifestyles

In this chapter we used two example lifestyles to see how to extract the necessary technologies for backcast lifestyles and how to find the technology seeds for their development in nature.

As explained in Section 6.2, any short paragraph describing a lifestyle will naturally elide over many details. In order to see the much broader image, it's necessary to develop the technologies every similar lifestyle requires. That means understanding and describing each of those lifestyles and picturing the details of people's actions and their situations. Here we have used the methods of conceptualization and action decomposition tree to do that. However, lifestyles can be broadened even further, and many more details can be uncovered. You are not restricted to just one of these methods, nor must you perform one before the other. Combine them in whichever way suits you. The more this is done, the more necessary technologies you will discover. Lifestyles contain a myriad of potential business seeds.

Acknowledgment

This work was supported by JSPS KAKENHI Grant number JP24120008 in Scientific Research on Innovative Areas: "Innovative Materials Engineering Based on Biological Diversity."

References

1. Fuji, S., Furukawa, R., Suto, Y., and Ishida, E. H. Development and verification of conceptualization method to raise social acceptability in lifestyle design. *Proceedings of the 28th Annual Meeting of the Japan Society for Science Policy and Research Management*, 67–70 (2013) (in Japanese).
2. Sasajima, M., Kitamura, Y., Ikeda, M., and Mizoguchi, R. FBRL: a function and behavior representation language. *Proc. IJCAI'95*, 1830–1836 (1995).
3. Kitamura, Y., Koji, Y., and Mizoguchi, R. An ontological model of device function: industrial deployment and lessons learned. *Journal of Applied Ontology* (special issue on "Formal Ontology Meets Industry"), **1**(3–4), 237–262 (2006).

Chapter 7

Bio-TRIZ

Takeshi Yamauchi,[a] Hidetoshi Kobayashi,[b] and Toru Kobayashi[c]

[a] *Department of Engineering, Materials Science Program, Niigata University, 8050 Ikarashi 2-no-cho, Nishi-ku, Niigata 950-2181, Japan*
[b] *Graduate School of Engineering Science, Osaka University, 1-3 Machikaneyama, Toyonaka, Osaka 560-8531, Japan*
[c] *Division of Electrical Engineering and Computer Science, Graduate School of Engineering, Nagasaki University, 1-14 Bunkyo, Nagasaki City 852-8521, Japan*
yamauchi@gs.niigata-u.ac.jp

7.1 Introduction

In recent years, researchers have been stepping up their work on biomimetic engineering (biomimetics), a field focused on introducing high-efficiency and high-performance biofunctions into material design [1]. More and more articles on biomimetic engineering have been reported every year, and expectations that the industry will develop practical applications for such are likewise on the rise. Numerous well-known applications of biomimetics may be cited such as self-cleaning paints based on lotus leaves, easy-to-peel-off tapes inspired by the microstructures in the soles of a

Lifestyle and Nature: Integrating Nature Technology to Sustainable Lifestyles
Edited by Ryuzo Furukawa

ISBN 978-981-4774-62-8 (Hardcover), 978-1-315-14687-4 (eBook)
www.panstanford.com

gecko's foot, nonreflective films structured like the compound eyes of a moth, shark skin-patterned high-speed swimwear, automobile designs that incorporate ideas taken from a boxfish's skeleton, and labels that use the structural colors of the morpho butterfly. The ranks of companies whose interest in developing materials based on biomimetic engineering principles sparked by news reports about such developments likewise has been increasing [1]. That said, there are more than 7.8 million species of living beings in the world, with an enormous number of different functions and behaviors. Whatever biofunction attracts our attention, it is unclear as to which ones will be useful toward developing new technologies and materials and lead to an optimal material design. In short, most engineers and researchers find themselves vexed by their inability to home in on a single target owing to the excess of options. Thus, case-by-case material design rules biomimetic engineering in the present day. Only a portion of the limitless number of biofunctions are being put to use, and there are no effective means for digging out those technological elements that may be necessary. Furthermore, with the International Standardization Organization (ISO) currently studying a variety of regulations regarding biomimetic engineering, there is demand for biomimetic products to be created that conform to international standards. According to ISO specifications, developing biomimetic biometric products requires they go through the following process: (i) identify issues with existing technologies and materials, (ii) search for biofunctions that can resolve those issues, (iii) extract and generalize the principles behind the biofunctions that have been discovered, and (iv) create and optimize new technologies and materials. Figure 7.1 presents this process schematically along with the main issues involved. The question also arises of the best approach to take for identifying functions among the 7.8 million living things said to exist and for optimizing them. In this chapter, we present a summary of how we have applied the Theory of Inventive Problem Solving (**Теориярешенияизобретательскихзадач** in the original Russian)—a prominent engineering problem-solving approach—and constructed a database that can transfer biofunctions to engineering patents as a way to resolve these issues.

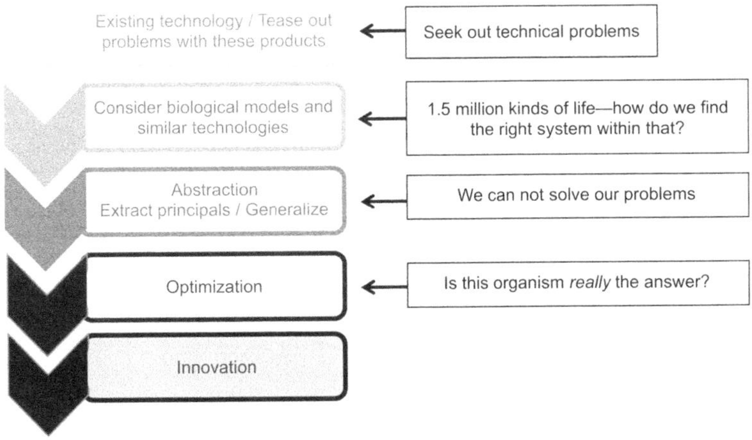

Figure 7.1 Standardization scheme for biomimetic products at the International Standards Organization and issues at each step.

7.2 Current Conditions Regarding Patents for Biomimetics

Why are we focusing on the functions and workings of living things as a means for solving problems? Let us begin our probe by reviewing the existing process for generating patent proposals. In a company, the first step is to look for a solution to an issue using in-house technology. If it can't be solved that way, the company will then solicit opinions from specialists. Furthermore, the company will also turn its eyes toward technologies at other companies in the same industry and try to get ideas about new ones from there. On this point—a key one for generating new, high-quality patents—there is in this the possibility of achieving a breakthrough on one's limitations in thinking by getting information from other lines of business. To wit, one might say that so long as the line of business is a different one, it may create opportunities to acquire information that is novel. In that respect, the sophistication and great efficiency of the structures of living things is likely why engineers and researchers have turned their focus on them (see Fig. 7.2).

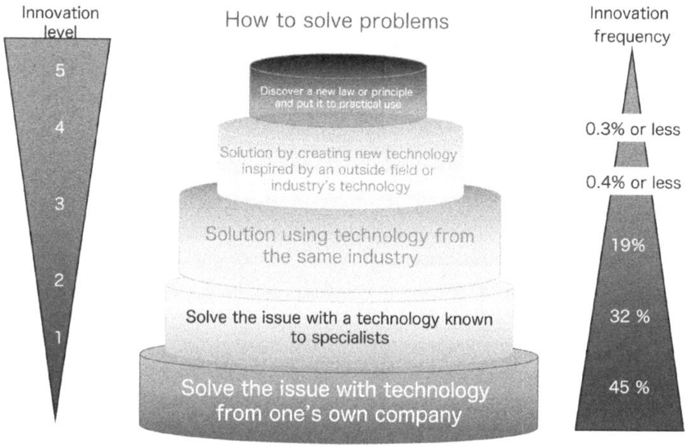

Figure 7.2 The relationship between various problem-solving methods and their degrees of difficulty.

The Japan Patent Office's *Survey Report on Technology Trends in Patent Applications* gives us a picture of current tendencies with regard to patents focused on biofunctions [2]. It can be inferred from a review of the data for those products that are mainly related to biomimetics that at present the number of instances in which a patent has been commoditized is limited regardless of country or region.

The main products can be broken down as belonging to one the following three broad categories.

(1) Products related to molecules and materials. Examples include hydrophilic and hydrophobic materials (e.g., water-repellent paints, tile-related building materials, and water-repellent glass), structural coloration materials (e.g., chemical fibers), optical materials (e.g., antireflective film and displays), adhesive and gummed materials (e.g., adhesive tape, carpet tile, and cyclone vacuum cleaners), medical and biocompatible materials (e.g., injection needles and cosmetics), low-value resistance and low-friction materials (e.g., competition swimwear, fuselage paint for airplanes, and ship-coating materials), and antifouling materials (e.g., antifouling coating for ship bottoms).

(2) Products meant to reduce the resistivity of a structure (e.g., cooling fans, washing machines, mixers, and the shape of the noses on bullet trains) or save weight (e.g., automobile wheels)
(3) Products related to robots (e.g., robots patterned after dragonflies, elephant noses) or machine controls (e.g., sensors and software for controlling smart grids)

In Japan, we find numerous products related to the molecular and materials fields such as hydrophilic and hydrophobic materials, optical materials, and adhesive and gummed materials. In the West, in contrast, we encounter products in the machine field. Much work was once being done in Japan in the areas of biomimetic chemistry and robotics, and the country was a leader in molecular-scale biomimetics. However, researchers and engineers have not been able to get a grasp on the new currents that emerged in material-related biomimetics in this century. Broadly speaking, this is because in Japan collaborations between different fields as biology and engineering never take place owing to preconceptions in Japanese research and academic settings about interdisciplinary interactions.

Additionally, it is believed that the materials field will become the primary destination for applications of biomimetics. If more products are to be brought to market in these areas, it will be necessary to develop even more technologies in the area of microstructure fabrication techniques in order to reduce manufacturing costs and improve durability. Furthermore, if the biomimetics market is to expand, would-be producers will need to adopt a "biomimetic" way of thinking in the control and processing fields. In particular, observers have highlighted the importance of autonomous distributed control systems, and the expectations of people involved in the field for advances in the development and practical application of such systems are high.

On the basis of this analysis, it is evident that to date biomimetic product development has concentrated on materials made with a focus on the structures of living things. No work is being done to find applications for the sophisticated process engineering evident in living things. The reason for this seems to be that even if engineers attempt to find uses for the data they get about living things, it is extremely difficult to come across data that provides the stimulus for inventions. We believe that with our database engineers in that position will be able to find data on living things that will point them in serendipitous directions and facilitate coming up with patents relevant to material design, development processes and systems, and a wide range of other problems.

7.3 The Job of a Bio-TRIZ Database

There are a variety of approaches that one can take to solving engineering problems. Researchers and engineers have found success in wielding such approaches as strengths, weaknesses, opportunities, and threats (SWOT) analysis, which applies a business framework to the problem, "logic" (or "issue") trees, Osborn's checklist, mindmaps, quality control methodology, and the Taguchi method. Among these methods is one that has been applied in a variety of fields since 1990 as a means for generating ideas for material design in mechanical engineering. This is the Theory of Inventive Problem Solving, usually known by its Russian acronym TRIZ (Teoriya Resheniya Izobreatatelskikh Zadach, Теориярешенияизобретательскихзадач). TRIZ was developed by Genrikh Saulovich Altshuller (1926–1998), who worked as a clerk in a Russian patent office. Altshuller came up with TRIZ as a set of "principles for performing creative problem solving" that put a system to the regularities he uncovered in his examinations of some 1.5 million patents. The distinguishing feature of this method is that it presents a set of rules and principles for solving issues in engineering technology. This it does by looking for the patterns in the problems for which solutions are being sought, thus making it possible to get ideas about how to solve those problems regardless of one's field of research [3, 4]. There are several problem-solving techniques in TRIZ, a technique to settle a problem from an invention principle of 40 by an inconsistent matrix way, a technique as the effect which makes a reverse lookup of a natural phenomenon from the function, a technique to recognize a target system composed as material and field, and a technique to think of an idea by grasping a trend of technological evolution [5, 6]. In this research work, technological contradiction matrix method was focused on because it is easy for engineers of various fields to have technical contradiction among their businesses. This method is a way of the conventional solving-problem technique by Altshuller [7]. With this theory, Altshuller in essence derived 40 principles for problem solving based on his analysis of those 1.5 million patents and put them in a treasure chest. You can then draw upon the principles collected in this book for leads on how to solve the problem with which you are faced. Altshuller's 40 principles are presented in Table 7.1. Consider the example of the AlphaGo computer program, which in April 2016 made news around the world for beating a human master of the board game Go. What made AlphaGo different from earlier types of artificial intelligence is that rather than searching out the best move from massive amounts of data-containing diagrams of Go games that have been played, the program was able to find the patterns in those games and then limit itself to those moves it should choose. In short, the method used was able to come up with new and original ideas by coming up with optimal solutions through restricting itself to a search for patterns within an enormous dataset rather than seeking out a specific solution within it.

In this chapter, we focus on a TRIZ offshoot known as Bio-TRIZ, a new concept put forth by former University of Bath professor Julian F. V. Vincent and his associates [8–11]. The aim of Bio-TRIZ is to bring high-efficiency biofunctions found in nature into TRIZ's 40 problem-solving principles and create systems from this idea. For example, in metallurgical and mechanical engineering, to open pores in a material means reducing its strength. However, living things create sophisticated structures that prevent cracks from forming when they open pores. This maintains the strength of a material while simultaneously saving weight. Biofunctions in this way are a gold mine for patentable new technologies. They imply ideas for the development of next-generation materials and can also be expected to do much in the area of education in new engineering design. Developing next-generation materials demands that their creators consider such factors as harmonizing with nature, environmental friendliness, and the use of biomass among others. The invention of materials that function effectively on low energy is also essential. Drawing inspiration from these ideas, the authors have created a

Table 7.1 The 40 problem-solving principles of TRIZ

Principles for performing creative problem solving		
(1) Segmentation	(16) Partial or excessive actions	(31) Porous materials
(2) Taking out	(17) Another dimension	(32) Colour changes
(3) Local quality	(18) Mechanical vibration	(33) Homogeneity
(4) Asymmetry	(19) Periodic action	(34) Discarding and recovering
(5) Merging	(20) Continuity of useful action	(35) Parameter changes
(6) Universality	(21) Skipping	(36) Phase transitions
(7) Russian dolls	(22) "Blessing in disguise"	(37) Thermal expansion
(8) Balance	(23) Feedback	(38) Strong oxidants
(9) Preliminary anti-action	(24) Intermediary	(39) Inert atmosphere
(10) Preliminary action	(25) Self-service	(40) Composite materials
(11) Beforehand cushioning	(26) Copying	
(12) Equi-potentiality	(27) Cheap short-lived objects	
(13) The other way round	(28) Mechanics substitution	
(14) Curvilinear surface	(29) Pneumatics and hydraulics	
(15) Dynamics	(30) Flexible shells and thin films	

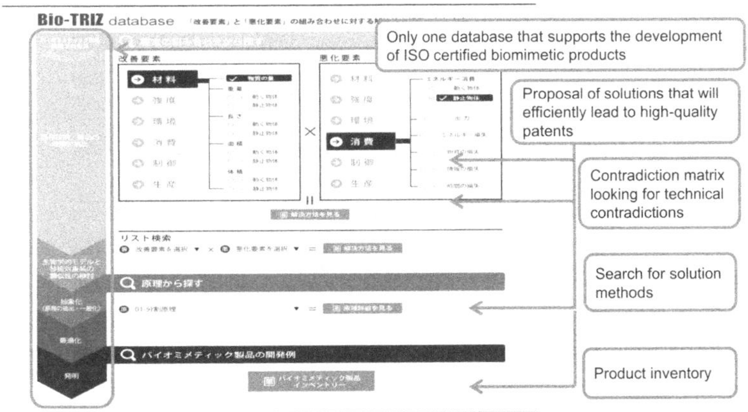

Figure 7.3 An outline of the Bio-TRIZ database.

database for resolving the variety of technological contradictions that arise when developing ISO-certified biomimetic products.

The strength of this database lies in the fact that when the user queries it over technological contradictions—the problem of improving one parameter having a negative impact on another—it has the framework in place to automatically display any number of biofunctions that might suggest a solution as it runs through the four processes shown earlier (see Fig. 7.3). Our database comprises carefully selected information—stored in files categorized into the 40 TRIZ problem-solving principles—about living things that could be used for coming up with patents. We describe it here as a Bio-TRIZ database in the sense that its data could contribute to decision making in the creation of biomimetic products. By merely condensing the technological problems that engineers face and putting them into a contradictions matrix, our database can supply the problem-solving principles that would seem appropriate for resolving those issues and furthermore will search out biofunctions for the user. The user can get ideas for efficiently developing biomimetic products in a short period of time by optimizing the biofunctions the base presents. Furthermore, the data contained can also contribute to product development based on existing work since the user can also view an inventory of biomimetic products. Let us turn next to review the structure of this database.

Table 7.2 The 39 features of Altshuller's Contradiction Matrix

Features you want to improve, and the features that may deteriorate as a result (39 feature parameters)		
(1) Weight of moving object	(16) Duration of action by stationary object	(30) External harm that affects the object
(2) Weight of stationary object	(17) Temperature	(31) Object-generated harmful factors
(3) Length of moving object	(18) Illumination intensity	(32) Ease of manufacture
(4) Length of stationary object	(19) Consumption of energy by moving object	(33) Ease of operation
(5) Area of moving object	(20) Consumption of energy by stationary object	(34) Ease of repair
(6) Area of stationary object	(21) Power	(35) Adaptability or versatility
(7) Volume of moving object	(22) Loss of energy	(36) Device complexity
(8) Volume of stationary object	(23) Loss of substance	(37) Difficulty of detecting and measuring
(9) Speed	(24) Loss of information	(38) Extent of automation
(10) Force	(25) Loss of time	(39) Productivity
(11) Stress or pressure	(26) Quantity of substance	
(12) Shape	(27) Reliability	
(13) Stability of the object's composition	(28) Measurement accuracy	
(14) Strength	(29) Manufacturing precision	
(15) Duration of action by moving object		

7.3.1 Extracting Issues from Existing Technology and Materials

The merit of the TRIZ method is that it presents a set of rules and principles for solving issues in engineering technology. This it does by looking for the patterns in the problems for which solutions are being sought, thus making it possible to get ideas about how to solve those problems regardless of one's field of research. There are any number of ways to apply it based, for example, on principles, predictions, or effects. Here, we will focus on problem-solving methods that make effective use of the 40 problem-solving principles the theory proposes. The principle-based approach entails taking the physical characteristics spanning the 39 parameters shown on Table 7.1 and using them to create a 39-by-39 matrix comprising "parameters wish to improve" and "problems that will arise" when improvement is made. The goal is to resolve the technological contradictions represented by the intersections between those parameters an inventor seeks to improve and those parameters that will change for the worse. The physical characteristics covered under those 39 parameters are shown in Table 7.2. The squares in the matrix show which of the 40 principles from Table 7.1 is the appropriate one to the problem at hand, and from this the inventor can get ideas for possible solutions. This 39-by-39 chart is called a "contradictions matrix." The database that we developed is equipped with a contradictions matrix as shown in Fig. 7.4, which gives it a framework for providing ideas about problem solving using TRIZ and Bio-TRIZ methods. As an example, let's consider the following problem: "I want to develop a pointer that I can use even from farther away, but this will result in a pointer that is quite bulky." The improvement parameter is "length of pointer," while the contradiction is "bulk increases." When we put this into a contradictions matrix, the suggestions for resolving this problem include—as shown in row 3, column 7 of Fig. 7.5—referring to the (7) Nested Doll and (17) Another Dimension principle. From this, the inventor can infer possible solutions—in the case of the former perhaps a pen-style pointer that extends like a fishing rod, or in the case of the latter a laser pointer that someone came up with.

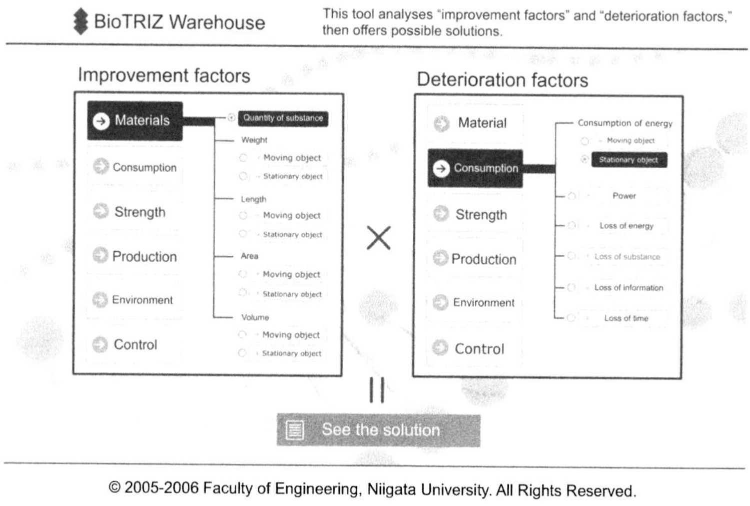

Figure 7.4 Using the Contradiction Matrix in the Bio-TRIZ database.

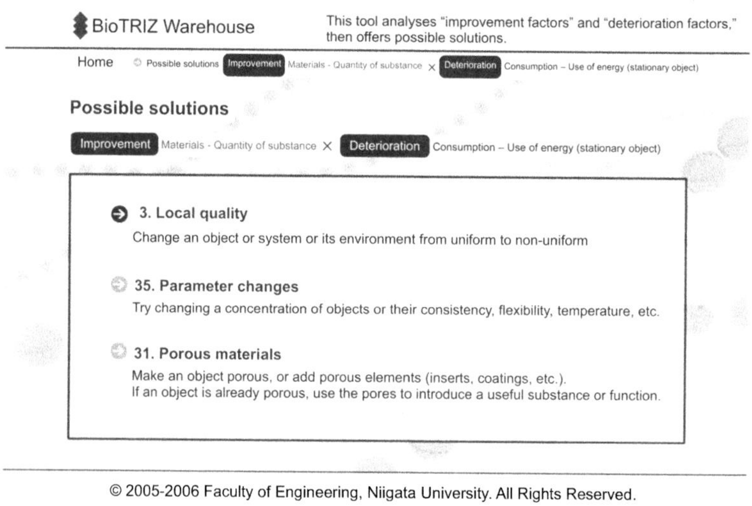

Figure 7.5 Example of TRIZ principles as shown in the Bio-TRIZ database.

7.3.2 Seeking Out Biofunctions to Resolve Problems

When you wish to improve the durability of an immobile object, but this will lead to an issue with the volume of matter, the matrix suggests that principle 31 on porous materials will be effective. From a mechanical engineering perspective, the uses of porous structures under TRIZ are limited since opening any holes in metal or inorganic material will degrade the material in question. On the other hand, porosity is used in various ways in the biological world. Plants transport water through netlike systems of ducts, while insects create sophisticated structures that prevent cracks from forming by opening pores in their casing. Porosity helps to maintain the strength of these materials while also saving weight. Furthermore, efficient use is made of flat surfaces in beehives with their arrangement of hexagonal openings. Triangular or quadrilateral shapes could also fill a flat surface without any gaps, but a hexagon offers superior mechanical strength and what's more creates openings that individually are more spacious than those formed by other shapes. Additionally, we can see from observing bees building their hives that creating these hexagonal shapes is easy and extremely efficient since they move in only three directions. One example of a patented invention developed using this structure is the adhesive bandage with a honeycomb structure, which serves as a means for maintaining the material strength of the bandage while improving its elasticity and increasing the volume of medicinal solution it contains [12].

7.3.3 Extracting and Generalizing the Principles of the Biofunctions Discovered

The data collected in the Bio-TRIZ database allows searching for the results of mathematical analyses of these frameworks in living things. The search is structured to once again show a list of the 40 principles and facilitate viewing of Bio-TRIZ data within each principle. Furthermore, it also allows the user to both search for and examine illustrative structural analyses. For example, for principle (3) Local quality as shown in Fig. 7.6 the results will show a

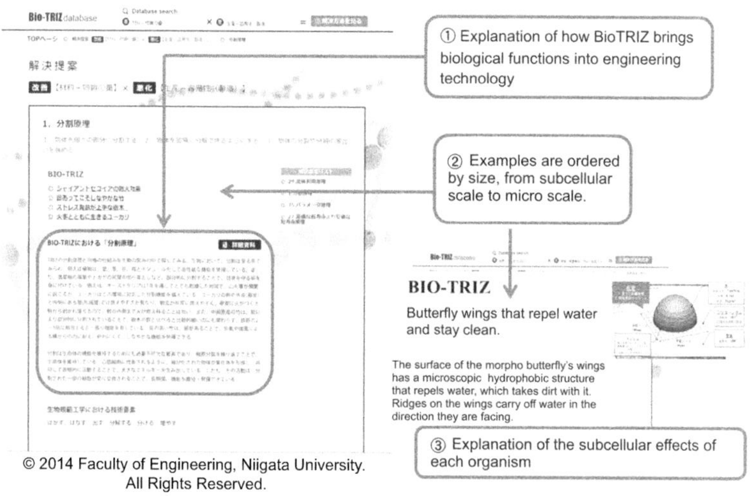

Figure 7.6 Sample display of Bio-TRIZ information from the Bio-TRIZ database.

Bio-TRIZ approach that makes the most local characteristics grounded in the microstructures of living things. The user can take this to search for ideas about how to transfer those biofunctions to an engineering context. Living things are arranged by sizes of functional sites, with local qualities presented for everything from the flower of a rose to the "yabukiri" shield-backed katydid (Tettigonia orientalis). For example, if the user searches out the structure of a lotus leaf, they will be presented with the results of an analysis on the relationship between the rod structures on the leaf blade surface and contact angle of water drops.

7.3.4 Creating Materials and Studying How to Optimize Them

When it comes to creating and optimizing new materials, it is useful to have information about existing products that were developed using patents inspired by biofunctions. The Bio-TRIZ database user can search for information about prior research as they develop their own product in order to optimize it. The user can also search the biomimetic product inventory across a variety of fields such as health, sports, and architecture.

7.4 Examples of Solutions to Technological Contradictions

Next, let us consider an example of how using the Bio-TRIZ database can assist even nonspecialist researchers with inventing next-generation materials that make the most of patents for existing engineering technologies that companies have developed while taking on board structures from living things. One technological contradiction common to materials development in a variety of fields is that of wanting to reduce the weight of a material but being faced with obstacles in terms of energy efficiency, power, and operating time. One example of how to resolve this using a TRIZ principle can be found in our own research into improving the functionality of polymeric actuators.

People working in a variety of fields have high hopes for employing conductive polymers as soft actuators that will provide drive at low voltages. However, when scaling this up the amounts of material required rises and the energy needed to provide drive grows dramatically. Reducing weight produces problems in terms of mechanical strength. Putting this technological contradiction into the TRIZ contradictions matrix reveals four inventive principles that are effective for solving it: (12) Equipotentiality, (18) Mechanical vibration, (31) Porous materials, and (36) Phase transitions. Examining existing patents that resolved this technological problem from an engineering perspective, we find Patent and Utility Model Number 2010-252422 for a voltage actuator disclosed by Seiko Epson Corporation that can enable high-efficiency driving. Principle 18 on mechanical vibration was used to develop an actuator equipped with piezoelectric elements that have the resonance frequencies of the lateral and antiresonance vibrations as the frequency for the drive signal.

Shifting our view to living things, we find that the porous structure of hedgehog's or porcupine's spines makes them an extremely lightweight and firm material. In short, they make skillful use of porous structures. This led to the development of an actuator that makes use of a porous structure. While there are many patents that are associated with structures similar to those of living things, there are few proposed solutions related to structural processes. Some further words of explanation are in order.

Given that opening pores in materials causes them to lose strength, there are only a limited number of examples of such in engineering.

Examples of materials with artificially opened holes would include those with foam or cell structures. Foams based on polymeric materials have widespread application. These include protective materials such as foamed polyurethane and styrene foam, as well as batteries, fuel cells, filters, lightweight structural materials, thermal insulation, and vibration-proofing materials. However, pores produce irregularities in the manufacturing process and are difficult to control. There are instances in which regular porous polymers have been developed through a method wherein silica crystal is as a mold and the silica removed after the polymer has been formed. This method requires consuming large amounts of thermal energy to create the silica. Removing the silica from the composite also requires using chemicals to melt the silica. High temperatures are required to achieve a phase transition in silica. However, plants of the gramineae produce silica internally at normal temperatures. They absorb silicates from the soil, which then crystallize. The reasons for this are believed to be the strength of the fibers and the lighting effects, i.e., the ease with which light is absorbed by this glass.

Underwater flora also make use of this sort of effective phase transition. For example, agar—a component of seaweed—uses the potassium in seawater to easily shift back and forth between sol and gel states and thus helps to affix the position of the seaweed in the water so it does not flow away. We studied making a porous conductive polymer by taking advantage of the phase transition phenomenon in agar to create gel beads, arranging the beads as a mold for synthesizing conductive polymer, and then melting the beads at a temperature of 80°C after compounding was complete. We can expect some researcher to develop an actuator

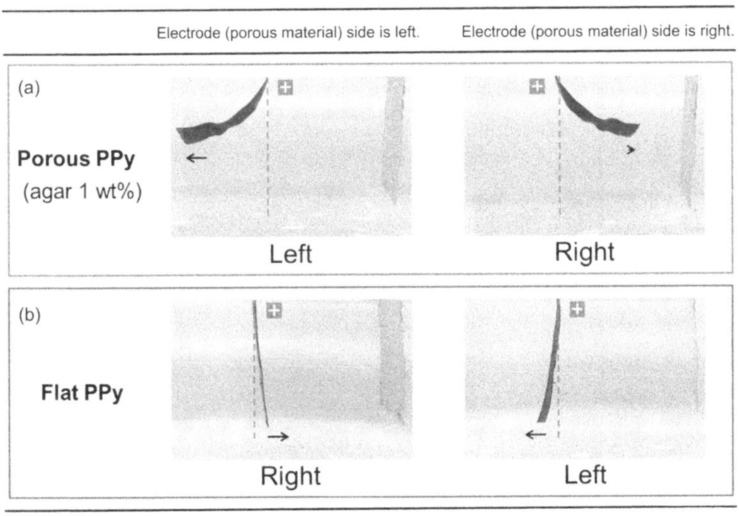

Figure 7.7 Bending motion of soft actuators (applied electric potential: 3V). (a) Porous material structure, (b) smooth, flat structure.

that uses the superb motor function like those seen on the surface of the sea. Specifically, some have used an ITO electrode coded in minute particles of agar as a work electrode and—taking into account the principle of phase transition—create a polypyrrole/agar microparticle membrane based on the electrochemical method of polymerization. The membrane obtained is then dipped in an ion exchanged water at 90°C and the microparticles of agar are removed, with the ultimate result of attaining a porous polypyrrole membrane. This porous conductive polymer membrane will have a porous membrane approximately 9 μm, with an average pore diameter of 108 nm. The actuation behaviors of the porous membrane (pore size: 108 nm) and the smooth and flat membrane are shown in Fig. 7.7. The porous membrane displays a relatively large bending motion compared to the smooth and flat one. Furthermore, we made porous membranes with a large average pore size (diameter: 293 nm) and induced a bending motion in the three types of conductive thin films (a smooth and flat membrane as well as porous membranes with average pore diameters of 108 nm and 293 nm). We then calculated the bending speeds of each to be

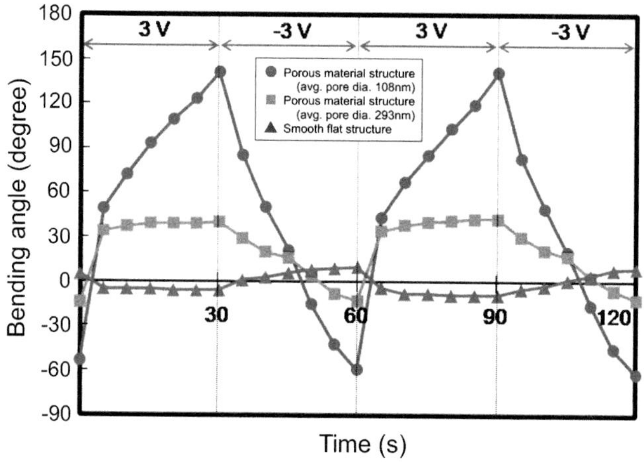

Figure 7.8 Functionality of curving motion of soft actuator (applied electric potential: 3V). (○) Porous material structure (avg. pore diameter 108 nm), (□) porous material structure (avg. pore diameter 293 nm), and (△) smooth, flat structure.

20.4 deg/s, 9.6 deg/s, and 2.4 deg/s, respectively (see Fig. 7.8). These results suggest that it is possible to control deformation movement and driving speed by applying a porous structure.

7.5 Applying Bio-TRIZ to Needs-Oriented Manufacturing

We can expect Bio-TRIZ to be applied to a variety of fields given how it can serendipitously bring to mind new ideas from chance encounters. The fact in particular that one may study the high-performance, highly efficient structures of living things means it should have a potent part to play in the creation of the technologies that will be key for a sustainable society. Here, as one application of Bio-TRIZ for needs-oriented manufacturing, we will focus on the technological contradictions that were seen in the action decomposition tree discussed in Chapter 6. Ishida and colleagues reported on the high levels of social recognition about the new lifestyles that are enveloping utility poles in noting that "wooden utility poles are serving as hubs for all-purpose, small power generators." The specific proposal is as follows:

Utility poles are often unpopular and criticized as unattractive. However, their potential importance as an infrastructure for small generators has been noticed. Windmills or solar panels are attached to them to generate electricity. They can also use rain to create hydropower. The electricity travels along wires to provide individual homes with power. Rechargeable batteries attached to poles also store such power, before being used for mobile phones and game consoles. Large utility poles are capable of powering electric cars, helping to promote this type of vehicle.

We have made note of human behaviors here, but nothing about technologies or functions. A variety of engineering technological contradictions were identified by the action decomposition tree. For example, technological elements that this lifestyle requires include flexible solar batteries that are unbreakable even by the strongest of winds and an electricity storage system that can allow for distribution from utility poles.

However, installing solar energy panels on utility poles will require an increase in the volume. This produces the technological contradiction of "large volume" versus "strength," as reflected by such concerns as damage due to mutual contact or wind and rain. Using the technological contradictions that arise when designing with these parameters in mind as a starting point, a search of the Bio-TRIZ database suggests that principles (14) Spheroidality, (15) Dynamicity, and (17) Transition into new dimension would be useful here. The data on living things made it possible to come up with designs for technologies that did not yet exist such as flexible solar batteries that used the spheroidality to draw lessons from the scales of a birdwing butterfly and road surface-installed rechargers that applied the Another Dimension principle in being modeled on the giant water lily. The spheroidality principle calls for methods that make use of curved shapes instead of linear components, surfaces, and patterns. Any number of biofunctions can be offered as examples such as the fins on bluegills that can be rounded to enable fast swimming, compound lobster eyes that offer a 360-

degree field of vision, the waterproof "tents" of lackey moths (their curved surfaces collect light and maintain temperatures), and the seeds of climbing gourds that can be scattered even without any wind. As for the dynamics principle, which suggests solving the problem by changing or optimizing the characteristics of a physical object, external environment, or process, the impact absorption method found in the woodpecker represents one expression of how to solve this problem. The Another Dimension principle, meanwhile, calls for arranging (2) The weight of a nonmoving object that moves in two or three dimensions across multiple layers rather than a single one. Examples of such unique characteristics include safe and reliable marine and air transport systems (magnetic) that draw lessons from the marine turtle, green-banded swallowtail butterflies dressed up with the latest in LEDs, new armors with ideas derived from ancient fish, the light but firm beaks of toucans, cuttlefish that can change their colors in a flash, and the secrets of the shell of an abalone that remain solid even if you hit them with a hammer or drive a car over them. Here, with regard to developing flexible solar batteries that draw upon the scales of birdwing butterflies, we will talk about how this has been applied to creating gels and to photoelectric conversion elements that bring together fullerenes and polyethylene glycols. In recent years, stimuli-responsive gels have attracted much interest as one of the most promising types of functional soft material. Stimuli-responsive gels have properties that respond to external stimuli such as pH values, heat, light, and force as well as changes in their external environment. Furthermore, they may also experience dramatic changes in such characteristics as volume, hydrophilia or hydrophobicity, color, and mechanical strength. One can expect that these properties will be wielded for use in devices and elements that require considerable levels of functionality such as flexible sensors, actuators, drug delivery systems, and artificial tissues. Accordingly, attempts are being made to manufacture high-performance gels from a variety of perspectives. There have been reports in recent years of one approach to manufacture high-performance gels that offer high-conductivity and other such characteristics by incorporating such carbonate ingredients as graphene and carbonate nanotubes into a network.

Focusing ourselves on the C_{60} fullerene, the authors have reported on forming a gel network (C_{60}–PEG gel) using polyethylene gycol (PEG) that employs C_{60} as a bridge. The C_{60} fullerene is a soccer ball-shaped cluster about 1 nm across formed from 60 carbon atoms. Its characteristics include electron acceptability, ultraviolet absorptivity, and radical complementarity. Work is already being done to make the most of these characteristics in a variety of fields including electronic materials, medicine, and cosmetics. In addition, research is also brisk on applying it as n-type semiconductor material for organic thin film solar batteries. Given that C_{60}–PEG gels have these properties, it is believed that compounding them with p-type semiconductors would make it possible to create a stimuli-responsive gel that presents an electrical response to light stimulus. By employing such gels to electron elements, we can anticipate the development of flexible ultraviolet sensors and solar batteries that have both flexibility and elasticity. The goal in our research has been to develop novel functional materials that possess the qualities of both fullerenes and composite materials by taking advantage of this gelatinous structure to compound it with a variety of materials. We selected poly(3,4-ethylenedioxythiophene)—a conductive polymer known as PEDOT for short—to make both C_{60}–PEG gel and C_{60}–PEG/PEDOT gel. In choosing a PEDOT polymerization method, we investigated a compounding method in which electrochemical polymerization occurs within the C_{60}–PEG gel and then evaluated the structure of obtained film. Furthermore, research is being done on applying such composite materials as fullerenes and conductive polymers toward organic thin film-type solar batteries. This is winning good reviews for the photoelectric characteristics of the gels. We can expect responsivity to improve and uses for new photoelectric conversion devices to be discovered through nanoscale network of the fullerenes and PEDOT distributed evenly around the gel. The C_{60}–PEG gel is formed after the Azo-PEG polymeric azo initiator has been caused to react with fullerenes and the excess solvent removed. You then dip the C_{60}–PEG gel into a monomer solution and use the electropolymerization method to force the PEDOT in the C_{60}–PEG to polymerize, a process that results in the formation of the C_{60}–PEG gel (see Fig. 7.9). We analyzed FT-IR, XRD, and SEM to make a structural evaluation of the gel we manufac-

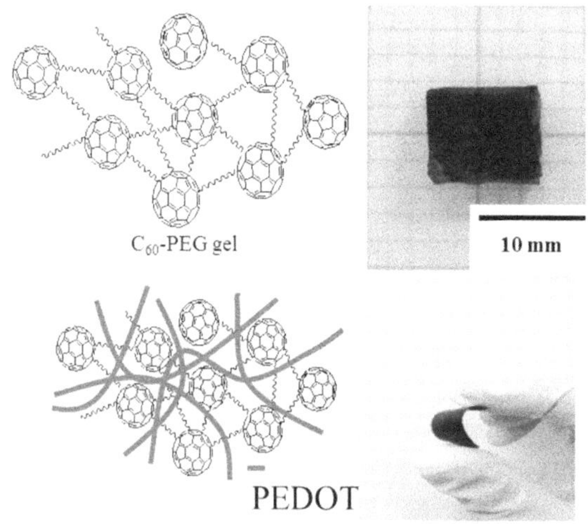

Figure 7.9 A solar battery made with C_{60} fullerenes.

tured. To evaluate its photoelectric characteristics, we investigated its ultraviolet response and also used false sunlight to evaluate its incident photo-to-current efficiency. The XRD measurements showed that the C_{60}–PEG we made as well as the simple PEDOT film displayed crystalline properties. By contrast, the C_{60}–PEG/PEDOT gel had an amorphous structure. Furthermore, SEM images revealed that PEDOT structures could be seen within the gel, while an energy-dispersive X-ray spectroscopy analysis showed that the fluorescence of PEDOT-derived sulfur could be observed throughout the gel. What's more, it was distributed in many locations near the surface. This suggests that the PEDOT within the C_{60}–PEG gel was polymerized in a dispersed state by use of electrochemical polymerization. We were able to use these results and the results of FT-IR measurements to manufacture both C_{60}–PEG and C_{60}–PEG/PEDOT gel. This method was also used successfully to bring dissimilar composite materials together into a nested doll structure (see Fig. 7.10). An evaluation of photoelectric characteristics shows that both the C_{60}–PEG and C_{60}–PEG/PEDOT gels present a responsiveness wherein the electroconductivity changes due to the irradiation of ultraviolet rays. Furthermore, compounding PEDOT

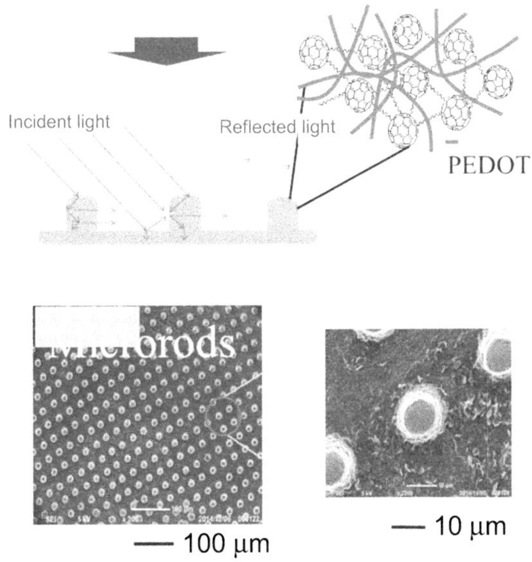

Figure 7.10 A solar battery made of C_{60} fullerenes incorporating a nesting structure.

led to dramatic improvements in the average conductivity and the amount of change in conductivity. Additionally, when shining false sunlight there were no changes in the current value when there was a single layer of either the C_{60}–PEG gel by itself or in combination with the PEDOT film. In contrast, a photocurrent was produced when only a layer of C_{60}–PEG/PEDOT gel was present, with an incident photo-to-current efficiency of approximately 21.6×10^{-3}%. On the basis of these results, we believe that the compounding PEDOT inside the C_{60}–PEG gel improved the contact surface with the fullerenes, and that responsiveness was improved due to the transfers among the electrons that excited by the light energy within.

All the foregoing suggests that it is possible to prepare C_{60}–PEG/PEDOT gel using the novel hybrid method of electropolymerization within the gel, and that light energy can be efficiently converted to electrical energy by compounding PEDOT.

7.6 Problem-Oriented Approach and Function-Oriented Approach

The outline of Bio-TRIZ database has been explained, and problem solution utilizing the traditional matrix method in Bio-TRIZ has been introduced from engineering technological problem-solving example to manufacturing example by a goal-oriented approach. This database is stocked separately in the state of which a resource of data about combination of an improvement factor and a deterioration factor, information about TRIZ, and function information from nature became independent, and each piece of information is not connected the other as respective information. When it's possible to put a mutual relation in order and follow a related chain about information on this database, you can get information from the various approaches. So we transform all information to the resource description framework (RDF) on a database. RDF is a standard model for data interchange on the Web and has features that facilitate data without premising on a specific application and knowledge range. It can rewrite data by triadic relations of the subject (resource), the predicate (property), and the object (object: the price of the property), we can search data easily and trace related chains to these data. Some engineers who have engaged in the product development have opinions that it is difficult to create a technological contradiction matrix at conventional TRIZ method. So the new searching system is developed this RDF framework applying TRIZ solution method as known as "effects." It was that data of a conventional Bio-TRIZ database transform using RDF, and we can search biomimetic information from a function parameter. This method can search for a natural principle extracted natural phenomenon and develop it to search biomimetic solution guide from 40 principles in TRIZ. Users can get idea for problem solving with function parameter thorough natural phenomenon and information on Bio-TRIZ by a different viewpoint. So we construct two way research process, one is "problem-oriented approach" from data of a conventional Bio-TRIZ database and the other is "function-oriented approach" which

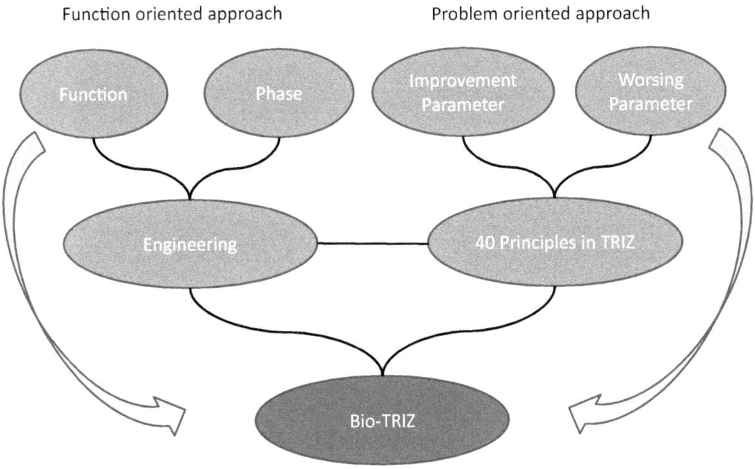

Figure 7.11 Concept of "problem-oriented approach" and "function-oriented approach" at Bio-TRIZ database.

can get Bio-TRIZ information from the function (Fig. 7.11). Moreover we can search optimized principle in Bio-TRIZ from "function-oriented approach" reversely using RDF. When it's input what kind of function you'd like to make act on it where of the solid, liquid and gas by a search of "function-oriented approach" as shown in Fig. 7.12, you can search for mechanism of a similar creature as well as the natural principle to improve its function. For example, when you search for a problem about contradiction of the large volume vs strength in the above-mentioned solar cell by function approach at Bio-TRIZ, "protection" of the material of the solar cell will be the necessary for new technology. What kind of "protection" does nature have? We can search the spread barrier, covering, chemisorption and gelation by "protection" × "the solid" through function oriented approach at Bio-TRIZ. A learned flexible solar cell was indicated by a scale of a birdwing like the above-mentioned problem orientation approach as Bio-TRIZ data about the spread barrier. Moreover, function orientation approach showed that the idea of new materials design of the solar cell for which flexibility of the bamboo was used and the cooperation gluing for which a micro space of ivy was used.

Search from functions

Function	Situation
O Rough O Change from concave to convex, v.v. O Solidify O Change hardness ◉ Separate O Divide	◉Solid OLiquid OGas

See solution in the case of searching from functions

Combination of elements that you chose

【Function】Separate 【Situation】Solid

Engineering	Biomimetics	TRIZ principal
Acoustic wave	-	-
Stiction	Octopus adsorbing by mussels	Local quality
Centrifuge	-	-
Coanda effect	-	-
Corona discharg	-	-
Electret	Reason that angleworm is not bemired (surface is taken by minus electric charge)	Equi-potentiality
Filter	-	-
Friction	Lotus leaf with repellent water and mud	Local quality
Inertia	-	-
Magnetic field	-	-
Magnus effect	-	-
Boundary layer momentum	-	-

Figure 7.12 Search system by "function-oriented approach" at Bio-TRIZ database.

7.7 Conclusion

In this chapter, we introduced readers to the Bio-TRIZ method and explored the possibilities of bringing the wonders of living things into engineering. Biofunctions comprise a gold mine of technologies to be patented. The researcher or inventor who can unravel those biofunctions, work out the regularities submerged within, and skillfully wield those findings like a winning chess piece should be able to help come up with new designs in engineering. It is our belief that by making the most of this Bio-TRIZ database, individuals involved in research at companies, whatever the field, will be able to come up with fresh ideas for resolving technological contradictions, unlike those that have already led to patenting efforts. It will also help them to take advantage of the engineering technologies at companies that in the past backed up the work of Japanese industry, and as a result of these various applications they will produce new technologies that will create the new industries for the future.

References

1. Among numerous works, we can single out Shimomura M. "Seibutsu no tayōsei ni manabu shinsedai: Baiomimetikku zairyō gijutsu no shinchōryō" [A new generation learns from the diversity of life: New currents in biomimetic materials technology]. *Science and Technology Trends*, 9–27 (2010).
2. Japan Patent Office (JPO). *Tokkyo shutsugan gijutsu dōkō chōsa hōkokusho* [Survey report on technology trends in patent applications]. *JPO*, 39 (2015).
3. Altshuller, G. S. *Creativity as an Exact Science: The Theory of the Solution of Inventive Problems* (trans. by Williams, A.), Gordon and Breach Science Publishers, New York (1984).
4. Sawaguchi, M. *VE to TRIZ* [VE and TRIZ], Dōyūkan, Japan, 56–153 (2002).
5. Kasai, H. *Kaihatsu sekkei no tame no TRIZ nyūmon* [An introduction to TRIZ for development design], JUSE Press, Japan, 35–175 (2006).

6. Kasuya, S. *Zukai—kore de tsukaeru: TRIZ/USIT* [Visual guides—With this you can use it: TRIZ-USIT], Japan Management Association Management Center, 15–74 (2006).
7. Hasebe, M., et al., *Hajimeyō kantan TRIZ* [Let's begin—TRIZ made simple], Nikkan Kōgyō Shinbun, Ltd., Japan (2007).
8. Vincent, J. F. V., and Mann, D. *Baiomimetikkusu handobukku* [Handbook of Biomimetics], trans. Osada, Y., NTS Inc., Tokyo, 3–14 (2000).
9. Vincent, J. F. V. *Baiomimetikkusu no shintenkai* [New developments in biomimetics], NTS Inc., Tokyo, 9–25 (2002).
10. Vincent, J. F. V., and Mann, D. TRIZ in biology teaching, *TRIZ Journal,* September 2000 (2000).
11. Mann, D. Integrating knowledge from biology into the TRIZ framework, *TRIZ Journal,* October 2001 (2001). (Note: As with the previous endnote.)
12. Ishida, Y. Patent and utility model number 2007-236831 (2007).

Chapter 8

Creating a Fountain of Future Lifestyle Ideas: Project Implementation

Ryuzo Furukawa[a] **and Masae Mitsuhashi**[b]

[a]*Faculty of Environmental Studies, Tokyo City University, 3-3-1 Ushikubonishi, Tsuzuki-ku, Yokohama, Kanagawa 224-8551, Japan*

[b]*Graduate School of Environmental Studies, Tohoku University, 468-1 Aoba, Aramaki, Aoba-ku, Sendai, Miyagi 980-0845, Japan*

ryuzofkw@tcu.ac.jp, masae.mitsuhashi.d7@tohoku.ac.jp

A number of local municipalities, including Toyooka City (Hyogo Prefecture), Kitakami City (Iwate Prefecture), Akita City (Akita Prefecture), Okinoerabu Island (Kagoshima Prefecture), and Shima City (Mie Prefecture), have started to implement proof-of-concept projects to change lifestyles, developed by interviews with 90-year-olds and backcasting. We would like to introduce the case of a food-related lifestyle in Toyooka City where a small district, inspired by its doubts towards a lifestyle that prioritizes convenience, has started to implement changes in lifestyle for future generations, including children. In Kitakami City, we'll be looking at a private sector approach to lifestyle design in Tenshochi, an area famous for its cherry trees. Also being implemented in Kuchinai, a district of Kitakami, is a lifestyle design based on the traditional Kuchinai Japanese umbrellas that make use of local resources.

Lifestyle and Nature: Integrating Nature Technology to Sustainable Lifestyles
Edited by Ryuzo Furukawa

ISBN 978-981-4774-62-8 (Hardcover), 978-1-315-14687-4 (eBook)
www.panstanford.com

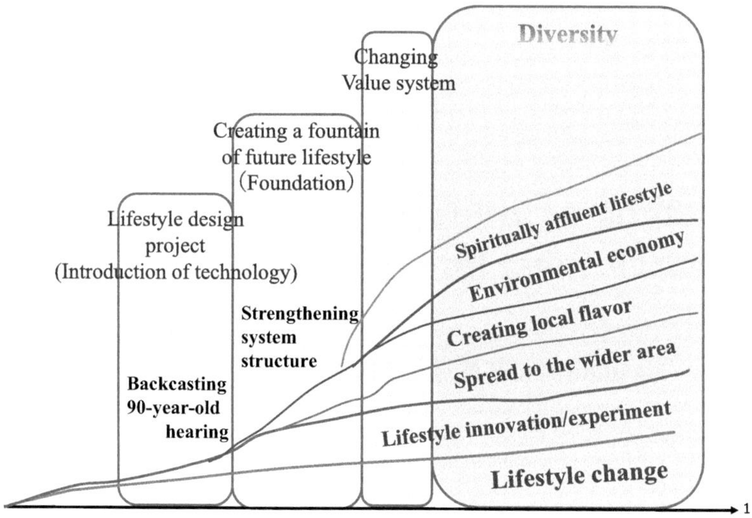

Figure 8.1 Process of the creating a fountain of future lifestyle ideas.

We will take an overall look at the different processes implemented by various regions and consider how and why changing lifestyles has become a growing trend across Japan.

8.1 Creating a Fountain of Ideas for Future Lifestyles: Project Implementation

In order to achieve sustainability under the harsh environmental constraints projected for the future, each region in Japan must give serious thought to the important values that need to be maintained and develop new businesses and policies that enable the creation of an independently sustainable and spiritually affluent lifestyle that makes the most of regional characteristics. In order to achieve this goal, the JST-RIXTEX decided to adopt *Creating a Fountain of Future Lifestyle Ideas* as one of its projects from 2015 to 2018.

This project is aimed at implementing ideas inspired by the lifestyle design methods introduced in Chapters 3 and 4, which are based on interviews with 90-year-olds and backcasting. In addition to changing peoples' mindsets, it focuses on how to make the most of regional characteristics; in other words, how to utilize natural regional resources to create a sustainable and spiritually affluent lifestyle for future generations.

Interviews with people around the age of 90 who remember life before the war have made us aware of values that are disappearing. But our aim is not to turn back the hands of the clock and restore old lifestyles and values. Instead, we are aiming to apply their words of wisdom to design new lifestyles that enable people to feel spiritually affluent despite the severe environmental and social constraints predicted for the future.

For example, before the war, people in the Tohoku region used to collect and share natural resources such as firewood, which was kept in the community woodshed. However, as we come to grips with future environmental constraints, rather than sharing firewood, we can apply the concept of sharing natural resources in a *different* way, for example by sharing the use of storage batteries charged by photovoltaic power generation. Before the war, people took great care of their belongings. This value was reflected in every aspect of daily life. Faced with all kinds of constraints, we could apply this value to emphasize the maintenance and repairs of community property and goods to make them last for a long time. Such applications are possible by analyzing lifestyle before the war.

Here are some examples of lifestyles inspired by the lifestyle design method. The city of Toyooka came up with the idea of using its local produce to bring people together (where several generations could enjoy cooking and eating local produce in season). Kitakami City envisaged a lifestyle where people could choose where to live according to the seasons and enjoy living in harmony with nature (by giving people an opportunity to experience living in other districts, the aim is to nurture understanding and empathy towards other regions and live in accordance with nature). The author collaborated with people in various areas to come up with these lifestyle designs through interviews with 90-year-olds and backcasting. Those aged 90 and over interviewed in Toyooka City talked about how they used to enjoy local produce in season. In Kitakami City, the 90-year-olds reminisced about how they used to adapt their lifestyles according to the seasons to survive in the region's harsh climate. These concepts were applied in designing new lifestyles.

Table 8.1 Specific problems, causes, and bottlenecks in target regions

Target region	**Toyooka City, Hyogo Prefecture**	**Kitakami City, Iwate Prefecture**	**Ise Shima District, Mie Prefecture**	**Okinoerabu Island, Kagoshima Prefecture**
Population (approx.)	82,000	93,000	234,000	13,000
Features of natural environment	–Rich natural surroundings (mountains, Maruyama River, Japan Sea) restoring wild stork population	–Rich natural surroundings (Kitakami River and basin) –Attracting businesses	–Rich natural surroundings –Pearl aquaculture, lobster fishing and other industries, Ise Shrine	–Isolated island –Mild climate –Water shortage
Features of lifestyle	–Urbanized, car–oriented –Tourism (4.7 MM/year)	–Urbanized, car-oriented –Tourism (1.27 MM/year)	–Nature-based industries –Tourism (12 MM/year)	–Farm-dependent –Tourism (40,000/year)
Focal challenges	–Shifting from environment conservation to environment industry –Lack of vision for future lifestyle values	–Lack of distinct local industries –No self-reliant form of entertainment	–Decline of nature-based industries from global warming –Compatibility of tourism, environment conservation, work and community living	–Damage to agricultural crops due to increase of heavy typhoons –Deterioration of regional circulation system
Regional bottlenecks	–Homogenous values of citizens who aspire to urban life –Seeking partially optimum solutions without creating new values (solutions to immediate problems only, no consideration towards future lifestyles)		–People are so used to living in harmony with nature that they are unaware of the importance of their wisdom and knowhow on living sustainably in harmony with nature.	

As a first step towards this new lifestyle, Toyooka City has already held more than six times experience-oriented events that drew participants across several generations. But we need to go beyond the organizing of events and consider a methodology for realizing a new way of life that would be both economically and spiritually affluent. In particular, we need to consider the inspiration, incentives, and constraints that are necessary for encouraging people to adopt different values and make the transition towards a sustainable way of life.

Medium-term plans should include the development of a collaborative relationship between existing or newly established nonprofits independent from local governments, laboratories, and Tohoku University. The methodologies developed in collaboration with researchers from local universities could then be introduced or handed down to local leaders who would serve as facilitators in organizing experience-oriented events or trials that would lead to new businesses and policy proposals. In the long term, this approach would be adopted in other local municipalities.

The target regions for this project are Toyooka City (Hyogo Prefecture), Kitakami City (Iwate Prefecture), Shima City, and other areas in the Ise Shima region (Mie Prefecture) and Okinoerabu Island (Kagoshima Prefecture).

This project is aimed at designing future lifestyles through the following process. First, organize a forum of local government officials or representatives from the private sector to share understanding towards global environmental constraints (population, global warming, energy, resources, water, food, biodiversity, etc.) predicted in 2030 based on data published by national or international institutions. After confirming the data, half a day should be spent in sharing opinions on the social situation in 2030. This should not be about predicting the future but about how certain conditions would bring about certain changes in society. This would make it easier to envisage life in 2030. The next step would then be to consider what aspects of current lifestyles would be impossible to maintain or would be problematic in that social context, and then work backwards to identify solutions to such problems and envisage a lifestyle that would be spiritually affluent (backcasting). In designing such a lifestyle, we would refer to the results of 90-

year-old hearing about the way of life before the war to highlight the special features of the region (values that need to be preserved). The problems revealed through this process would then be relevant to future global environmental constraints, and the envisaged lifestyle would be a lifestyle that would be spiritually affluent despite such constraints. A lifestyle that ignores global environmental constraints will not reflect any of the new values created through this process. The envisaged lifestyle should provide solutions to the problems that are likely to occur in the future.

The next step would be to consider the technologies, businesses and policies necessary to realize the envisaged lifestyles. Rather than attempting to realize such lifestyles in a single leap, people who hold current values would create a roadmap for transitioning towards these future lifestyles (a methodology for designing a lifestyle in stages by analyzing the so-called *slope of convenience,* clarifying the change in values, the inspiration and incentives for change, the constraints needed to maintaining a self-contained lifestyle, and the spiritual affluence gained by making the transition from a dependent lifestyle to a self-contained lifestyle) and consider the necessary technologies, businesses, and policies for achieving this goal. Ontology engineering (implementing available software) would be used to come up with matching technologies.

After selecting a district within the municipality, the first step would be to organize an event to experience new lifestyles, which would help determine the challenges in realizing such lifestyles. The second step would be to consider the businesses and policies this would necessitate. We have already carried out these steps in four target regions, each with different issues. The objective of this project is to develop a project for *Creating a Fountain of Future Lifestyle Ideas* for each of these four regions.

In the long term, we are hoping to apply the above system to municipalities that are not targeted by this project. In other words, we aim to establish similar projects for *Creating a Fountain of Future Lifestyle Ideas,* which would then be implemented according to similar processes. We hope this system will give rise to collaborative research with local businesses which would then lead to the development of lifestyle proposing products and services or local university ventures. Once we succeed in developing a methodology for creating new businesses based on lifestyle design, research leaders could then collaborate with major corporations interested in the lifestyle design methodology to develop new businesses, which we hope will lead to the creation of new products and services that would promote changes in urban lifestyles other than in the regions targeted by this project.

On March 14, 2016, we invited lifestyle design project collaborators in Toyooka, Kitakami, Okinoerabu Island and the Ise Shima area to a joint symposium in the Ise Shima region to report on activities conducted in 2015 and share information on lifestyle design related developments in each region. The symposium signified the full-scale launch of the project.

In 2016, we set up five laboratories in Toyooka City, Kitakami City, Okinoerabu Island, Shima City, and Akita City to conduct research led by Dr. Furukawa of the Graduate School of Environmental Studies, Tohoku University, allowing us to conduct cutting edge research on transforming lifestyles in collaboration with local municipalities. The laboratories are open to local residents interested in the research and serve as a forum for discussion.

We will now look at the progress made on several precedent cases.

8.2 Toyooka City

8.2.1 Launch of Toyooka Lifestyle Design Project

The city of Toyooka in Hyogo Prefecture is aiming to realize a lifestyle that is in harmony with nature, exemplified by its efforts to restore its wild stork population. By 2030, the impact of global warming and the shortage of fossil fuels will have given rise to various environmental constraints. How then can we achieve a lifestyle for Toyooka that provides a sense of spiritual affluence with a local flavor?

In 2013, a lifestyle design project was launched in collaboration with Toyooka City. A project team was set up, comprising several staff members from the local government employee and several members from local businesses. The team organized workshops every two months or so to learn about global environment constraints and discuss how this would affect society in the future. They then began to design various lifestyles for 2030 using the backcasting method. The local government employee members and local businesses members came up with a total of 71 lifestyles. Interviews with Toyooka's 90-year-olds were conducted at the same time to gather information about what lifestyle was like before the war to see how people managed to find spiritual affluence in Toyooka.

In 2014, several designs were selected and work began on developing strategies to realize these lifestyles. One of the lifestyles designed by the local government employee members was similar to a design proposed by the local businesses members and was selected as the main focus of development. It involves bringing people together through local produce, a lifestyle where several generations could gather and learn about traditional cooking methods and enjoy cooking and eating local produce in season. The following is a description of the lifestyle design.

"In 2030, surplus agricultural products etc. are collected from local home gardens and farms and sold at traveling night markets in each district. The market is crowded with people on their way home from work. Producers can convert losses into profit, and consumers can obtain information on production and shipment from the website and place orders in advance, as well as exchange information with producers. Electric or hydrogen vehicles are used to collect, deliver and sell the produce. Nature technology is used to preserve vegetables in a storage box. Meanwhile, local senior residents bring the produce from their home gardens to the Toyooka Kitchen, where they cook and eat together. The kitchen operates on a membership fee system and is held regularly at the city hall. It also organizes cooking classes and contests. In the future, it may be able to play a role in taking care of local children, providing the elderly with a sense of community participation and accomplishment."

To realize this lifestyle, we created a relationship chart to clarify the future issues that would be solved through this lifestyle, the system through which new values would be created, what the values were, the challenges in realizing the lifestyle, the necessary policies and technologies, and a roadmap for implementation.

At present, the level of environmental constraints on the community is not high enough to motivate local residents to adopt the "Gather Around Local Toyooka Food" lifestyle right away. We therefore decided to organize events that would trigger a change in people's thinking as a precursor to achieving this goal, events aimed at "enjoying foods in season."

As a first step towards realizing the lifestyle design, the Nakasuji district in Toyooka City was selected as a model district to hold events that would give people the opportunity to experience the lifestyle. A total of five events to "enjoy Nakasuji's foods in season" were held over the year. Let's look at each of the events.

8.2.2 Enjoying Nakasuji's Foods in Season Event

This event was the result of numerous discussions with leading members of the community about how to reconfirm traditional values. In order to realize the lifestyle designed by backcasting in which local produce would be the focal point in bringing people together, we realized we would need to change people's values regarding local production and consumption, the richness of the local area and meeting with people across generations—all of which were in decline—and implement lifestyle changes in stages. We decided to start off by with a simple reminder of how delicious foods were in season and how good local vegetables were, and how important it was to listen to what the elderly had to say and make use of their wisdom. These were things that could be achieved right away and would enable people to reconfirm traditional values. Feedback from the event was very positive and included comments on how enjoyable it was to spend time and effort on cooking, how it was the first time for them to experience the richness of local produce and the wisdom of traditional methods, and how they were able to effortlessly communicate with their children. The "Enjoying Nakasuji's Foods in Season" event helped reconfirm eroding values.

Enjoy the Bounties of Autumn

On October 16, 2014, the Nakasuji District Community, the Nakasuji District Public Hall and the city of Toyooka organized a joint event to "Enjoy the Bounties of Autumn" at the Nakasuji District Public Hall. Around 100 people, including fifth and sixth graders from Nakasuji Elementary School, took part in the event.

Suriyaki Cooking Experience

Suriyaki is a pancake made with flour, water, and carbonated water. It's a traditional children's snack that was commonly eaten in Nakasuji District, a major producer of wheat until about 50 years ago. The participants were able to compare plain suriyaki made with just wheat and soda water, and a richer version with eggs and sugar that was enjoyed by children of wealthy families or prepared for special occasions such as ceremonies and annual events.

Photo 8.1 Suriyaki cooking experience.

Vegetable Quiz

Children were asked to name the 52 different crops harvested in the Nakasuji District which were laid out on a table. The quiz helped to reconfirm the huge variety of vegetables produced within the community. Since the food prepared during the event was made entirely with vegetables from the area, the children were able to appreciate the bounties of nature and realize that this was another measure of affluence.

Photo 8.2 Vegetable quiz.

Enjoying Local Foods in Season

Residents of the Nakasuji District served a variety of dishes made with local produce in season to the fifth and sixth graders, their parents, and people associated with Tohoku University. The menu included rice cooked with sweet potatoes, canned mackerel curry, simmered green bell peppers, boiled vegetables with tofu dressing, mixed vegetables simmered in a savory sauce, grilled chicken with green herb sauce, mixed vegetable soup, and deep-fried sweet potatoes coated with brown sugar. Most of the children said they didn't like vegetables, but they devoured the food that was served, even though it was mostly vegetables. Some children said they never had such a variety of vegetable dishes at home.

Photo 8.3 Enjoying local foods in season.

Photo 8.4 Enjoying local foods in season.

Professor emeritus of Tohoku University, Emile H. Ishida gave a lecture titled "Thinking about the Earth and Poop!" where children learned about the limited availability of natural resources and how all matter is circulated throughout the ecosystem.

Photo 8.5 Class taught by Emile H. Ishida, Professor Emeritus, Tohoku University.

Photo 8.6 Workshop with children.

The children were divided up into groups and were asked to share what they'd learned or felt from taking part in these seasonal food events and think of ways to ensure the production and consumption of local vegetables.

Photo 8.7 Workshop with children.

The seasonal food event attracted many people from in and around the Nakasuji District. Everyone had a great time, providing food for thought on what constitutes a spiritually affluent lifestyle. Similar events are being held regularly.

Photo 8.8 Participating people.

In addition to holding seasonal food events in Nakasuji throughout the year, we also interviewed 90-year-old people in the district to learn about life in the old days when resources and energy and were limited, especially about traditional foods and cooking methods. Through the process of gathering basic data, local residents (parents and their children, the elderly) became more aware of what foods were in season, which in turn encouraged them to think of how to enjoy good food despite various constraints.

In order to build upon these events and preserve important values that would enable people to lead a spiritually affluent lifestyle despite various constraints, we needed to give serious thought on how to pass on these values to future generations. Children as well as the elderly came up with various proposals, inspiring the entire community to think about the future. Children can be quite astute and more often than not are able to focus on essential matters. For example, when asked about what they thought should be done to make sure that people would be able to continue to enjoy local produce, the children gave astute answers that included: adopting the use of local products in school lunch programs; encouraging mothers to buy local produce; and cooking more meals at home and sharing the surplus with people who are unable to do so. What are the issues hampering the realization of such proposals? One has to do with the school lunch center's emphasis on work efficiency. Rather than buying in small volumes from local farmers, it opts for mass procurement. Then there's the fact that local supermarkets only stock small amounts of locally grown vegetables and rely heavily on vegetables from the Kinki region to secure a stable amount of stock. To tackle these issues and achieve a lifestyle that connects people through local produce, it was decided to implement the proposal to use of local ingredients in school lunches and provide children with more opportunities to eat local crops.

8.2.3 Implementation of a Lifestyle Using Snow Cellars in Nakasuji District, Toyooka City

Having studied the model project to promote a lifestyle in which children are provided with school lunches that make use of local ingredients and are given the opportunity to grow and consume safe and delicious produce, we agreed that snow storage facilities would be an essential technology for this. The idea of introducing snow cellars was suggested by residents of Nakasuji upon hearing how their children wanted to have local vegetables served in school lunches. It's an example of a multi-generation joint effort led by the

Photo 8.9 Snow cellar opening event.

Photo 8.10 Children discuss what to store in the cellar.

Nakasuji District Community Center. Proof of concept studies began in fiscal 2016 and school lunch center started supplying the school lunch center with onions and potatoes from 2016.

The technology makes use of a natural resource that is plentiful because of Toyooka's heavy snowfalls. The snow cellars will allow potatoes and onions to stay fresh for a long time, providing school lunch centers with a stable supply over a long period. In addition to recording the temperature, humidity and state of preservation in the snow cellar once a week, we will also monitor the behavior and level of spiritual affluence experienced by people who live near the snow cellar. This case study shows that it is possible for the community to take the initiative in making a gradual transition, towards the envisaged lifestyle design, in this case, coming up with the idea of using snow cellars to make a transition towards a lifestyle of producing and consuming food locally.

Snow Cellar Opening Ceremony

Local people are coming up with ideas to make a gradual but steady transition towards a lifestyle in which people gather around local produce. It's easy to be swept up by the pace of everyday living, but here, people across generations have realized the need to stop every now and then to think about life and reconfirm and implement local values that have been handed down through generations. This case set a precedent, proving that once local residents understand the significance of backcasting and interviews with 90-year-olds, they can play a major role in designing and implementing lifestyles.

8.2.4 Integrated Studies at Nakasuji's Elementary School

In 2015, based on basic data on the local food culture that would enable the creation of a spiritually affluent lifestyle for the Toyooka City Nakasuji district despite future environmental constraints, we examined the possibility of providing elementary school children with the opportunity to learn about local traditions at school. This led to the development of educational materials in 2016. We believed it was important to pass on precious local knowledge.

We worked together with the principal of Nakasuji Elementary School to develop the teaching tools and came up with a booklet titled "My Hometown, Nakasuji," and a DVD introducing the interviews with Nakasuji's 90-year-old people. These materials were put to use in the six graders integrated studies program to envisage a future lifestyle for Nakasuji through backcasting.

In the DVD, the elderly residents talk about prewar life, about Toyooka's natural landscape, its wildlife, people, and work. They share their memories of daily life, about how they used to gather firewood for fuel; how boats were the main means of transportation other than walking; how there existed a spirit of give-and-take that allowed people to walk freely through paddy fields without having to ask for permission; how, because bathtubs in private homes were only filled and heated once a week, it was customary for people in the neighborhood to go from one house to another to share baths; how they did the washing in the river; how all the villagers worked together to shovel snow; how they used to raise silkworms as a source of income; and how they would burn straw to ward off mosquitoes. They also reminisced about how much fun they had as children, making the most of their rich natural surroundings, playing

Photo 8.11 Lecture at Nakasuji Elementary School, Toyooka.

and enjoying the delicious foods, how they used to catch cicadas and enjoy skiing on homemade bamboo skies, how they would carry pieces of basalt as weights so that they could walk along the river bed of the Maruyama River. The DVD conveys facets of prewar life unknown to most of the younger generation. After watching the DVD, the children will be asked to think about the differences between past and current lifestyles and consider what brought about such changes. They become aware of the spiritual affluence that used to exist and think of what sort of life they would like to lead in Nakasuji in the future. Without using technical expressions such as backcasting, the educational materials will enable children to apply a similar approach to envisaging the future.

In October 2016, Nakasuji Elementary School began using these educational materials in the integrated studies program for sixth graders.

Photo 8.12 Lecture at Nakasuji Elementary School, Toyooka.

The example of "lifestyle experience" events in Toyooka's Nakasuji District shows that once we start reviewing lifestyles, local residents begin to take the initiative in developing new projects and ideas that seem to inspire even more ideas. Ideally, it should be the

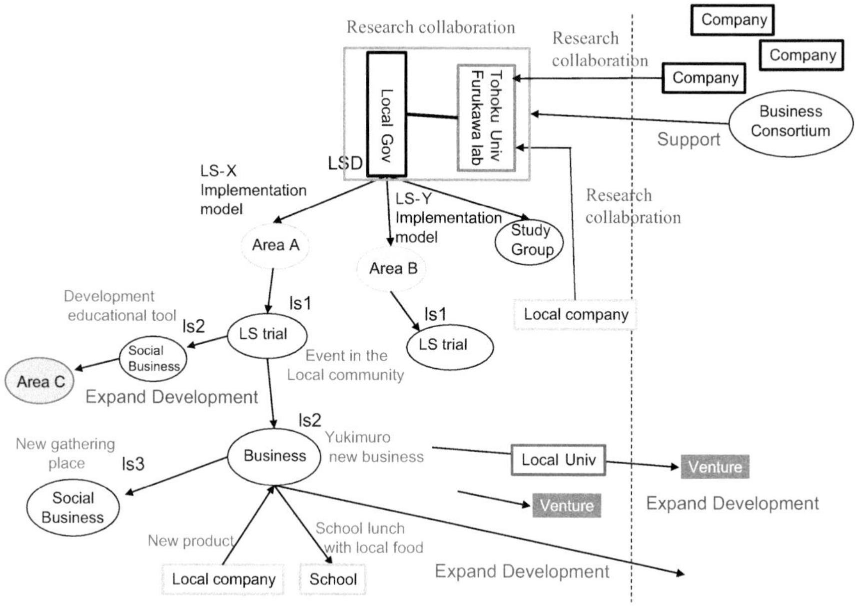

Figure 8.2 Expand of lifestyle implementation.

local community that takes the initiative in seeking and transitioning towards a new lifestyle—it should not be something forced on them. In fact, this is probably the only way to nurture a lifestyle for the future.

8.3 Kitakami City

The city of Kitakami in Iwate Prefecture proposes a hydrangea-style city for sustainable living. A hydrangea-style city is a network of local communities that collaborate and coexist with each other while functioning as one comprehensive city. Each community is self-sustaining in the sense that all functions that support daily life are concentrated within walking distance. The local communities comprising the center are compared to hydrangea blossoms. Each community is self-sustaining, flourishing by making use of local resources. Through mutual collaboration among communities, the city will be able to pursue sustainable development and provide a spiritually affluent lifestyle to its residents despite its declining and super-aging population.

8.3.1 Lifestyle Design in Kitakami City

Based on this concept, the Kitakami Lifestyle Design Project was launched in collaboration with the Kitakami City government in 2014. The first year was spent working together with about 10 members from the local government employee to, in the same way as occurred in Toyooka City, come up with a total of 108 lifestyle designs using the backcasting method, and interview the city's 90-year-old people about what life was like before the war. We would now like to introduce the three most popular designs for implementation selected by vote by the project members.

"Choosing Where to Live According to the Seasons"

Kitakami City's population continues to shrink from its peak of 94,000, with aging and a decline in the working population causing a rapid decrease in tax revenue. The city's budget is also falling. It's a large city and it's becoming difficult to manage the overall infrastructure in town and in the surrounding hills and mountains. Bottlenecks in infrastructure management include delays in clearing snow, causing problems in winter months in some districts. In such areas, an increasing number of people have come to spend the winter months in town, renting vacant houses. People are becoming quite rational in terms of living environments, opting to spend the winter in town where the infrastructure is well managed, regarding their own homes as a villa that offers a retreat from the summer heat. This style of living where people move house according to the seasons is somewhat reminiscent of the way a hermit crab upgrades its shell. Since uninhabited houses are easily dilapidated, the idea of making effective, seasonal use of vacant houses is attracting attention not only among the elderly but families with small children.

"Buying Home-cooked Meals at a *Sharing Convenience Store*"

In the past, people would often share surplus food with neighbors. Now, this will be done via the local convenience stores. Local housewives bring in the excess food they've prepared to a nearby store. The store posts "Today's Share Menu" on its website. Businessmen who have been transferred to the region and are living away from their families can look at what's available and buy home-cooked food at the store. Housewives can also choose to buy home-cooked food to add to their menu. From the supplier's point of view, housewives can look at the "share menu" and prepare foods that they feel are missing from the menu and might sell well. Since the underlying spirit is sharing, there're always be an extra something for people who come to buy the food. The meals offered by the sharing convenience stores reflect the local food culture, providing employees living away from their families, or families who are frequently uprooted because of a parent's job, with a chance to connect with the community.

"New Mothers Raising Children with Confidence"

The increase in nuclear families has led to an increase in houses for mothers and newborn babies. These are houses where mothers with newborn babies can stay after leaving the hospital. Old houses have been refurbished to create a comfortable space for both the mother and baby. They can stay here until the mother has gained enough confidence to care for her baby. When the mother is tired from caring for her baby who seems to be crying all the time, she can rely on help from a volunteer who acts as a surrogate grandmother. Each mother and baby have a private room, but there's also a lobby where the mothers can enjoy each other's company. The fathers can pop by on their way home from work. Since it's an old house, it requires some maintenance and occasional repairs. Unless the work calls for a professional, the fathers do it themselves.

8.3.2 Private Sector Designed Lifestyles for Limited Areas

In 2015, we launched a lifestyle design project for the Kitakami area using the backcasting method, inviting eight people from the private sector to take part. Generally speaking, involving the private sector or businesses in lifestyle design can be quite problematic. This is due to the very nature of backcasting. Even if an attempt is made to develop products and services for the future by applying the technology and knowhow of private sector businesses and corporations, it's difficult to come up with products and services that will be of use in a society under environmental constraints. It's impossible to guarantee immediate returns to the private sector, or to companies applying their business domains or technologies. In order to effectively involve the private sector in lifestyle designing, we therefore decided to target a limited area. By targeting a small, limited area, it's possible to come up with a clear vision for an optimum lifestyle, providing an incentive for relevant corporations in the private sector to take part in the lifestyle design project.

That's why we invited eight people in their late 20s to 40s from the private sector with ties to Kitakami, including several companies, the restaurant or shops at Tenshochi and the Kitakami Municipal Museum Kitakami City, to join us in designing a new lifestyle for the area. Although the young project members had a heavy workload, they were also concerned about the future and willingly took part in workshops to create a spiritual affluent lifestyle for the future. In order for businesses, nonprofits, and other private sector organizations to effectively design future lifestyles, it's crucial to have an appropriate theme, otherwise it can be very difficult. By limiting the target area to Tenshochi, we were able to focus the strengths of the private sector in designing a lifestyle specifically for the area.

We would now like to look at one of the eight lifestyles designed for the next generation. We measured the level of social acceptability by conducting an internet survey. The survey was sent out to research monitors registered with the marketing research company Macromill. To compare the differences in perspective towards lifestyles in different areas, the survey was sent out to 200 monitors in both the Tokyo Metropolitan area and the Tohoku region. Equal numbers of men and women were selected from each of the equally proportioned age groups (20s, 30s, 40s, 50s, and 60s).

The monitors were asked to read the following paragraph that describes a future lifestyle and select the attributes they felt would enhance the sense of spiritual affluence. The graph shows the mean value for each attribute. Overall results show that monitors in the Tohoku region found more factors that would bring them spiritual affluence than the monitors in the Metropolitan region. In other words, the spiritually affluent lifestyle designed by the private sector group in Kitakami City was more appealing to the monitors in the Tohoku region. This indicates that lifestyle designs by people who are familiar with the region are likely to have a more positive effect on spiritual affluence.

"Tenshochi as the Center for Promoting a Sense of Goodwill through Effective Use of Vegetables": Social acceptability, 56.2%

With growing interest in a healthy and self-sufficient way of life, an increasing number of people are growing their own vegetables. When they've harvested more seasonal vegetables than they can eat, they take the surplus to the Tenshochi Vegetable Station for other people to take home. This way, instead of being wasted, the vegetables can be put to good use. In exchange for donating their surplus crops, the home gardeners can take home other crops from the station (bartering system).

Restaurant owners and people without home gardens can go to the vegetable station to obtain local vegetables for free. The vegetables are safe, having been grown chemical free or with limited use of agrochemicals. Restaurants will use the vegetables in their menus, promoting Kitakami by emphasizing the use of locally grown vegetables that can only be enjoyed in the area—an "only found here" menu. Local residents who receive vegetables for free will express their gratitude to the vegetable growers in their own way. Some may comment about the vegetables in their blogs or on social media, or upload recipes, while others will leave thank you cards at the station, which will be put up for display. The Vegetable Station is also popular with people who have relocated here from other areas, and businessmen who are living away from their families. The exchange that takes place through local vegetables allows them to communicate with local residents. The home vegetable growers are motivated to grow more vegetables because of the positive feedback they receive from station users. News of the Vegetable Station and the restaurants that use the station to prepare local specialties will spread via Facebook, attracting people from outside the community. Tenshochi becomes a magnet that brings people together.

8.3.3 Implementing Lifestyles in Kitakami City

The Kitakami district of Kuchinai, with its rapidly aging population, was chosen as the model district for implementing events that would enable people to experience the lifestyles designed for Kitakami City. In 2015, a kick-off meeting titled "Seminar for Creating Future Lifestyles in Kuchinai" was held to discuss the contents of the experience-oriented events with community leaders in Kuchinai. Before implementing a lifestyle, it's important to promote understanding throughout the community. Life in rural areas is different from life in the city. People need time to understand the concept behind the initiatives aimed at changing their lifestyle. The concept needs to be explained repeatedly at community gatherings along with values or wisdom in the old lifestyles gained from interviews with 90-year-olds. It may take time, but it's an essential process.

On February 2, 2016, we organized a joint meeting with members of the Smart Community Promotion Model Project to report on the lifestyles for Kitakami City envisaged in 2015. The meeting enabled us to share the significance of lifestyle design with the participants, including technically savvy members of the community. Feedback was positive. Participants noted that the meeting provided a comprehensive viewpoint on creating a smart community from the perspective of both hardware (facilities, etc.) and software (approach to designing lifestyles). Too much emphasis on the hardware, and the software will be neglected and vice versa.

8.3.4 Designing Lifestyles Based on Regional Resources

With the help of the Nature Technology Research Consortium's Monozukuri Working Group, set up by the Monozukuri Nippon Conference, we decided to apply the backcasting method to create

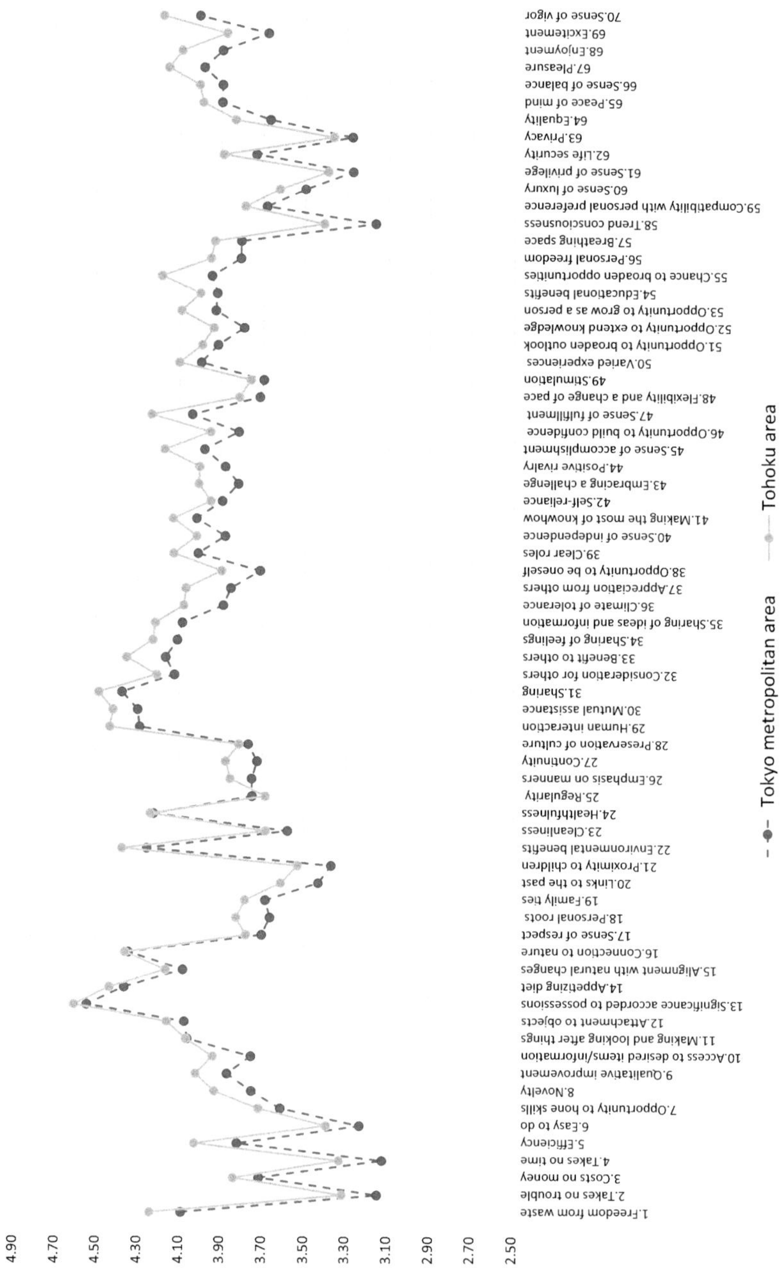

Figure 8.3 Differences in perception of lifestyle between people in the Metropolitan area and the Tohoku area (degree of importance of 70 lifestyle evaluation factor in each area).

Photo 8.13 Kuchinai workshop.

a lifestyle design based on the use of regional resources. We began by studying the case of Kitakami City. A lifestyle design methodology based on regional resources requires the utmost caution. This is because of the tendency to emphasize the importance of reviving traditional crafts. Rather than applying the backcasting method to designing a lifestyle that enhances spiritual affluence despite the environmental constraints that are predicted in the future, the focus is on forecasting what sort of training and technologies are needed for preserving and reviving traditional crafts that are no longer needed. Having said that, it's also true that many traditional crafts made with local materials provide people with a sense of spiritual affluence. If we make sure to take into consideration future environmental constraints, we believe that a wealth of ideas could be found in traditional crafts and technologies. This is why we decided to develop a methodology that involves focusing on regional resources to backcast and design future lifestyles.

In cooperation with the Nature Technology Research Consortium's Monozukuri working group, we conducted research on the traditional crafts and regional resources throughout the Kitakami

Photo 8.14 Having fun with nature.

region. This led us to discover the traditional Japanese umbrellas produced in Kuchinai. These historic umbrellas used to be made entirely with materials produced in the Kuchinai district of Kitakami City and provided people with the income to survive famine. We realized that focusing on the umbrella making technology might be a way to discover some of Kitakami's distinctive features, which could then serve as a foundation to design a lifestyle using the backcasting method. Towards this goal, we organized several workshops over a period of six months.

The workshops led to the creation of three lifestyle designs, one of which will be explained in the following section. In addition to reviving the Kuchinai umbrellas, we believe that we have managed to depict a new lifestyle with new values.

Adding Color to Life with Umbrellas and Lanterns Made with Seasonal Washi Paper

"Scenery"

Excitement is building with the approach of the annual Kuchinai Umbrella Festival. People get together to make Japanese umbrellas,

Figure 8.4 Image of lifestyle.

using washi paper that they've designed or made themselves. The organizers meet with artists beforehand to decide on this year's theme and its duration. Since the participants will need to design and create their own washi paper, they turn to the senior residents for ideas on what sort of colors and patterns would reflect the season. They then head to the mountains to gather the branches and flowers they will need to make beautiful sheets of wash paper which will then be used to make umbrellas and lanterns.

The umbrellas and lanterns decorating the streets are in harmony with the surroundings, creating a dazzling sight. At night, the lanterns are lit up to create a totally different atmosphere.

After the festival, the umbrellas are rented out to tourists and are put to good use. The participants who made the washi paper used for making the umbrellas and lanterns are filled with a sense of achievement. People using the umbrellas appreciate the way they blend into the surroundings, which gives them a good feeling. The umbrella makers are delighted to see how everyone seems to be enjoying their work. Visitors to the city are so impressed that they come back, this time with their families.

8.4 Other Case Studies

Similar methods have been applied in designing lifestyles for Okinoerabu Island and Shima City, both of which are model districts targeted by this project. The project's methodology has been attracting interest from non-targeted areas as well. We organized the seminar for creating future lifestyles in Hiroshima, Takayama City, and Suginami-ku, Tokyo.

In Akita City, members of the Seminar for Creating Future Lifestyles and representatives from Kamishinjo district, including university students, got together to discuss the possibility of setting up a Share Kitchen in Akita. The first of several meetings was held on February 21, 2016, to consider the district's future development.

Photo 8.15 Workshop "Share Kitchen" in Akita City.

We also made a point of showing the following picture to as many people as possible. It depicts the lifestyle we envisaged for Akita City on the basis of interviews with 90-year-olds about life before the war.

Photo 8.16 Thinking about Akita's future with students.

Our methodology for designing lifestyles is also attracting the attention of businesses. In 2015, a number of major companies adopted the backcasting method of lifestyle design towards the development of new products and services for 2020. These companies included gas companies, electric appliance manufacturers, automobile industry companies, railway companies, and IT companies.

To promote the methodology of reflecting the insights into prewar life gleaned from the interviews with 90-year-olds, and the concept of lifestyle design, we commissioned the rakugo storyteller Katsura Sanshiro to create a rakugo story titled "Summer homework," "University at the small island," "Diary of a Crane" and "Demon" and perform it. People who find it too difficult to grasp the methodology explained in academic terms have commented on how the rakugo made it easy for them to understand what it was about. This has made us realize that lowering the barrier between academia and ordinary life can be quite effective in implementing lifestyles.

Figure 8.5 Future Akita City that continues to live in harmony with nature.

With the help of Sendai City, we also produced a short educational film based on interviews with three Sendai residents who were about 90 years old. Many of the people who provided us with feedback commented on how the film helped them understand the project. Some members of the audience said the film made them cry because it reminded them of their grandparents. We believe it will be worth our while make more films like it as they can be very effective in promoting the project.

8.5 The Spread of Lifestyle Reforms and the Challenges Involved

Japan is seeing an increase in collaborative research between municipal governments and businesses based on lifestyle design and the methodology of listening to 90-year-old people. While this chapter focused primarily on the cases of Toyooka City and Kitakami City, we have in fact received numerous requests from other municipal governments and corporations to develop lifestyle-changing policies, products, and services. They seem to have undergone a major change in mindset somewhere around 2014. Previously, the pursuit of lifestyle design was limited to a handful of forward-thinking municipal governments and corporations. But since 2014, lifestyle design has been a focus of major corporate research institutes, with key industries such as the automobile industry incorporating lifestyle design in future strategies. One reason for this change in mindset is that statistics were showing how the population was declining and was on the fast track towards a super-aging society. It was becoming obvious that this would have an effect on corporate strategy. With just five years to go till 2020, the timing coincided with the customary revision of long-term strategies carried out by corporate research institutes every five years. Meanwhile, municipal governments were already being confronted with serious issues before 2014. The rate of population decline and aging was causing businesses to shut down and was making it difficult to provide and maintain public services. Japan's birth rate and population is steadily declining. These municipalities have to operate on a tight budget and are short of staff. Apprehension towards the future must be rising.

The Great East Japan Earthquake that occurred on March 11, 2011, sent shockwaves throughout the country. It brought about a major change in values placed on material possessions. A comparison between people seriously impacted by the disaster and people who were not shows a difference in how values towards material possessions changed because of the disaster. People who were seriously impacted were more aware of the value of material possessions, whereas those who were not came to place less value on them. In other words, those who really suffered from the lack of goods came to take greater care of their possessions. However, the results of a similar survey conducted in 2016 showed that people's appreciation of material goods had returned to the pre-disaster level. Five years after the disaster, life had returned to normal and they no longer felt its effects, leading to a return to pre-disaster values. This goes to show how difficult it is to change people's values and establish new lifestyles.

The key to promoting new lifestyles lies in being able to evaluate the implemented lifestyles. An effective evaluation tool would be the action decomposition tree introduced in Chapter 6 that applies ontological engineering. By breaking down lifestyles into goals, actions, ways, challenges, and the sense of spiritual affluence to create an action decomposition tree makes it possible to evaluate the lifestyle as explicit knowledge. However, with so many different variables to spiritual affluence, and numerous external factors that influence lifestyles, further research is required. If ordinary people can be encouraged to try out these new lifestyles, and they then see the value in them, the spread of lifestyle changes will accelerate.

Acknowledgment

This research was supported by JST, RISTEX.

Chapter 9

Creating Local Flavor and Passing It Down

Ryuzo Furukawa
Faculty of Environmental Studies, Tokyo City University, 3-3-1 Ushikubonishi, Tsuzuki-ku, Yokohama, Kanagawa 224-8551, Japan
ryuzofkw@tcu.ac.jp

In the future, when we face harsh environmental constraints, where will "local flavor"—the things that make a given place special—come from? Regional Japan is looking at the future and thinking hard about how it is going to pass down the best of its local culture to future generations. If you examine ways of life in prewar Japan, you find remarkable examples of local flavor. By contrast, in a way of life with no constraints, it becomes difficult to preserve an area's unique appeal. If we do nothing, the local flavor of certain areas is going to disappear. In a future with harsh environmental constraints, how can a region generate local flavor? This question is vitally important. We believe that, to create truly local flavor that no other area can imitate, each region must take as its foundation its unique natural environment and local resources, which generate spiritual affluence. Moreover, in order to pass this local flavor down to future generations, it is essential to consider the constraints faced by, and the spiritual affluence of, an area's people.

Lifestyle and Nature: Integrating Nature Technology to Sustainable Lifestyles
Edited by Ryuzo Furukawa

ISBN 978-981-4774-62-8 (Hardcover), 978-1-315-14687-4 (eBook)
www.panstanford.com

9.1 Local Flavor and Value Systems in Japan's Prewar Way of Life

Before the Second World War, the Japanese way of life put great importance on finding common ground and existing in harmony with others. Chapter 3 introduced a list of "44 values from disappearing from a way of life." Drawing on that list, we hoped to compare, in as quantitative a way as possible, what makes different regions of Japan special.

To that end, we took the interviews with 90-year-olds that we had previously conducted, and converted that footage (video and audio) into text (transcription). We then carried out a quantitative analysis of the prewar way of life in five areas: Toyooka City, Kitakami City, Okinoerabu Island, Akita City, Miyagi Prefecture [1]. Taking the list of 44 values, which we developed using data drawn from the interviews with 90-year-olds conducted in Miyagi Prefecture and elsewhere, we counted the number of times each value came up in the interviews with people from Toyooka City ($N = 28$), Kitakami City ($N = 18$), Okinoerabu Island ($N = 13$), Akita City ($N = 20$), and Miyagi Prefecture ($N = 40$), and also compared how often each value came up among people in different areas. Our "44 values from disappearing from a way of life" list was based on 40 of our interviews with 90-year-olds in Miyagi Prefecture that were analyzed over the course of the year by four researchers. From that analysis, the researchers identified 70 concepts associated with practices disappearing from contemporary life, and then examined data from the interviews with 90-year-olds from places other than Miyagi Prefecture. Ultimately, they condensed the list down to its current total of 44 values.

Each 90-year-old hearing record has roughly 30,000 Japanese characters when transcribed. The interviews were two hours long and touched on a wide range of topics concerning daily life. In this study, some areas comprise much smaller sample sizes than others, so we must be careful about any conclusions we draw, but we can identify general trends. For example, Okinoerabu Island has a warm climate and mountains that host an abundance of food. The interviewees from there spoke about "treasuring

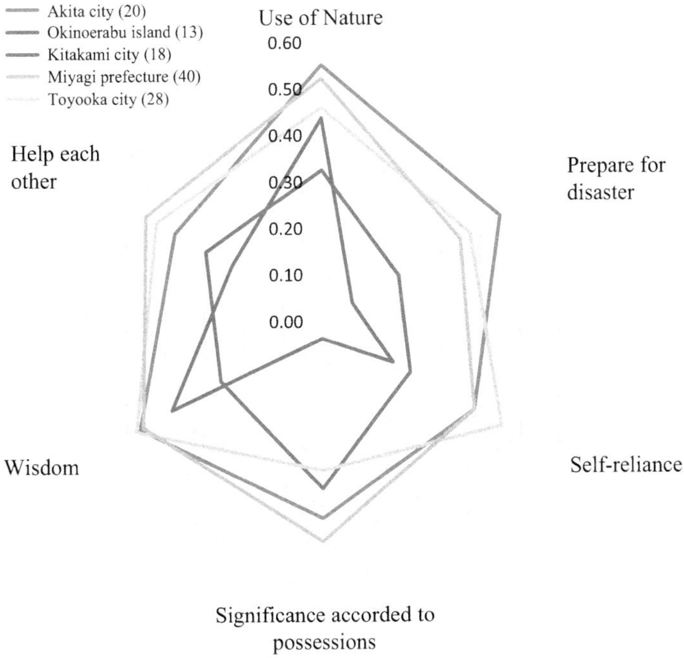

Figure 9.1 Comparison of local ways of life in prewar Japan.

things" and "stocking up" for tough times much less frequently than interviewees from snowy northern regions. On the other hand, if you compare Akita City with Miyagi Prefecture, you find extremely strong similarities (Fig. 9.1). In prewar Japan, an era when people pretty much traveled on foot, or perhaps by train, different places truly did have local flavor—a way of life particular to that given area. Our "comparison" methodology is able to demonstrate that.

However, when it comes to neighboring areas with similar natural environments, comparison alone cannot pinpoint exactly what the features of an area's local flavor are. The meeting point of what makes a region special in comparison with others, and the specific environment and resources that the region possesses, is where you will find the region's local flavor. For example, in Ogatsucho, part of Ishinomaki City, Miyagi Prefecture, for centuries people have mined a type of slate rock called ogatsuishi. As a result, the area became a hub for the production of slate tablets, inkstones, and tiles. This industry had an influence on the local way of life—the area became home to a variety of craftsmen. The things that make a region what it is arise from these sorts of local resources.

On the basis of the results of our analysis, in order to clearly determine the local flavor of each region of Japan, rather than trying to specify through comparison, it would be better to examine the facets of a region's specific natural environment, local resources, and so on. If we do that, then even in a future with environmental constraints, we will be able to pinpoint a given region's local flavor.

9.2 Local Flavor Comes from the Natural Environment and Local Resources

We can tease out a region's local flavor by comparing it with other regions, but we can also tease out that local flavor by looking at the special qualities that emerge out of that region's particular natural environment and resources. Let us use Toyooka as an example to demonstrate this idea. Toyooka has the large, languidly flowing Maruyama River, and agriculture thrives there. Kinosaki Onsen near the Sea of Japan is a bustling resort town. Storks glide overhead, contributing to the sense that time passes slowly here. Willow trees live along the river. Once, there was a thriving industry of wicker luggage made using the willows. The particular customs that had their origins in the local resources of Toyooka were in evidence everywhere in the prewar way of life. In old surviving documents from the area, we find expressions like "co-existence, co-prosperity," and "practical philosophy." It's easy to surmise how the natural environment and local resources of Toyooka made it what it is. Now, let us look at what some local residents who took part in the interviews with 90-year-olds said about the prewar way of life in Toyooka, about the natural environment and local resources, as well as these ideas of "co-existence, co-prosperity," and "practical philosophy." From what these Toyooka people said, we have tried to tease out what prewar life in the area was like, and, by extension, what Toyooka's local flavor was and is.

(1) Natural Resources

Example 1: The Maruyama River

There's a river down there. The Maruyama River. It's got this wide riverbed. When I got home from school, I'd put my bag down. And then I'd undress and go down to the riverbed in my underwear. The river was just in front of it. Usually all the kids would be down there.

*The Maruyama River runs through Toyooka City

Example 2: Pit Vipers

We would put pit vipers in alcohol to pickle them. Someone gave me a piece of a pit viper, peeled the skin off for me—I took and it dried it. One of these vipers. Then, one time, I cut my hand badly. When something like that happened, you would stick the viper skin on the cut. You'd wrap it in the viper skin. You'd get the viper skin wet, to soften it, then wrap it right around. It smelled really bad! But even without going to the doctor, the cut healed.

*Pit vipers were found in Toyooka

Example 3: Willow Trees

Lots of families near mine made baskets from the willow. We did it in our own house. We would collect white, dry willow wood. Then, when it was time to work with the wood, we'd bring water to moisten it, and then bend it into the right shapes. Mainly we made baskets, things like that. Back in those days, the old men who were good with their hands would make these cute baskets for kids to take to preschool. I remember I got a beautiful basket like that—it even had these decorations on it. But now everyone who took part in that tradition is gone. But back then everyone was good with their hands, even the old people.

*Toyooka City has a large population of willow trees, and the area is famous for wicker luggage made from willow wood. It is still a prominent center of bag-making in Japan.

(2) Co-existence, co-prosperity

Example 1: Storks

There were pine trees here in the old days, and the storks would breed up in these pine trees. So around March, you'd always see, up in the pine trees, the storks taking care of their eggs, and then the baby storks would be born. They'd get bigger in April and May, and then in the summer they'd migrate north. They typically nested up in the pine trees. That was just what they did.

Example 2: Kinosaki Onsen

The baths were open 365 days a year. You'd see your friends there. Sometimes people would teach each other things—sometimes they'd get into arguments! I think that's a good thing. But that group bathing culture is gone now. It was like going to an old-fashioned variety show. I mean, not exactly, but it was always interesting. People would teach you, share their knowledge.

(3) Practical Philosophy

Example 1: Co-existence and co-prosperity with living things

We take care of our ancestors. And everything has life. So it's important that we treat out possessions in a way that they, too, can live full lives. We have to do that. In other words, we have to use things in a way that they can fulfill their purpose. So take care of your things. It's the same thing as taking care of your ancestors.

The example we just provided shows what differentiates Toyooka from other areas (Fig. 9.2). The factors that differentiated Toyooka from other places, and the factors of Toyooka's natural environment that gave rise to its unique local flavor, basically corresponded. However, after the Second World War, as information and transportation technology advanced, people stopped drawing on the natural environment in daily life, and the things that made Toyooka different from other places became less apparent.

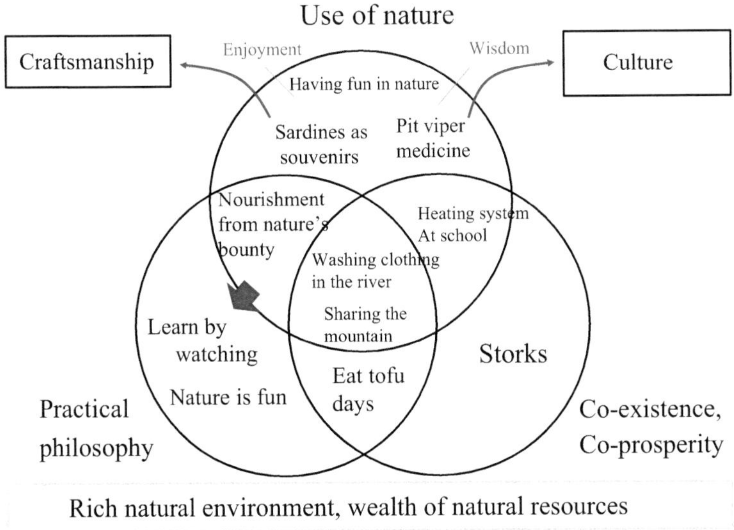

Figure 9.2 Features of the prewar way of life in Toyooka City.

In a future with harsh environmental constraints, information technology, transportation, and so on will become more expensive. The natural environment and local resources which create local flavor will become powerful assets. And they will give rise to things that no other region will be able to imitate. Accordingly, each region must now work to create its own unique appeal, and that involves looking closely at the natural environment and local resources. The starting point is creating local flavor and appeal. If a region can manage to do that, then even in a world of harsh environmental constraints, the people there can pursue spiritually affluent lifestyle.

9.3 A Growing Trend of "Passing Things Down"

What encourages the creation of local flavor, and what hinders it? In Chapter 4, we presented a list of 70 "lifestyle evaluation factor." On that list, the item most connected with the idea of passing things on

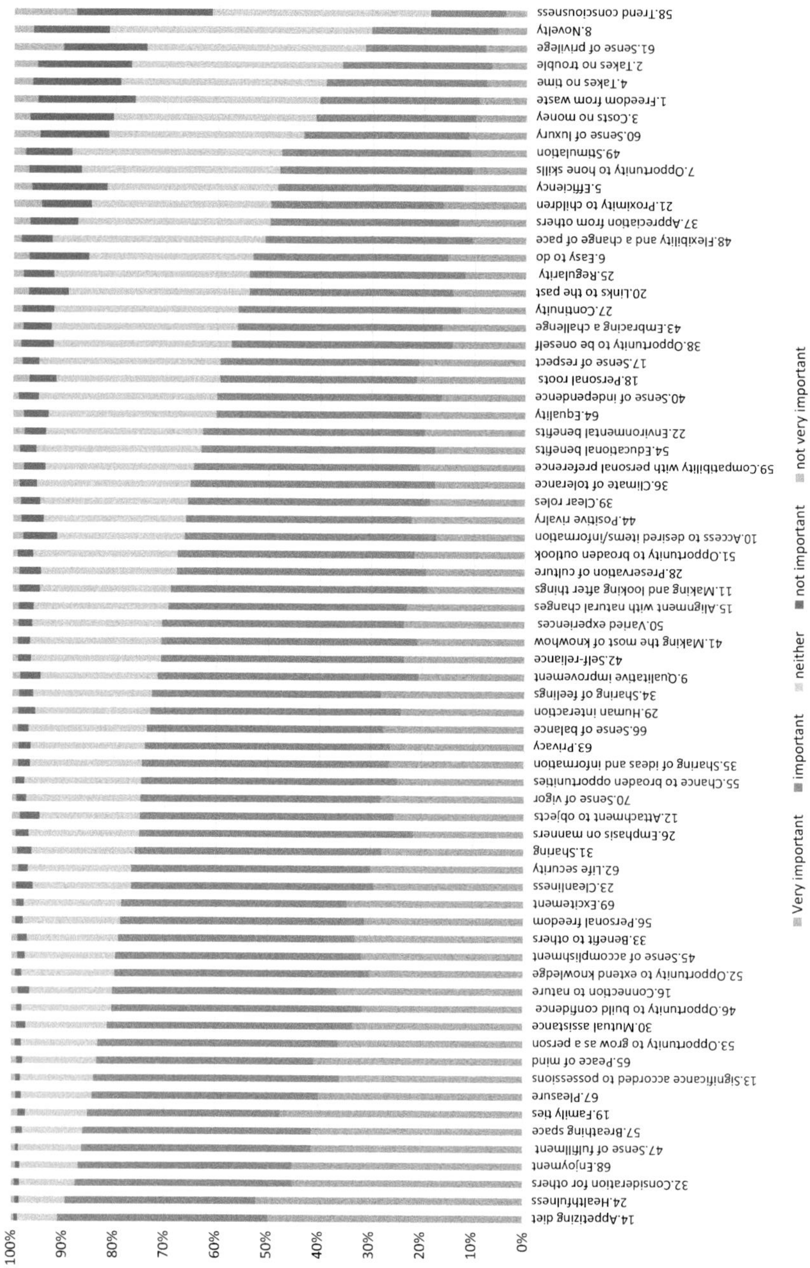

Figure 9.3 70 lifestyle evaluation factors in order of their average “significance” score on the survey for spiritual affluence.

was "continuity." We used a questionnaire survey, and subsequent analysis of it, to investigate the factors that might have an influence on how much a community values continuity [2, 3].

In order to investigate which of these 70 factors people valued, and to what extent, we used an internet survey, "Rakuten Research" to conduct the survey. Their monitors answered this survey were divided by age (29 and under, 30–39, 40–49, 50–59, 60 and older), gender (2 categories), and annual household income (under 3 million yen, 3–5 million yen, 5–7 million yen, 7–10 million yen, 10–15 million yen, over 15 million yen). With five age groups, two gender options, and six income groups, the monitors were thus divided into a total of 60 groups. The aim was to have roughly equal numbers of participants from each group. Over 2000 people were surveyed over three days (January 19–21, 2015). The survey asked the following question: "At some point in your life, you've thought, 'I feel spiritual affluence.' Think of what kind of lifestyle evaluation factor means to your spiritual affluent lifestyle. Now, how significant are the following factors in your spiritually affluent lifestyle?" Monitors looked at each of the 70 "lifestyle evaluation factors," and ranked them from 1 to 5 in the following way:

1 = Highly significant
2 = Quite significant
3 = Neither significant nor insignificant
4 = Not very significant
5 = Not at all significant

The Fig. 9.3 shows the 70 lifestyle evaluation factors, in order of their average "significance" score on the survey. The highest average score went to #14, "appetizing diet." The lowest average score went to #58, "trend consciousness." We can see that the importance people gave to these 70 lifestyle values, as they related to a spiritually affluent lifestyle, varied significantly from item to item. As for #27, "continuity," the people who chose either "quite significant" or "highly significant" outnumbered those who chose "neither significant nor insignificant," "not very significant," or "not at all significant," so it seems fair to say that people do consider continuity a vital part of a spiritually affluent lifestyle. However, it ranked only 53 out of the 70 lifestyle values. For people in Japan today, it would appear that this idea of continuity, of passing things down, is relatively unimportant to feeling spiritual affluence.

Did certain attributes of these respondents affect in some way how significant they considered "continuity" to be for a spiritually affluent lifestyle? To investigate that question, we grouped the respondents based on certain attributes, and then examined the differences between those groups with regard to how valuable they perceived continuity to be [3].

To see if differences in the respondents' social or natural environment had a connection with the degree to which they valued continuity, we began by taking three factors in the areas where the respondents lived: the ratio of elderly people, the population density, and the ratio of forest cover. As for the "differences" in social and natural environment between respondents, we classified the prefectures of Japan where the respondents lived as either "urban" or "rural" with respect to these three categories. We determined the ratio of elderly people, the population density, and the ratio of forest cover in the areas where the respondents lived, and then compared the mean score of the significance of "#27 continuity" of respondents from urban areas to those from rural areas. We also set the means of the three factors—ratio of elderly people, population density, and ratio of forest cover—as the "borderlines" between urban and rural, and then categorized each area as either "urban" or "rural" in each category depending on which side of the mean it fell on. We took our data from the following sources. Data on ratios of elderly people was taken from population estimates done in 2013 by the Ministry of Internal Affairs and Communications; these estimates contain information on the proportion and total population of people 65 and older in each prefecture of Japan. Data on population density for each prefecture was taken from the Japan Statistical Yearbook, 2014 (published in 2013), also done by the Ministry of Internal Affairs and Communications. Data on forest cover ratio for each prefecture was taken from 2014 forest cover statistics compiled by the Ministry of Agriculture, Forestry and Fisheries.

The "natural increasing/decreasing" category is based on the number of births and deaths within a given area, while the "social

Table 9.1 Population increase/decrease in areas where "population is socially increase"

		Name of the prefecture	Number
Population increasing	Nature increasing, socially increasing	Tokyo, Kanagawa, Aichi, Shiga, Okinawa	5
	Nature increasing, socially decreasing		0
	Nature decreasing, socially increasing	Miyagi, Saitama, Fukuoka	3
Population decreasing	Nature increasing, socially decreasing		0
	Nature decreasing, socially increasing	Chiba, Oosaka	2
	Nature decreasing, socially decreasing	Other 37 prefectures	37

increasing/decreasing" category is based on the number of people moving to or away from a given area (Table 9.1). Areas where "the population is growing society" are areas where there are a relatively high number of people moving in; we therefore expected these areas to have a diverse mix of value systems and ways of thinking among their populations. For our analysis, we have labeled the Japanese prefectures where social increasing is the β (beta) group (Table 9.2). For each of the three categories—ratio of elderly people, population density, and ratio of forest cover—we analyzed both the "group including β (47 areas)" and the "group excluding β (37 areas)." Going by the population estimates done in 2013 by the Ministry of Internal Affairs and Communications, β includes 10 prefectures: Miyagi, Chiba, Saitama, Tokyo, Kanagawa, Aichi, Shiga, Osaka, Fukuoka, Okinawa.

Table 9.2 Areas where the population is socially increasing (β)

Areas where the population is socially increase	Miyagi, Chiba, Saitama, Tokyo, Kanagawa, Aichi, Shiga, Osaka, Fukuoka, Okinawa (10 areas)

If you look at the ten areas in Table 9.2, you will note that most of the 10 areas in β, in terms of the ratio of elderly people, the population density, and the ratio of forest cover, are quite urban. Seven of the areas—Chiba, Saitama, Tokyo, Kanagawa, Aichi, Osaka, and Fukuoka—are indisputably urban (Chiba, Saitama, Tokyo, and Kanagawa all contain parts of the Tokyo metropolitan area; Aichi is home to the large city of Nagoya; and Osaka and Fukuoka prefectures are home to the major cities of the same names). In other words, when we look at the "including β" group, we can already see a trend as to what kind of areas have socially increasing, and what kinds of areas do not. Moreover, because we imagine that the β areas have populations with a diverse mix of values, in our analysis of the "including β" group, we did not expect to identify many clear, significant differences between urban (the ratio of elderly people and ratio of forest cover is lower than the mean and population density is higher than the mean) and rural areas (the ratio of elderly people and ratio of forest cover is higher than the mean and population density is lower than the mean) in terms of what people value.

First we performed t-tests comparing urban and rural areas for both the "including β" group and the "excluding β" group. We found that, in the group including β, people who lived in areas with lower percentages of forest cover valued continuity more highly than people who lived in areas with higher percentages of forest cover. We also found that, in the group excluding β, areas with higher percentages of elderly people valued continuity more highly than areas with lower percentages of elderly people.

The number of lifestyle evaluation factors where we found a significant difference was, depending on the category, anywhere between 0 and 5, out of a total of 70. From the beginning, we expected that in the group including β we would not find many significant differences, but in fact, the group including β had a slightly greater number of significant differences than the group excluding β. This leads us to believe that the areas in the β group may have a certain degree of shared values. However, between the group including β and the group excluding β, the number of items showing a significant difference only differed by a few, out of a total of 70—we cannot say there was a significant difference in the number of lifestyle evaluation factors that were valued differently between the group with β and the group without β.

As for what people thought of the value of "#27 continuity," the results are as previously noted. In the group including β, people who lived in areas with lower percentages of forest cover valued continuity more highly than people who lived in areas with higher percentages of forest cover, and in the group excluding β, areas with higher percentages of elderly people valued continuity more highly than areas with lower percentages of elderly people. Out of the 70 lifestyle evaluation factors, #27 continuity was one of the few where we did see a significant difference. It would seem that, among this wide range of factors that make up a spiritually affluent lifestyle, how much you value "continuity" is particularly affected by whether you live in an urban or rural area. However, the fact that we received different results for the group including β and excluding β meant that we were not able to clearly say what kinds of living conditions influence the extent to which people value continuity. To the end, we undertook a more detailed analysis of Japan's 47 prefectures.

Our results suggested that the areas in the β group had a certain degree of shared values. They also seemed to suggest that whether you live in an urban or rural area influences how much you value "continuity." However, our results did not clearly tell us what specific factors influenced the value a person placed on continuity. We therefore divided the 47 prefectures of Japan into smaller groups, based on whether an area had a growing society, and on the extent to which it was urban or rural. We continued to categorize the prefectures where society was growing as the β group. We then split the other prefectures into three categories. Areas where the ratio of elderly people, the population density, and the forest cover ratio were all "rural," we put into the α group; areas where all three of these things were "urban," we put into the $\gamma 1$ group; all the other areas, we put into the $\gamma 2$ group. We now had four categories in total. We compared the mean scores of survey respondents in these four groups with regards to "continuity" as well as all the other lifestyle items. The groups and the areas that comprise them are listed in Table 9.3.

We investigated whether an area was a social increasing area, and to what extent it was urban or rural, made a difference in how its

Table 9.3 Areas in each of the four groups

Areas where socially increasing	Areas where socially decreasing		
β (10 areas)	$\gamma 1$ (7 areas) All categories urban	$\gamma 2$ (16 areas) Others	α (14 areas) All categories rural

people perceived the value of continuity. To do this, we divided the 47 prefectures of Japan into the aforementioned four groups (α, β, $\gamma 1$, $\gamma 2$), and carried out t-tests between the groups. We found that people living in α areas valued continuity more highly than people living in $\gamma 1$ areas. We also found people living in β areas valued continuity more highly than people living in $\gamma 1$ areas. From this, it would seem we can say that β areas have a certain measure of shared values. There were other lifestyle values for which we found similar trends.

In the Table 9.4, the mean scores of the groups in the vertical column are subtracted from the mean scores of the groups in the horizontal column. The differences of the mean score of 70 lifestyle evaluation factors are the numbers you see in the boxes. The numbers in parentheses in the left-hand column are the mean scores for "#27 continuity" for each group. Comparison of the groups shows

Table 9.4 Comparison of the importance of certain lifestyle evaluation factors among the four groups

	α	β	$\gamma 2$	$\gamma 1$
α (2.3557)	–			
β (2.3713)	Personal roots −0.030**	–		
$\gamma 2$ (2.4425)	Clear roles −0.164* Varied experienced −0.150*	Appreciation from others −0.127*	–	
$\gamma 1$ (2.5520)	Personal roots – 0.198* Continuity −0.1963*	Continuity −0.1806** Compatibility with personal preference −0.128*	none	–

anywhere from zero to two significant differences among the 70 lifestyle values. "#27 continuity" is among them. This suggests that, among this wide range of factors that potentially make up a spiritual affluence, how much you value "#27 continuity" is particularly affected by whether you live in an urban or rural area, and whether you live in an area where society is growing.

Among the three groups of social increasing area, the group that considered the most of these 70 lifestyle evaluation factors "significant" to a spiritually affluent lifestyle was α. On the other hand, the group that considered the most items not significant was $\gamma 1$, with $\gamma 2$ just one item behind. We can thus identify a trend among these groups: the more rural an area is, the more it values these 70 lifestyle evaluation factors for living a spiritually affluent lifestyle, whereas the more urban an area is, the less it values these things. β, the group of social increasing, valued these 70 lifestyle evaluation factors second-most, just after α, the most rural group. The groups followed the same order when it came to how much they valued "#27 continuity" specifically: α the most, then β, $\gamma 2$, and finally $\gamma 1$. We think it's fair to say that α and β fall into the same category when it comes to how much they value continuity.

We suspect that what is generating these results could be a relatively high proportion of people originally from α areas either moving to β areas permanently, or living in β areas for a time, and bringing their values with them. The places where local flavor is in the greatest danger of not being passed down are urban areas which is social increasing.

9.4 Passing Down Local Flavor: A Case Study from Shinmei, Ago, Shima City, Mie Prefecture

We will now introduce a case study carried out in Shinmei, a village in the town of Ago, which is part of the city of Shima in Mie Prefecture [4]. Our aim was to collect data through a specific case study of what factors encourage, and what factors hinder, the passing down of local flavor.

Shima City, Mie Prefecture, sits in the southern part of the Shima Peninsula area. In September 2004, the towns of Hamajima, Daio, Shima, Ago, and Isobe merged to create Shima City (before that, the towns were grouped into an area called Shima District). The northern part of Shima borders multiple bustling cities, including Ise, home to the famous shrine Ise Jingu, as well as Toba, a thriving castle town that was originally part of the feudal Toba Domain. Many people in Shima City commute to these areas for study and work. Our study was conducted in the town of Ago, which has a population of 22,350 (Shima City figures, 2015). Ago is the most populous of the five towns that make up Shima City. Ago is further divided into seven villages: Ugata, Shinmei, Tategami, Shijima, Koka, Ko, and Anori. Ugata and Shinmei are home to stations on the Kintestu Railway's Shima Line. Shinmei is further divided into eight districts: Naka, Nishi, Higashi, Minami, Maekata, Kashiko, Urajiro, and Miyanishi. These districts often compete against each other in fun events like relay races. Shinmei is on the coast and features inland seas. In the inland waters, there used to be a thriving pearl farming industry. However, in recent years, there has been an influx of cheaper pearls from abroad, and fewer and fewer people are able to survive on pearl farming alone. Many people will now do pearl farming and also cultivate sea lettuce, or they'll farm the land. There are also more people working at companies outside Shima City.

In this area, the local word for shinji (Shinto rites) is "jinji." The specific subject of this study is an aspect of religious festivals in the Shinmei and Tategami areas of Ago known as "toya." The toya are a collection of solemn duties, and the people who take them on are chosen each year from the local community. People are said to be "taking up the toya," and the people who take the toya up are known as "tonin." The most important rites in Shinmei are held at Shinmei Shrine: religious festivals, and ceremonies to drive out evil spirits, bad luck, and other misfortunes. The tonin, who must play a central role in the festivals at the shrine, are men in their 42nd year (counting the old Japanese way, where you are one year old at birth). This year of life is traditionally considered a year of bad luck and misfortune for men. In recent years, between four and 10 people have been chosen as tonin each year. The religious festivals and the ceremonies to cast out misfortune are separate events, but because the people "taking up the toya" are men in their year of misfortune, even locals will often confuse the two if they are not particularly well informed about toya. The year after fulfilling the duties of toya, the tonin become "koto" (lit. "old tonin"). They teach the next generation of tonin all the duties involved in carrying out the toya. These duties are wide-ranging and include cleaning the shrine once a month, growing the rice that will be used to make mochi offered at the shrine, alternating being in charge of the keys to the main shrine building, and participating in shrine festivals, which are held 16 times a year. The work of toya requires at least four tonin, who are known as the first to, second to, third to, and fourth to. At festivals, the "first to" has the most important role; this is the person who is able to participate in all the rites and rituals to the greatest extent. In the event that there are fewer than four tonin present, someone who has been a tonin before will step in—usually a koto.

At the biggest festivals, there are other roles besides those related to toya. There are lantern carriers, "servers," and female dancers. The lantern carriers are former tonin, and they carry lanterns in front of the current tonin. The servers are responsible for serving food and alcohol—typically, they are the sons of tonin. The female dancers do what their name suggests. They are girls who have not yet had their first period, and they are typically the daughters of tonin. If servers and dancers do not come from a tonin's immediate family, then another relative may fulfill these roles. Also, at periodic intervals, Shinmei Shrine holds a ritual where Shinmei's resident deity is installed in a new shrine building, similar to the famous installing ritual held at Ise Jingu. (Shinmei Shrine's ritual is staggered a few years apart from Ise Jingu's.) Men who are tonin in the years when this ritual takes place have the additional responsibility of helping to carry it out.

On January 18, 2016, a "jutosai" (lit. a "teach the toya festival") was held. At this festival, the duties of the Shinmei toya pass from one group to the next, in a ritual called the "jutoshiki." The locals also call the festival the "uketo," literally the "taking" of duties by the new crop of tonin. One part of the jutosai is the "hankishiki" (lit. rice container ritual). The tonin line up as the servers who have been providing food and drink each pour alcohol in a metallic teapot-like vessel called a "hanki," and then pour the mixture into the tonin's cups. Finally, the tonin drink.

The Tategami area has five major ritual events. July 14th is Tenno-sai (lit. the Emperor's Festival). Originally this festival was held to pray for the warding off of famine, disease, and so on. The festival includes a special dedication ritual, where elementary school students perform a "child's dance," but with the decreasing number of children in the area, people fear that this tradition is going to fade away. On August 13th and 15th, Tategami holds its annual Bon dance, a traditional summer event all across Japan. On August 14th, there is a "dainenbutsu," a gathering where people pray together to the Buddha. At this dainenbutsu, the local community chants along to the beat of taiko drums, praying for families who have lost members in the last year.

During the dainenbutsu, a special group called "the nine" perform a chant that serves as a memorial service. Traditional doctors join in with a prayer featuring gongs and drums./Gongs and drums are added to the performance "The nine" is actually a group of four people—a chair, vice-chair, treasurer, and secretary-general—and they are responsible for organizing Tategami's festivals. They are chosen by election from among the citizens of the local community, and the term of office is four years. Tategami is broken up into nine numbered subdivisions, and it is believed that the term "the nine" comes from the fact that, before the mid-19th century, one person would be elected from each of the nine subdivisions. It is pretty much decided among the candidates before the election who will be elected to the group. However, in the past, after some people felt that this was not really a proper voting system, an election was held where the result was not fixed beforehand. In the event, some of the people chosen declined the position, because they were too busy with work and so on. So Tategami returned to the old way of doing elections.

On August 14th, the "bon-okuri" is performed. Two people, chosen from families who have lost members recently, hold torches. Led by the nine, a procession of straw horses (which are stuck through with a bamboo rod held by two people, one on each end), gongs, taiko drums, and torch bearers march through the area. At the end, the straw horses are burned at the riverside, and a bonfire called an okuribi, which sends off the spirits of the community's ancestors, is lit. On January 2nd, at Ukihi Shrine, the "hipporo rite" is performed. Traditional flutes are played during this ritual; the name "hipporo" comes from the sound a traditional flute is said to make ("hippori-ryo-ryo"). Differently from toya in Shinmei, the toya duties in Tategami rotate among local citizens, not just men in their 42nd year.

*Sengu . . . shikinensengu: a new shrine building is constructed after a fixed number of years and the resident deity is moved from the old shrine to the new one. This is among the most important rituals performed at Ise Jingu.

Now we will present an overview of the interviews we did with the citizens of Ago, Shima City, Mie Prefecture, about their religious rituals and customs (January 16–18, 2016).

[Shinmei, Ago, 47-year-old man, Mr. C]

Mr. C is a 47-year-old man who lives with his wife, two sons, and daughter (five-person family). He is originally from Hiroshima City, Hiroshima Prefecture, and he currently lives in the Shinmei area of Ago. He "took up the toya" in 2009. Mr. C's wife (Ms. D, introduced below) is originally from Shinmei.

2009, the year Mr. C took up the toya, was a year when the deity of Shinmei Shrine was installed in a new shrine building, and so he helped with this process. As Mr. C is not from the area, he didn't feel any particular pressure or sense of duty when he first took up the toya, but after he started, he said he felt that he "had to preserve the toya tradition." He is eager to continue his involvement with the toya—at the very least, he wants his name to remain on the temple's list of people who helped to continue the tradition. To be a tonin for one year costs about a million yen, and he said that expense had been tough, but that he hadn't wanted to quit his duties.

He says that even now, when he enters the grounds of the shrine, he feels as if his body has been purified.

[Shinmei, Ago, 44-year-old woman, Ms. D]

Figure 9.4 Jinji of Shinmei Shrine.

Figure 9.5 Jinji of Shinmei Shrine.

Ms. D is Mr. C's wife. She is a 44-year-old woman. Originally from Shinmei in Ago, she lives there now. Ms. D grew up with the toya tradition. She longed to be the wife of a tonin; when she was younger, she absolutely wanted the man she eventually married to become one. She said that since the only people who can serve the deities as tonin are men in their "bad luck" 42nd year, if they have the opportunity, they should take it. She says she is happy the she and the rest of her family were able to support her husband while he was a tonin. Her husband is not originally from Shinmei, but the fact that he took up the toya is a symbol to her that he has truly put down roots there.

[Shinmei, Ago, 89-year-old man, Mr. E]

Mr. E lives with his wife, son, and daughter-in-law (four-person family). He is originally from Shinmei in Ago and lives there today. He is an 89-year-old man. In 1969, his bad luck 42nd year, he took up the toya.

Until two years before Mr. E's turn at the toya, the tradition that tonin had to be men in their 42nd year was not set, but there was a worry that there would not be enough successors to carry on the tradition. Based on the results of a survey taken by local citizens, the current system was established. When Mr. E looks back on his time performing the toya, he says that because it was a holy duty, all the tonin worked together peacefully, without a single fight, giving everything they had to fulfill their duties. Originally, Mr. E didn't have any particular interest in religious customs. But the year he took up the toya, when the koto taught him what he needed to know, he began to feel respect for the gods. Ever since, he has prayed every single morning at his household altar.

[Shinmei, Ago, 40-year-old man, Mr. F]

Mr. F lives with his parents, wife, and three sons (seven-person family). He is originally from Ago and lives there now. He is a 40-year-old man. He became a tonin on January 18, 2016. He is in the same year of tonin as Mr. G, whom we will meet next.

Mr. F and his fellow tonin practiced the jutoshiki ritual morning, noon and night up until the day of jutosai (the festival where the toya are passed from one group of tonin to the next). Mr. F thought the considerable expense of becoming a tonin would be a burden, and he wasn't really thinking about doing it. However, he knew that the local community would take note of what he did, and so he decided to become a tonin.

If, during the course of his year taking up the toya, the duties have a significant negative impact on his life, he does not intend to continue.

[Shinmei, Ago, 40-year-old man, Mr. G]

Mr. G lives with his grandmother, parents, wife, and son (six-person family). He is originally from Shinmei in Ago and lives there now. He is a 40-year-old man. He became a tonin on January 18, 2016.

Both Mr. G's father and grandfather took up the toya, and it is a local tradition—he feels he cannot let it die. If it fades away, he thinks that would be very sad.

[Shinmei, Ago, 41-year-old man, Mr. H]

Mr. H is originally from Shinmei in Ago and lives there now. He is a 41-year-old man. On January 18, 2016, he went from being a tonin to being a koto.

Having taken up the toya, he understands well what it involves. His understanding of what it means changed greatly after, on the day before the jutosai, the tonin from Mr. H's year met and talked about the future of the toya all night.

[Shinmei, Ago, 42-year-old man, Mr. I]

Mr. I lives with his parents, wife, and two sons (six-person family). He is originally from Shinmei in Ago and lives there currently. He is a 42-year-old man. On January 18, 2016, he finished his year of service as a koto.

Before he took up the toya, he said he felt like it was "a duty" and that he "didn't not want to do it, but it seemed like a burden." When it was time to begin the work of the toya, he said, he realized he had to do it, and it was a long road, so he might as well think of it not as something he was forced to do, but as something he was taking on voluntarily, of his own volition. He would think of doing the toya as his proper role; this allowed him to do the necessary work without any angst. He said there was a lot of tough aspects of taking up the toya, but that, actually, that's what made it fun. Everyone working as one, and receiving the thanks of the local community—things like that were great.

When it became Mr. I's turn to pass on the traditions to the next generation, he told them about how to overcome the difficulties they would face while taking up the toya, and he also wrote down many teachings that, until then, had only been handed down orally. He has done a lot to support the tonin of the future and ensure the tradition will endure for decades to come. There has been a great effort over the years to hand this tradition down from generation to generation, and he wants to continue it. He said he didn't want the tradition to end with his generation.

[Tategami, Ago, 64-year-old man, Mr. J]

Mr. J is originally from Tategami, a neighboring community of Shinmei, and lives there currently. He is a 64-year-old man. He was a junior high school teacher and is currently the chair of the Tategami Local Council.

Ever since he was a boy, he had seen the hipporo rite at the festival at Ukihi Shrine, but back then he didn't have much of an interest in it. When he was a student, it was Japan's postwar era of rapid economic growth, and he saw traditions and old culture gradually fading away. That was when he took an interest in passing down traditions, but he had just started working as a junior high school teacher, so he was very busy—he hardly took part in Tategami's religious rituals at all. Around the age of 55, he took part in the hipporo rite for the first time as a tonin. He also became the head of the local council. He said taking part in the hipporo rite made him realize how fun and interesting it was.

He is also glad he took part in the toya. He was able to meet and converse with local people he didn't know before. Also, he said that this is an area where lots of people—whether it's younger people in the local youth association or pearl farmers—work to support the Tategami community . . . but that he had hardly taken part in the local customs at all. By participating in this unique local hipporo rite as a tonin, Mr. J said he felt like he became a "full-fledged member" of the community.

By serving as a tonin, and heading the local council, Mr. J says he has developed a deep interest in the customs and events of Tategami. He wants to learn, of his own volition, more about local traditions. He wants the traditional customs and events of Tategami to be passed down, starting with the hipporo rite.

These days, fewer people in the area run their own businesses, and more work for companies. Also, tonin must basically cover the costs of the traditional dress worn at festivals, which amounts to roughly 180,000 yen, a big expense. These are two big reasons why a lot of people decline to become tonin, and why it has become difficult to find enough people each year. In Tategami, there are nine areas called "seats," and one tonin is selected from each of the nine seats. The role rotates between the families of each seat. Because the number of households differs from seat to seat, it has become necessary in some seats to rotate tonin every few years rather than every one year; some people have complained that this is unfair. Mr. J says that at the local council, they try to share information with local citizens to get them interested in local customs, and they also collect donations to lessen the financial burden on the people putting together the festivals. But he says the situation is difficult. He says that if it comes his turn to do the toya again, he would accept, feeling a sense of duty to protect the local festivals.

On the basis of these interviews, we can make the following implication.

Increasing the Desire to Carry on Traditions

If passing down a tradition offers enough "spiritual affluence" in relation to the constraints or burdens a community faces, then the desire to pass down that traditional will increase. Even if there is not enough spiritual affluence in relation to constraints, there are several other factors that can lead to a strong desire to pass on traditions: perhaps helping to pass down a tradition creates a powerful sense of community, or passing down a tradition is simply seen to be a matter of course, or if there are certain constraints/special circumstances as to how a tradition can be passed down, or the person passing down the tradition feels the weight of history, or the tradition feels like a vital part of one's hometown. The implication is that even if the spiritual affluence attached to a tradition does not outweigh the strain it places on the community—indeed, even if the burden outweighs the spiritual affluence—there are still various factors that will lead to people feeling a strong desire to pass on a tradition.

Increasing Spiritual Affluence

In terms of the factors that will amplify the spiritual affluence derived from passing on traditions, some include: the admiration you receive for being the one to pass on a tradition; the way the spiritual affluence of passing on a tradition can connect you to the past; if there is a special time within which a tradition must be passed down; the cooperation needed to pass on a tradition; understanding the meaning and significance of a tradition; the feeling that what you are doing is worthwhile; the constraints/special circumstances as to how a tradition can be passed down; if there is a special place where a tradition must be passed down; or if only certain people can pass on the tradition. Things like this can overcome constraints and other challenges.

Increasing the Desire to Pass Things Down

We think that the following factors are some of the things that will increase the desire to pass things down: if the meaning and significance of passing down a tradition is understood; if passing down a tradition offers people a role to play; if people feel the weight of history in passing down the tradition, if a tradition is at risk of being suspended or dying out; if, after people fulfill their role in passing down the tradition, they truly begin to understand it; if passing down a tradition is a vital part of one's hometown; or if passing on a tradition is a way of nurturing the next generation. Also, nurturing the next generation is also connected with increasing things like a community's sense of individuality and sense of its role. Accordingly, if things can be found that have a connection with raising these senses of individuality or role, they will also likely be things that will raise desire to pass things down. Moreover, understanding the meaning and significance of a tradition will both increase a person's desire to pass on that tradition and increase that person's spiritual affluence. On another note, desiring strongly to pass down a tradition because you feel it will help to nurture the next generation is something that should be felt not just by older generations, but by younger generations as well—this is one approach to encouraging the passing down of traditions.

Ensuring the Importance Placed on Passing Things Down Is Secure

One man said since he took up the toya over 40 years ago, he has prayed every single morning at his household altar. Another man lives far from his hometown, but he uses his vacation to go home for festivals. It seems fair to say that people like this—people who treat the passing down of traditions as a habit, who see it as part of the yearly cycle of life—have an unshakable belief in the value of passing things down.

9.5 How to Bring Back Local Flavor

Truly discovering your area's local flavor is about more than simply passing on or reviving old rituals and crafts and performing arts. Regions must seek out ways of life that will work in a future that is going to have harsh environmental constraints. It is evident that the value placed on passing things down is closely related to a place's way of life. The environmental constraints faced by the people of the past are different from the ones people are going to face in the future, and the way to lead to a spiritual affluence existence will also be different. Especially in areas that become more urban, it is essential for communities to maintain their sense of individuality, and a sense of the role they play.

We once established ways of life that were in harmony with nature. The roots of that value system were the ways of thinking and knowhow that were necessary to find common ground with nature. Within the values that make up the old Japanese way of life, we find ways to achieve lives rich with spiritual affluence even amid constraints. However, the feeling that these values are necessary is rapidly fading away, and getting lost. This chapter gave specific examples of this, but Japan faces a low birthrate and an aging population—customs and communities are losing richness and complexity, and many places have simply given up on passing down their traditions. "If there are no people to pass down the tradition, then it won't get passed down. It's finished." In some parts of Japan, you will hear comments like this. Communities must come up with and introduce new measures that will make sure they do not lose their sense of individuality, and that they continue to maintain a sense of the role they play. In communities facing severe degradation—where the foundations are crumbling, where the systems on top of those foundations are crumbling, where the regional environment is feeling the effect of environmental constraints—simply feeling a sense of crisis is not enough. These regions must act before it is truly too late. That much is certain.

Acknowledgment

This research was conducted in the research project of 'Creating a Fountain of Future Lifestyle Ideas' supported by JST-RISTEX.

References

1. Furukawa, R. Chikyu kankyo seiyakuka ni okeru kokoro yutakana kurashikata-shizennkannkyo ga kibishii tohoku chiho no sennzenn no kurashi wo bunsekishite-, NETT No. 94, 38–41 (2016).
2. Ogawa, K., and Furukawa, R. Kokoro yutakana kurashi no katachi no kouzo bunseki-hyouka grid ho wo mochiite-, Kenkyu Gijutsukeikaku gakkai dai 29kai nenji gakujutsu taikai ippan koen Yoshi shu, 205–208 (2014).
3. Matsuda, Y., and Furukawa, R. Kokoro yutakana kurashikata no densho wo sokushinnsuru youin ni kannsuru kenkyu, kenkyu gijutsu keikaku gakkai dai 30 kai nenjigakujutsutaikai koen youshishu, 986–988 (2015).
4. Matsuda, Y. Kokoroyutakana kurashikata ni motomerareru densho to iu kachi wo takameru youin ni kansuru kenkyu, Master Thesis, Graduate School of Environmental Studies, Tohoku University (2016).

Chapter 10

Developing Needs-Oriented Technologies with Bio-TRIZ: An Example of a New Lifestyle in Kitakami Achieved through Japanese Umbrellas

Takeshi Yamauchi,[a] Hidetoshi Kobayashi,[b] and Toru Kobayashi[c]

[a] *Department of Materials Science and Technology, Faculty of Engineering, Niigata University, 8050 Igarashi Ninomachi, Nishiku, Niigata 950-2181, Japan*
[b] *Department of Mechanical Science and Bioengineering, Graduate School of Engineering Science, Osaka University, 1-3 Machikaneyama, Toyonaka, Osaka 560-8531, Japan*
[c] *Division of Electrical Engineering and Computer Science, Graduate School of Science and Technology, Nagasaki University, 1-14 Bunkyocho, Nagasaki 852-8521, Japan*
yamauchi@gs.niigata-u.ac.jp, hkoba@me.es.osaka-u.ac.jp, toru@cis.nagasaki-u.ac.jp

10.1 Introduction

In order to enjoy an emotionally satisfying lifestyle despite the harsh environmental constraints that lie ahead, we need to envisage a sustainable society and consider the necessary technological components that would support such a society. This lifestyle-oriented approach in developing innovative technologies can be

Lifestyle and Nature: Integrating Nature Technology to Sustainable Lifestyles
Edited by Ryuzo Furukawa

ISBN 978-981-4774-62-8 (Hardcover), 978-1-315-14687-4 (eBook)
www.panstanford.com

expected to be highly effective. Since we cannot expect to resolve the many challenges that confront us with existing technologies alone, the development of innovative, eco-friendly technologies is crucial for the realization of a rich and fulfilling lifestyle. This chapter looks at Japanese umbrellas in Kuchinai, Iwate Prefecture, as an example of how to extract technology elements and apply Bio-TRIZ to the development of key technologies that support an emotionally rich society. To be more specific, we will look at how the folding and unfolding of a morning glory flower inspired a new style of Japanese umbrella that folds inwards so that the water droplets are enclosed inside. We'll also look at how the liquid crystalline film secreted by insects inspired the development of liquid crystal washi paper. Finally, we will consider how the application of such innovative technologies inspired by nature can help to realize spiritually affluent lifestyles.

10.2 Supporting Lifestyle Design by Taking a Needs-Oriented Approach

In Chapter 7, Section 7.5, we explained at the application of Bio-TRIZ to needs-oriented manufacturing and the development of biomimetic products based on a needs-oriented approach to engineering. In this chapter, we demonstrate how the use of the Bio-TRIZ Database can support the application of serendipitously discovered biological functions to engineering technology, solve various technological contradictions, and help design a sustainable lifestyle. Specifically, we extracted the technology components necessary to create the Kuchinai umbrellas of Kitakami, Iwate Prefecture, and considered how the innovative designs inspired by Bio-TRIZ applications are helping to create a new lifestyle. We conducted an ontological analysis of peoples' hopes and actions in relation to lifestyles to determine the desired functions and technologies. We then verified the means to solve the resulting technological contradictions.

In particular, we highlighted the elements of a new lifestyle that included the wide use of Japanese umbrellas, umbrellas that were in harmony with the scenery, a cityscape dotted with lanterns made with washi paper, and an arcade of Japanese umbrellas. In order to solve the technological issues hampering the realization of such elements, we came up with transparent Japanese umbrellas using nanocellulose produced from recycled waste; lanterns that repelled bugs; an arcade of Japanese umbrellas that could serve as a rainwater tank; and a Japanese umbrella inspired by a morning glory flower that will not drip when folded. We then examined the action decomposition tree introduced in Chapter 6 and looked at that tree for achieving a lifestyle featuring Japanese umbrellas. In this chapter, we will explain how the technological contradictions revealed in the action breakdown tree were resolved from the perspectives of materials and mechanical engineering.

10.3 Designing and Developing Japanese Umbrellas from the Perspective of Material Engineering

Using the Bio-TRIZ Database

Japanese umbrellas are made primarily with natural materials that have been used for centuries, such as washi (Japanese paper) and bamboo. Unlike Western umbrellas that only have several ribs, Japanese umbrellas can have as many as several dozen ribs, depending on the size. This is because of the difference in the opening mechanism. Unlike Western umbrellas that rely on tension to stretch the umbrella canopy, Japanese umbrellas use bamboo to support the ribs and the washi covering. Japanese umbrellas are highly waterproof but are not very durable. Also, because they're made mostly of natural materials, they tend to be heavier than Western umbrellas. In fact, they are too heavy to point up and open. To protect the rings and ribs from excessive force, Japanese umbrellas are usually opened with the tip pointing down and then held up. A Western umbrella is put in a stand or stored with the tip facing down, whereas a Japanese umbrella has to be hung from the ceiling or the space under the eaves with the tip pointing up, to prevent rainwater from accumulating in the tip and causing it to tear.

Because of the predominant use of washi paper, Japanese umbrellas are endangered by moths and erosion due to humidity. They also do not offer adequate protection in heavy rains. Then, as with most papers, the colors tend to fade over time.

If an umbrella has not been used for a long time, the water-repellant oil on the washi paper may cause the umbrella to stick, making it difficult to open. These shortcomings coupled with the spread of Western umbrellas and fewer opportunities to wear kimonos have led to a sharp decline in the use of Japanese umbrellas. In recent years, it's become rare to see people using Japanese umbrellas even when wearing a kimono. There are only a limited number of occasions that call for the use of a Japanese umbrella.

If the use of eco-friendly Japanese umbrellas were to become widespread, more people would start wearing kimonos to match their umbrellas. The increase in people using Japanese umbrellas would bring about diverse changes to the environment and help to achieve a new lifestyle design. It's important to emphasize the fact that Japanese umbrellas are simply a beautiful sight. But in order to emphasize the colors, the umbrellas would have to be given multiple coatings, which would destroy the translucency of washi, which allows the colors of the umbrella to change with the intensity of light—one of the most striking features of Japanese umbrellas.

If we were to enter the above characteristics in the contradiction matrix, the resulting technology contradiction would be "Stability of composition" vs. "Illumination intensity." We propose to resolve this contradiction by applying the following four TRIZ principles.

Principle 32. Color changes

(1) Change the color of an object or its external environment.
(2) Change the transparency of an object or its external environment.

Principle 3. Local quality

(1) Change an object's structure from uniform to non-uniform.
(2) Make each part of an object fulfill a different and useful function.

Principle 27. Cheap short-living objects
Replace an expensive object with multiple inexpensive objects, compromising certain qualities such as service life.

Principle 16. Partial or excessive actions
Try using "slightly less" or "slightly more" to adjust to the right (operational) balance.
A further Bio-TRIZ search results in the following biological functions:

Principle 32. Color changes
The Argiope spider's nest attracting its prey (using ultraviolet rays)
The skeletal structure of a weevil that appears green from every angle
The light but tough beak of a Toco Toucan
The cuttlefish's ability to change color instantaneously

Principle 3. Local quality
A painless needle that mimics the mechanism of a mosquito's stinger
The powerful legs of *Tettigonia orientalis*, a locust found in the savanna, enables it to cling to objects
A moth's eye, which does not reflect light (moth-eye structure)
The secret of the morpho butterfly's blue wings that glow in the dark
New wing shapes inspired by dragonflies
Lotus leaves that repel water and stay clean
The burs of wild burdock that are impossible to remove
The osclum of knife fish fry
The muscular grip of the North Pacific giant octopus
The ability of dolphins to swim fast by wrinkling their skin
The energy-saving fins of humpback whales

Principle 27. Cheap short-living objects
Living organisms are basically able to adapt and function by changing their structure within limited components. In that sense, they can be said to be making effective use of inexpensive objects.

Principle 16. Partial or excessive actions
A stylish crab that is good at hiding (using its hook-like hair to cover its body with sponges and other materials to disguise itself)
New wings that change shape—morphing wings
The superbly adjustable eyes of penguins

Problem Solving from the Perspective of Materials Engineering

So far, the synthesis and optimization of materials with appropriate performance has been of utmost importance in materials engineering. However, living things have a mechanism that enables them to achieve high performance and effectively accumulate information in a short time by changing the structure of limited materials. We can apply TRIZ principles to this. For instance, the application of Principle 32, color changes (the skeletal structure of weevils that appear green from every angle) has inspired lamination, and the application of Principle 3, local quality, to how a moth's eye does not reflect light (moth-eye structure) has led to the development of microrod structures.

Observing how lotus leaves repel water and stay clean leads us to believe we can develop a super-water-repelling spray that creates an irregular surface by coating it with a nanocellulose solution. The surface structure of crab that uses its hook-like hair to snag and cover its body with sponges can also be of reference (Fig. 10.1). In Fig. 10.2, we describe how the skeletal structure of weevils led to the development of structural colors. The main component of washi paper is cellulose, which is structured from bundles of nano-size cellulose. Separating the bundles extracts nanocellulose. Since the nanocellulose fibers used to make a film are shorter than the wavelength of light, the resulting washi paper is translucent. If the nanocellulose can be chemically modified, it may well lead to development of iridescent washi paper. By focusing on such subcellular structures, we can inspire the development of innovative state-of-the-art materials with a low environmental impact.

LS	Improvement factor	Deterioration factors	The 40 Principles	BioTRIZ vocabulary	BioTRIZ examples	BioTRIZ patents
Japanese umbrellas are simply a beautiful sight	Stability of the object's composition	Illumination intensity	32. Color changes 1. Change the color of an object or its external environment 2. Change the transparency of an object or its external environment	Replace	A paramecium that appears green from every angle	Development of nano-cellulose using cellulose derivatives
			3. Local quality 1. Change an object's structure from uniform to non-uniform 2. Make each part of an object function in suitable conditions	Destroy, Solidify Change surface Change hardness Separate, Divide	A lotus leaf not getting wet	Development of nano-compound solution in spray form
			27. Cheap short-lived objects Replace an expensive object with multiple inexpensive objects, compromising certain qualities, such as service life	Replace		Ability to recycle abandoned plastic umbrellas
			16. Partial or excessive actions Try using "slightly less" or "slightly more" to adjust to the right (operational) balance	None	A stylish crab that's good at hiding	Jap. umbrellas that make use of seasonal pressed flowers

Figure 10.1 New technology for Japanese umbrellas—a Bio-TRIZ innovation.

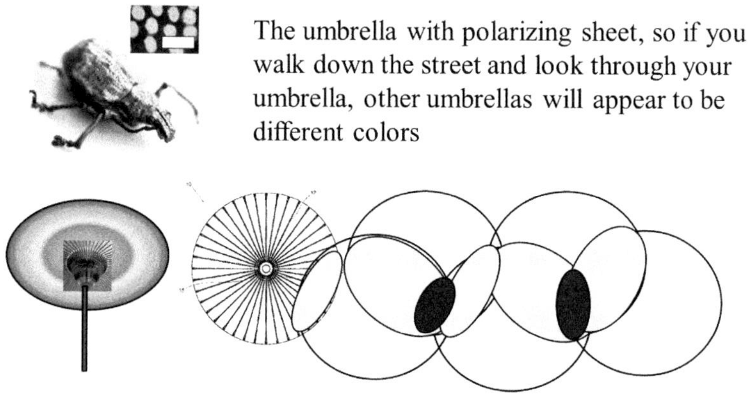

Figure 10.2 Development of nanocellulose washi paper inspired by the structural color of the elephant beetle. If you look up at an arcade of Japanese umbrellas, you'll notice that the colors of the overlapping sections will be different due to the use of polarized washi paper.

10.4 Designing and Developing Japanese Umbrellas from the Perspective of Mechanical Engineering

Using the Bio-TRIZ Database

Figure 10.3a is an ontologically engineered action decomposition tree for enjoying a spiritually affluent lifestyle on the theme of Kuchinai umbrellas. This tree contains numerous ideas for revitalizing the town of Kuchinai, introducing new lifestyles, and realizing a sustainable society through its traditional Japanese umbrellas. Arrows indicate the relationship between strategic words and phrases. For example, at the top of the section framed in red is the phrase "Preserving traditional craft techniques." A closeup of the section shown in Fig. 10.3b indicates an arrow pointing to "Encountering traditional craft techniques," which in turn links to "Using traditional goods" → "Use craft objects" → "Having a personal Japanese umbrella," and finally, "Using a Japanese umbrella." From the phrase "Using a multifunctional Japanese umbrella," you see arrows pointing out to the phrases "Carry the umbrella," "Open the umbrella," "Use the umbrella," and "Close the umbrella." Each phrase then inspires other images of a new lifestyle featuring Japanese umbrellas, including "a picturesque scene of brightly colored Japanese umbrellas," "Fragrant umbrellas, "My personal Japanese Umbrella," and "Japanese umbrella rentals."

The green nodes indicate positive aspects of using Japanese umbrellas in daily life. For instance, "assert one's personality," "a sense of nature," and "camaraderie with others." The yellow nodes, on the other hand, are negative aspects of Japanese umbrellas that include "heavy," "oily smell," and "expensive." These negative words and phrases are very important, serving as guideposts on the problems that need to be resolved in order to achieve a spiritually affluent lifestyle in which Japanese umbrellas play a part. The phrase "closing the Japanese umbrella" in Fig. 10.3 is connected to "Closing the Japanese umbrella without getting other people wet." A related negative aspect, "Hands become wet when closing the umbrella," is written nearby, along with the solution "Make the umbrella super-water-resistant". If the negative aspect of umbrellas—"hands get wet when closing the umbrella"—can be resolved, people will no longer be hesitant about carrying an umbrella with them on a crowded train. Umbrellas will become much easier to handle.

What can be done to solve the problem? Let's use the Bio-TRIZ Databaseihttp://com. sixcore.jp/bio-TRIZ/index.phpjto search for bio-functions that may provide possible solutions. To do this we must first determine what needs to be improved, and the deterioration that may occur as a result of an improvement. The desired improvement is to be able to close an umbrella without getting your hands wet. Since this will make it easier for you to carry an umbrella in a crowd, this will improve the adaptability or versatility of an umbrella. So, for Improvement Factor, you should select "Control"—"Adaptability or Versatility." Since improvements in these areas will probably lead to a more frequent use of umbrellas in situations impacted by a large external force or use in a harsh environment, you should select "Environment"—"External harm affects the object" as the Deterioration Factor. The search results for possible solutions will display the following four principles which are listed in the order of effectiveness:

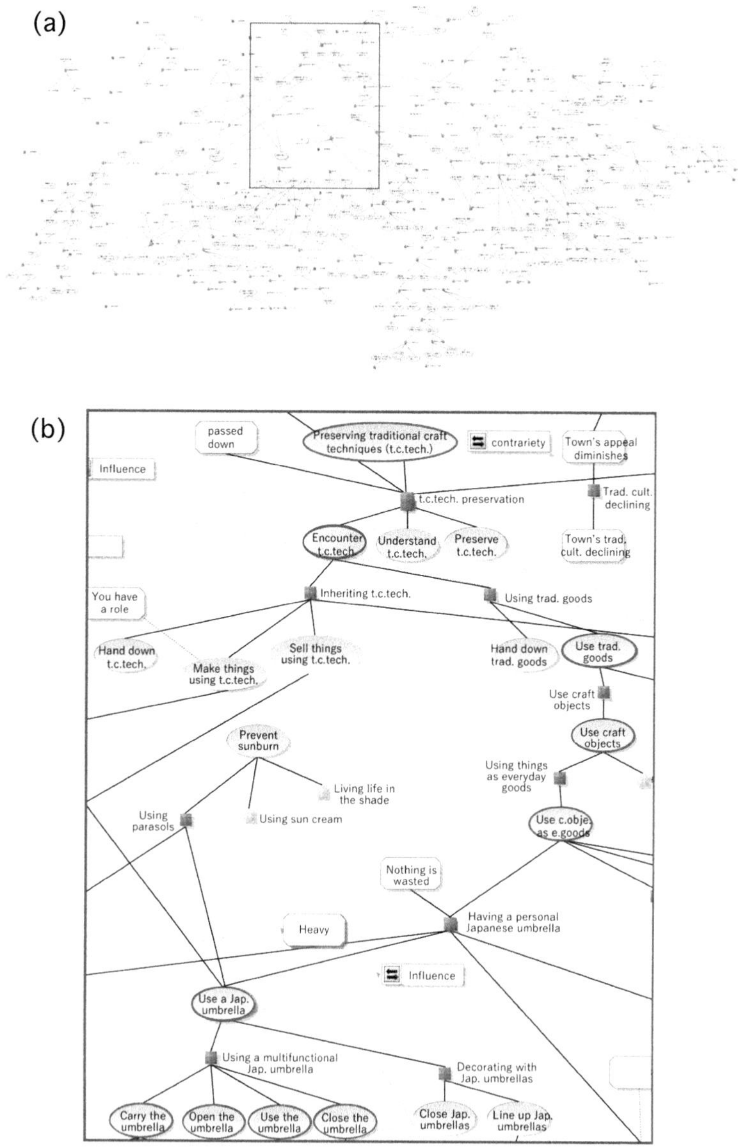

Figure 10.3 Action breakdown tree on the theme of "kuchinaigasa": (a) Full diagram and (b) zoom in the red rectangular area.

Table 10.1 Example of living things that are already using "Beforehand cushioning principles"

- Wasps' parasitic relationship with aphids
- The propolis (bee glue) that protects beehives
- The furanones that keep seaweed clean and healthy
- Kangaroo rats' self-preservation skills
- Frogs' antibacterial peptides (temporins)
- Desert lizards' temperature regulation and hydration system
- Bloodsucking Triatoma rubrofasciata's ability to treat a stroke
- Folding and unfolding of the morning glory flower

Principle 35. Parameter changes: Change an object's physical state (e.g., to a gas, liquid, or solid), or change the concentration or density, temperature etc.

Principle 11. Beforehand cushioning: Prepare emergency means beforehand to compensate for the relatively low reliability of an object.

Principle 32. Color changes: Change the color or transparency of an object or its external environment.

Principle 31. Porous materials principle: Make an object porous, or add porous elements (inserts, coatings, etc.). When the object becomes porous, use the pores to introduce a useful substance or function.

You can choose any one of these four principles. If, for example, you were to choose Principle 11, Beforehand cushioning, a list of living things that are already using this principle will appear (Table 10.1).

If you were to click, for example, on the third listing ("The furanones that keep seaweed clean and healthy"), you'll be given the following explanation: "Seaweeds in the Bonnemaisoniales family secrete a chemical substance called furanone to prevent them from being besieged by biofilm forming bacteria." You'll also be given links to websites for reference. In this way, you can refer to biological functions that may provide a clue to solving practical problems.

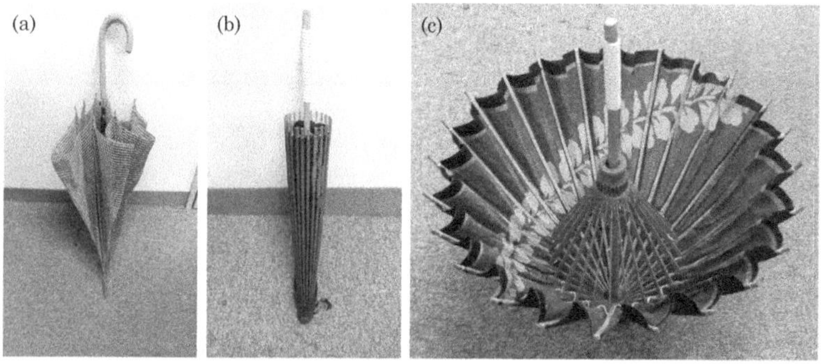

Figure 10.4 Folding of Western umbrella vs. Japanese umbrella: (a) Closed Western umbrella, (b) closed Japanese umbrella, and (c) partially opened Japanese umbrella.

Problem Solving from the Perspective of Mechanical Engineering

Figure 10.4a,b are both photos of folded umbrellas. Figure 10.4a shows a common Western umbrella that has become popular since its introduction in the Meiji era. Figure 10.4b is a Japanese umbrella. (Normally, a folded Japanese umbrella is placed the other way round, so that the tip is facing upwards, but it has been placed with the tip down to compare with the Western umbrella.)

One of the main differences between the two umbrellas is in the number of ribs. Western umbrellas usually have eight (some of the new umbrellas have 6, 16, or 24), whereas Japanese umbrellas have as many as 30 to 70. This means that the size of the panels, the triangular-shaped fabric (or in the case of a Japanese umbrella, the oiled washi paper) in between the ribs, is also quite different. But the biggest difference is in the way the panels are folded away. In Western umbrellas, the panels fold outwards, so that when wrapped around the shaft, the rain covered surface is facing outwards. With Japanese umbrellas, however, as you can see in Fig. 10.4c, as it is closed, the panels are slightly twisted and folded inward so that they are tucked away inside the umbrella. The only parts of the umbrella in sight are the backs of the ribs. Therefore, if the backs of the ribs were made super-waterproof (hydrophobic) by applying a lotus-effect coating or by creating a rough nanostructure on the surface

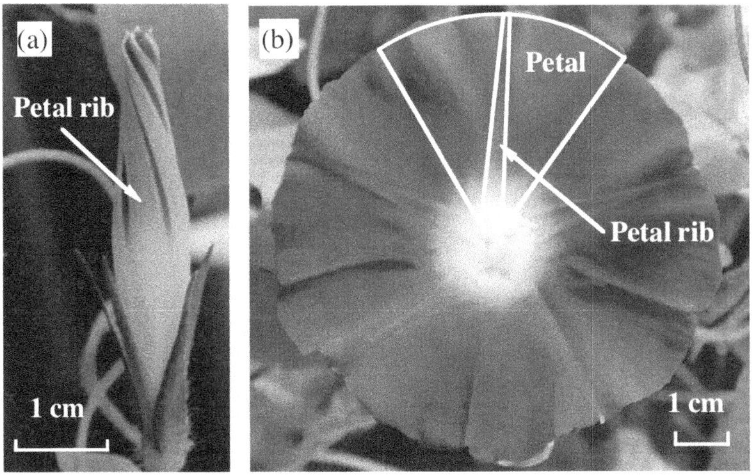

Figure 10.5 Morning glory flower: (a) Flower bud and (b) unfolded flower.

like that of lotus leaves, it would easily become a super Japanese umbrella that wouldn't get your hands wet when closing it! Since a Japanese umbrella has many ribs, when closed, the washi panels in between the ribs are folded into very long, flat triangles. A slight twist will be enough to fold the triangles away inside the cylinder formed by the collapsed ribs.

The folding mechanism behind the Japanese umbrella is basically the same as that of the folding and unfolding of the morning glory flower, at the bottom of Table 10.1. Figure 10.5 shows the morning glory flower bud, before and after it flowers. Each fan-shaped section indicated by the white lines in Fig. 10.5b is a petal. At the center is the floral axis. The bordering sections between the petals fold in half, allowing the petals to be folded inside the spindle-shaped bud formed when the axis rotates and twists, resulting in Fig. 10.5a. To confirm this, we cut open the bud and as shown in Fig. 10.6, the pink petals are curled around the white floral axis. This goes to show that the folding and unfolding mechanism of Japanese umbrellas mirrors that of the morning glory flower. When a Japanese umbrella is closed, the thin panels between the radiating ribs are folded inwards and twisted so that they are folded inside the cylinder formed by the closed ribs. This remarkable mechanism is testament to the way

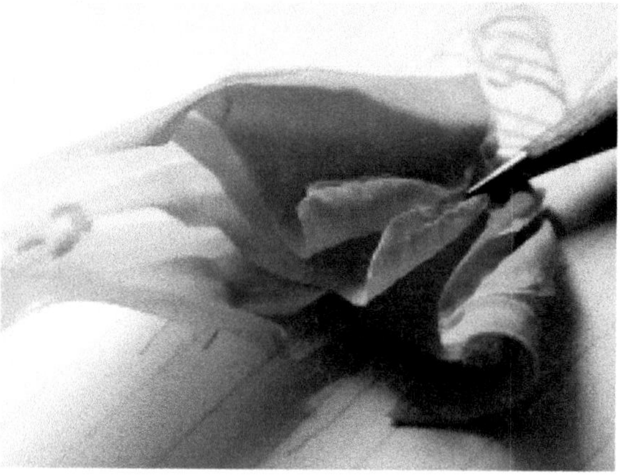

Figure 10.6 Morning glory bud cut open.

in which Japanese people have long lived in harmony with nature, closely observing and learning from natural phenomena.

Japanese umbrellas are made of washi panels treated with a waterproof oil coating. The panels between the ribs have a very long and thin triangular shape so that they can be easily folded inside the cylinder formed by the large number of ribs when closed. The morning glory flower, however, has far fewer petals than the number of panels in Japanese umbrellas. The number of petal is usually less than 8, which is the number of panels used in a standard Western umbrella. Despite the few number of petal ribs, the thin petals are intricately tucked inside the spindle-shaped bud formed around the floral axis. If the morning glory mechanism can be applied to Western umbrellas, it'll be possible to develop a new type of Western umbrella that mimics the mechanism behind a Japanese umbrella, and one that you will be able to close without getting your hands wet. In order to recreate the opening and folding mechanism of a morning glory, the engineering designed must make sure that

(1) the panels between the limited numbers of ribs are folded inwards so that they can be contained within the structure formed by the ribs when the umbrella is closed.
(2) at the same time, the panels are twisted around the central shaft so that they can be fully contained within the ribs.

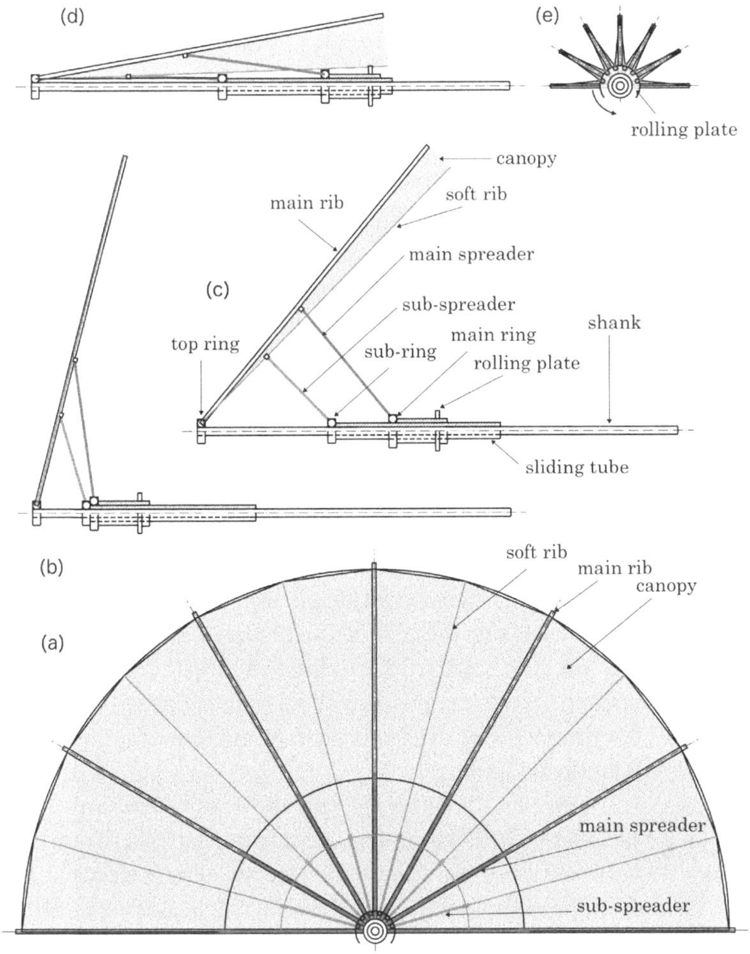

Figure 10.7 Western umbrella that incorporates Japanese umbrella/morning glory style of closing up. (a) Completely open Western umbrella, (b) main ribs and spreaders while completely open, (c) ribs being closed, (d) umbrella with the soft ribs closed up, and (e) almost closed umbrella, seen from the handle.

Figure 10.7 is an example of a Western umbrella that incorporates such an engineering design. Figure 10.7a is a diagram of the top half of a fully open umbrella seen from the inside. Flexible glass fiber sub-ribs have been placed between the main ribs, with spreaders attached to both types of ribs. In addition to having a main ring that controls the main ribs, there is also a sub-ring for the sub-ribs. Because of this double ring mechanism, when folding the umbrella, the sub-ribs close before the main-ribs. To make it easier to see how the positions of the main ribs, main spreaders and main ring change when the umbrella is being closed, close-ups of the relative positions of the main rib, sub-rib, main spreader, sub-spreader, main and sub-rings, and shaft are shown successively in 7b,c, and d. When the umbrella is fully open as in Fig. 10.7b, the main ring is right above the sub-ring. To close the umbrella, the sliding tube connected to the sub-ring has to be pulled towards the handle, forcing the main ring to move appropriately along the tube as well (Fig. 10.7c). This causes the sub-rib to collapse against the shaft before the main rib, as shown in Fig. 10.7d, with the tip of the sub-rib falling into one of the indentations around the rolling plate shown in Fig. 10.7e. Finally, we turn the rolling plate and twist the sub-ribs, causing them to wrap around the shaft. This causes the main-ribs to be pulled towards the shaft as well, creating a cylindrical shape. As a result, the panels are twisted and tucked inside the cylinder formed by the main ribs. Since the outer surface of the main ribs are sure to be treated with a super-waterproof coating, the wet panels will all tucked inside this super-waterproof cylinder so we can expect to have no rain droplets on the surface of the closed umbrella.

Western umbrellas that incorporate the traditional craft techniques of Japanese umbrellas and the folding mechanism of morning glory flowers will be more functional and will be easier and fun to use. We hope that the development of such new conceptual umbrellas will help create a lifestyle that supports the realization of a sustainable society.

10.5 Revitalizing the Regions with New Technologies

In the previous section, we considered the application of needs-oriented manufacturing in helping to create Japanese umbrellas and contribute towards a more spiritually affluent lifestyle in

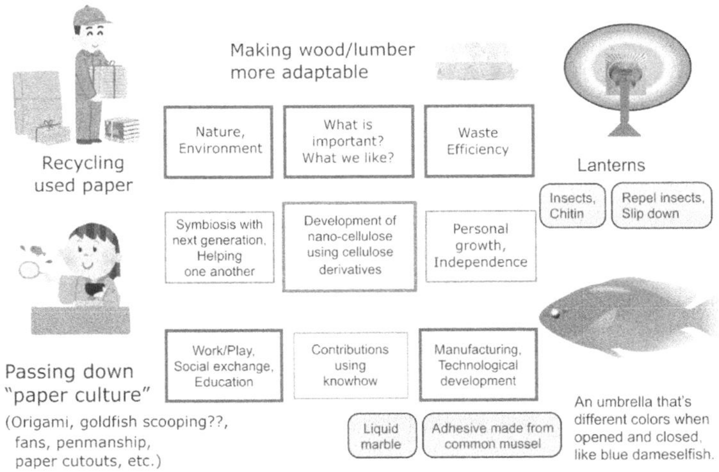

Figure 10.8 How nanocellulose, an idea that comes from the Bio-TRIZ process, is being put to use.

Kitakami. We used Bio-TRIZ to solve engineering problems from the perspective of material science and mechanical engineering. Although these technology seeds were found by analyzing the actions that are part of spiritually affluent lifestyle, it is important to verify whether these seeds are really effective in contributing to a spiritually affluent lifestyle. How should these seeds be developed? An effective solution to this problem can be found in the 70 standard vocabulary map created ontologically and explained in Chapter 6. This map comprises 70 actions, categorized into groups, that an individual would pursue in an emotionally fulfilling society. We decided to use these groupings as a framework for developing technology seeds.

Figure 10.8 indicates the framework for creating colorful Japanese umbrellas using nanocellulose. For instance, "nature/ global environment and nanocellulose" inspires the reuse of nanocellulose by recycling used paper. "Symbiosis with the next generation and nanocellulose" leads to the idea of having children make washi from nanocellulose at the public hall. The words "play and nanocellulose" can inspire origami, picture-story shows, silhouette cutting and other forms of play using nanocellulose paper.

"Technology development and nanocellulose" can lead to ideas such as extracting nanocellulose from pulp, and a spray-type water repellant and deodorizer.

"Reducing waste and effectiveness" can inspire the making of lanterns in the same workshop. Associated outcomes pertaining to the development of such technologies could be, for example, moth-repelling lanterns, to which it would be added by Bio-TRIZ, insect-repellent inspired by cicada wings. We can also apply the method in which the fish Chrysipteracyanea changes color to developing an umbrella that changes membrane color when opened and closed because of the tension change in membrane. Adding the morning glory to this framework will help inspire a whole range of new ideas such as hanging Japanese umbrellas from a helical stand so that anyone can use them—an idea stemming from "symbiosis with the next generation and helping one another." If we integrate these ideas, we can develop the technology seeds to create a lifestyle such as the one shown below.

10.6 Community Development Project: A Circus of Japanese Umbrellas

In designing a prototype for community development, we focused on reviving a local festival and invigorating it with a circus-like atmosphere with a profusion of colorful Japanese umbrellas. From these two technology seeds shown above, we can expect to achieve the following lifestyle.

Japanese umbrella craftsmen and their nanocellulose washi-making techniques will play a central role in this way of life. Existing technologies will be used to extract nanocellulose from used paper. The liquid nanocellulose solution will be used to make iridescent colorful washi. Children gather around the demonstration, their eyes sparkling with excitement. They want to have a go at it. They can hardly wait to make their own original washi. In addition to offering children the chance to learn about the culture of Japanese washi, which is an important part of their heritage, the experience enables them to further their understanding of nanomaterials, the cutting edge in materials engineering. They also get to handle products made from nanomaterials. The children are filled with a sense of accomplishment, and are dying to use the washi they made. After thanking the elderly craftsmen, the children rush off to ask their grandmothers to teach them origami, picture-story shows, silhouette cutting and other traditional techniques for washi. The elderly craftsmen have no time to rest, bombarded by requests from enthusiastic children who want to learn how to make colorful Japanese umbrellas. The craftsmen are all smiles as they watch over the children. Outside are festival booths offering various games and attractions, including goldfish scooping and picture-story shows. The children enjoy scooping goldfish with their handmade washi as they try to win a pet goldfish. They also get to use their handmade paper folding fans to hit a target and win a prize.

Each booth is set up under a Japanese umbrella, which provides shade. The umbrellas mimic the structural color of insects and are highly transparent, allowing sunlight to project colorful patterns on the ground. The umbrella shades form a corridor that fluctuates in the wind. The movement and the changes in light create a kaleidoscope of colors in the overlapping sections. Although projection mapping has made, it possible to project images on Japanese umbrellas, here, natural forces are at play, projecting colorful patterns on both the umbrellas and the ground.

By the river, a lantern floating ceremony is held to offer solace to ancestral spirits. The lanterns have been made out of Japanese umbrellas that were once prized possessions of the deceased. The lanterns are treated with a biomimetic coating inspired by cicada wings that repel harmful insects by slippery surface, so they are quite practical and can be used in everyday life, bringing back fond memories of the deceased. The festivities are abruptly halted by a sudden evening shower. Everybody rushes to borrow a community umbrella from the Japanese umbrella tree. Morning glory-inspired umbrellas are hanging from the spiral tree stand. Thanks to these community shared umbrellas, everyone can head home without getting wet. The next morning, the umbrellas are hung out to dry on the tree stand, transforming it into an object of art, with the umbrellas opened up like morning glory flowers. It's a dazzling sight that brings smiles to peoples' faces as they make their way to

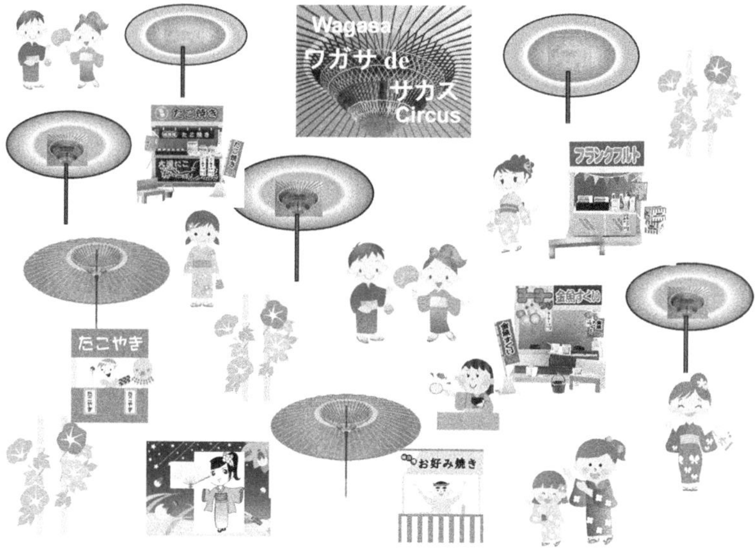

Figure 10.9 The revival of Japanese umbrellas in Kuchinai—one example of a great lifestyle change.

work. Festivals like this make you want to dress in a kimono rather than wear your customary Western clothing. The kimono industry is struggling to convey the appeal of traditional Japanese clothing. But creating an environment that seems right for kimonos may help to revive the industry. Figure 10.9 illustrates the above lifestyle. It's still a prototype but it contains numerous ideas for the development of an emotionally fulfilling community.

10.7 Conclusion

We have seen how Bio-TRIZ, the concept of applying serendipitous discoveries inspired by living nature towards invention, enable us to quickly and easily develop innovative technologies and materials necessary for achieving a spiritually affluent lifestyle following a development process that meets ISO standards. This also suggests that functions discovered through ontology engineering can be applied to the creation of technologies with Bio-TRIZ to achieve needs-oriented manufacturing. Furthermore, we have demonstrated how a vocabulary map that describes a spiritually affluent lifestyle can be used to extract and develop technology seeds which can be applied to community development and social infrastructure.

Chapter 11

Trends and Characteristics of Biomimetic Technology: Towards Social Implementation

Yoshinori Fujihira,[a] Yuta Uchiyama,[b] and Ryo Kohsaka[b]

[a]*Graduate School of Engineering, Muroran Institute of Technology, Hokkaido, Japan*
[b]*Graduate School of Environmental Studies, Tohoku University, Miyagi, Japan*
yfuji@mmm.muroran-it.ac.jp, yuta.uchiyama.c1@tohoku.ac.jp, kohsaka@hotmail.com

11.1 Introduction

11.1.1 Ecosystem Service and Biomimetic

Biomimetics is a process where we learn from the results of the long history of evolution in the natural world. In the past, chemistry and pharmaceutics have played an active role in research and development (R&D) by using the materials deriving from biological resources. In recent years, the field has expanded to material engineering or deign of constructions and cities. The benefits deriving from biodiversity can be categorized largely into four services: supporting, provisioning, regulating, and cultural, as shown in Fig. 11.1. Biomimetics falls into the supply services, mainly

Lifestyle and Nature: Integrating Nature Technology to Sustainable Lifestyles
Edited by Ryuzo Furukawa

ISBN 978-981-4774-62-8 (Hardcover), 978-1-315-14687-4 (eBook)
www.panstanford.com

through the supply of genetic resources of the biological diversity. For ideas and R&D of biomimetics, it is not the volume of biomass that matters but the existence of the diverse genetic resources, which are critical even in limited amount. This is a rather unique relationship when we consider the other supply services of water, foods, and timbers that are directly consumable and whose volume is also critical. In other words, the aspect of "intellectual property" deriving from genetic resources with billions of years of history comes into focus in the context of biomimetics.

The surge of biomimetics-related R&D can enhance the motivation to conserve biological diversity. The enhancement of the motivation is an urgent task in different regions of the world (Kohsaka 2010; Uchiyama et al. 2015). Simultaneously, the spread of technology that abstracted laws and principles by learning from nature can contribute to realizing spiritually affluent lifestyles under severe environmental constraints. This technology includes energy-saving technology or structure with less environmental impacts. In summary, technologies learning from nature are an integral part of a sustainable society in two folds: first, ideas or structures are found that are unique from the evolution and difficult for researchers to think of; second, the technology that learns from nature can potentially have environmental benefits with less impacts.

11.1.2 Current Status and Challenges for Biomimetic

There are a number of technologies that became products in use, yet for the majority of the sector, technology remains at the R&D or the conceptual phase, and mainstreaming and implementation of biomimetics remains a future challenge Kohsaka et al. (2017). It is necessary to identify what hinders the mainstreaming. In doing so, we need to consider not only technology but also its social implications and contexts of use. For this purpose, we focused on whether the number of published scientific articles have correlations with the patent applications in the same themes. We regarded the number of journal scientific articles as an indicator of "seeds" and examined whether they are linked with patent applications, which are a rough indicator of possible "needs" in society. In other words, such linkages are processes of (re-)bridging

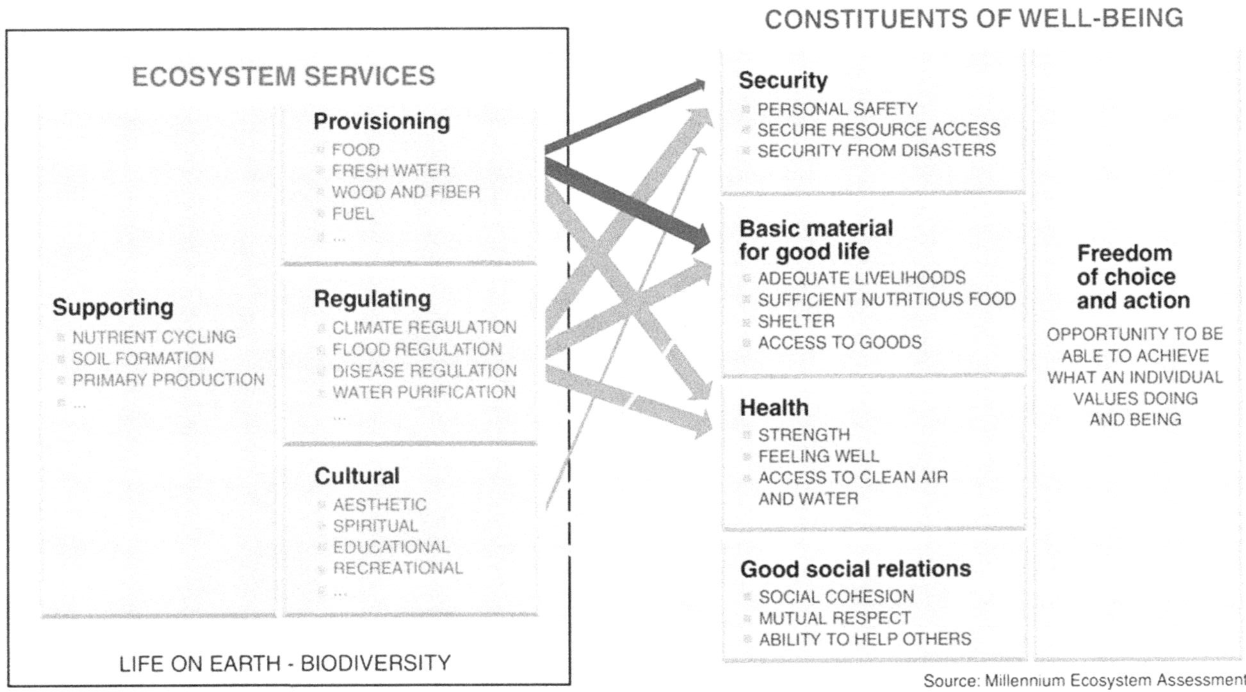

Figure 11.1 Linkages between ecosystem services and human well-being. *Source*: MA Conceptual Framework in the Millennium Assessment Reports, p. 28 (http://www.sustainablescale.org/ConceptualFramework/UnderstandingScale/MeasuringScale/MilleniumEcosystemAssessment.aspx).

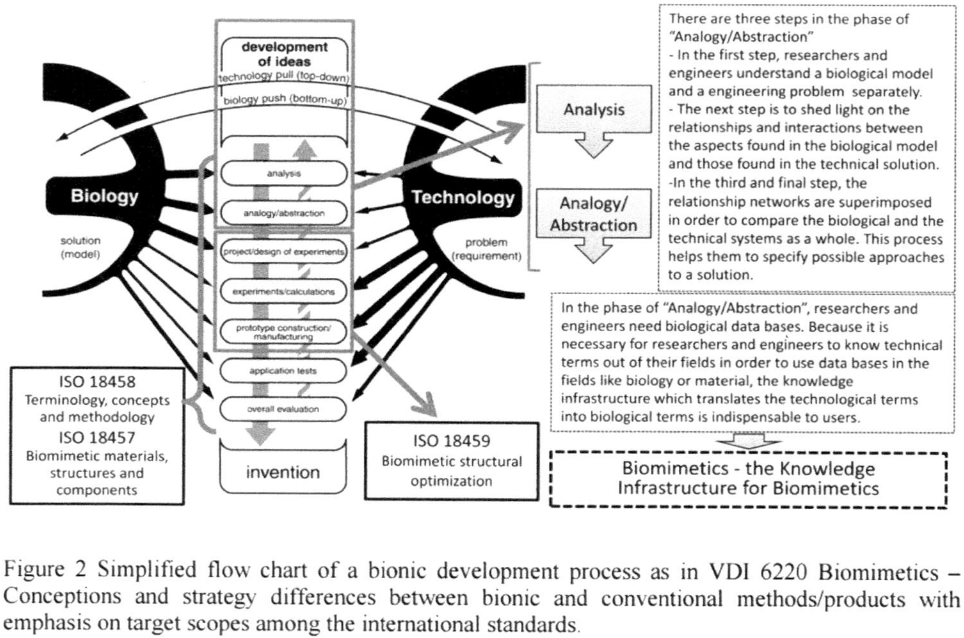

Figure 11.2 Role of the knowledge infrastructure for biomimetics. *Source*: VDI 6220, VDI-Fachbereich Bionik. Reproduced with permission of the Verein Deutscher Ingenieure e. V.

dialogue between scientists and engineers. Similar schemes are to be found in the biomimetics ISO standards (cf. ISO/TC 266 N 144). In these processes, it is necessary to involve various stakeholders and disciplines. These process will also include bridging dialogue with engineers and biologists, as shown in Fig. 11.2.

At national level, it is necessary to overcome these challenges related to stakeholder involvement. At international level, it is necessary for the stakeholders to adapt and engage in the development and harmonization of the ISO standardization by governments and private entities. The involvement of stakeholders, capturing social needs, and clarification of social needs and issues are crucial in this process as well. Developing effective R&D strategies and giving feedback to engineers are integral. It is critical to absorb knowledge from different disciplines.

11.1.3 Approach of Biomimetics in Japan

Given the domestic and international challenges in the realm of biomimetics, the Ministry of Education, Culture, Sports, Science and Technology (MEXT) sponsored a project with Kaken titled "Innovative Materials Engineering Based on Biological Diversity" (Engineering Neo-biomimetics). In the project, ontology is applied to bridge the language or concepts between biology and engineering in a systematic manner. In the existing studies, we identify that the challenges for Japan include collaboration between the industry and science as well as networking the interdisciplinary approach (Kohsaka et al., 2016a). As an overall framework of the project, the informatics serve as the basis of collaboration for biology, engineering, and social science, where "biomimetic engineering" is built on the basis of biomimetic technology. From the viewpoint of social science, various approaches are proposed. Communications, backcasting, and transitional management are proposed to form the basis of communication and a sustainable society, including biodiversity conservation and sustainable use. For this purpose, research institutions, private entities, and consumers are involved. Certain approaches such as backcasting are introduced and illustrated in this book as well (cf. Chapter 4). Transitional management is an approach to manage the changes in the future through participatory process. Social science plays a role in analyzing the impacts and significance of the technology and products beyond disciplines in trans-disciplinary manner by involving stakeholders. In the field of *biomimetics*, research institutes are the major players. Yet, implementations are conducted by private entities. In this chapter, we analyze the trends and status of the private entities involved in biomimetics, including their role and tasks. One point of focus in the chapter is intellectual property, which is highlighted through the analysis of patent applications and published articles related to biomimetics. We further compare them with the patent applications and articles concerning other technologies. In biomimetics, private entities play a major role in the process of the R&D and release of products into the market in collaboration with research institutes. They are also expected to mediate between the technologies and the consumers. The purpose here is to clarify how the private entities develop biomimetics-related technology and their strategies and measures to bring it into the mainstream. Based on the empirical data, it was suggested that databases such as Bio-TRIZ are instrumental in involving stakeholders and addressing the challenges for stakeholders and the end users. Identifying the nature-oriented technology system is an important process for environmental management of technology. As mentioned earlier, we examine patent applications and published papers as two indicators of the trends in patents and R&D, which are related to social implementations. We define "social implementation" as the phase where products or services with certain technology are mainstreamed. Before this phase, there are the phases of R&D, patent application, and product development. The phase in focus for this chapter comes before social implementation and the phases of patent application and R&D where dialogue between engineers and scientists is necessary. Thus, patent applications and published papers are examined as two indicators of the trends in patents and R&D that will lead to social implementations in the future. In addition to chemistry, biomimetics increasingly gained importance in material sciences. We compare the trends with other scientific and industrial domains at international level to gain insights into the characteristics of the themes. For concrete materials, we based our data on the Patent Application Trends Reports 2014 produced by the Japan Patent Office (JPO). Cross-region comparisons are conducted based on the numbers of patent applications and published papers. By doing so, we highlight the trends and focus areas of biomimetics-related technologies in each region.

11.2 A Comparative View of Research and Patents in Biomimetics and Other Themes

11.2.1 Background

In this section, we review the positions and characteristics of the themes in each region from patent applications and published scientific papers as indicators of invention and R&D. The areas where biomimetics is applied are rapidly growing. By reviewing the themes and trends of other countries and regions, we gain insights into both cross-region and cross-theme comparisons for biomimetics. These insights are instrumental for patent application, R&D, and social implementation of biomimetics. The areas, technologies, and positions related to biomimetics are highlighted on the basis of region and country.

11.2.2 Methodology

We compare biomimetics and other technology themes by examining patent applications and published papers as two indicators. Cross-regional comparisons are conducted among Japan, the United States, and Europe. We based our data on the Patent Application Trends Reports 2014 (JPO 2015) for the numbers of patent applications and published papers. In the report, the Agency highlights certain technologies with high social attention and reviews the general trend (cf. Table 11.1 for the individual highlighted technology). In 2014, biomimetics was chosen as a topic. We compare the trends in biomimetics with other themes at cross-regional level. The patents and papers in the report of the JPO are based on the definitions and concepts of ISO/TC266 biomimetics standards. The compared themes and durations are illustrated in Table 11.1.

For the report of the reports on patent application trends, there are two categories of normal and China-specific themes. We

Table 11.1 Research themes and duration for comparison

ID	Emerging technology theme	Period for patent review (~2012)	Period for paper review (~2013)	Nationality surveyed
1	Endoscope	1992	2005	JP, US, EU, CH, KR, TW, BR, RU, IN, A*, IL
2	Fitness Equipment	1996	1996	JP, US, EU, CH, KR, TW, AU, CA
3	Cultivation Technology	1993	1993	JP, US, EU, CH, KR
4	Breeding Technology	1993	1993	JP, US, EU, CH, KR
5	Ocean Industry	1998	1998	JP, US, EU, CH, KR
6	Minimally Invasive Medical Devices	2003	2003	JP, US, EU, CH, KR, BR, IN, A, IL, ME, AU
7	Automotive Engine Technology	2001	2001	JP, US, EU, CH, KR
8	Therapeutic Antibodies	1971	1971	JP, US, EU, CH, KR
9	Superabsorbent Polymer	1980	1980	JP, US, EU, CH, KR
10	Biomimetics	2001	2001	JP, US, EU, CH, KR
11	Transmission Technologies of Wireless LAN	2007	2007	JP, US, EU, CH, KR, TW, IN
12	Contactless Power Transmission	1995	2000	JP, US, EU, CH, KR
13	Artificial Intelligence	2008	2008	JP, US, EU, CH, KR

Note: Themes without datasets of published papers are excluded from the table.
*A stands for ASEAN.
Source: Fujihira et al., 2016

have chosen the theme from the normal theme groups. The focus areas are Japan, the United States, Europe, China, and Korea. In the report, the figures and annual trends of nationality of the patent applicants and affiliated institutions are aggregated for the specific theme. In order to conduct the comparisons for the themes and nationalities, we set a common duration of comparisons. We then compared the number of patent applications and published papers during the period. As a following step, we transformed the proportion that the individual regions occupy for the theme within the set time frame. The period is set from the year 2001, when the review of biomimetic was conducted. For the reviewed themes that were conducted after 2001, we applied the review period (Artificial Intelligence). For the theme where the patent and published papers had different time frames, as was the case with the Endoscope, we used the period that was later than the other (the year 2005 for the Endoscope). From these comparisons, we identify the trends which themes and regions are conducting patent applications and R&D. Furthermore, we identify the difference between the two indicators of patents and published papers, which potentially are the gap between scientific research and patent applications. By doing so, the characteristics of biomimetics are highlighted compared to other technological themes. We compare the themes listed in Table 11.1 for the regions of Japan, the United States, Europe, China, and Korea (hereafter the five regions). In a strict analysis, the applicants of a country or region should be subdivided into academic institutes and private entities where the priorities of applicants and publishing papers differ largely (as is the case with Japan). Such detailed analysis remains a future task. For now, we will focus our analysis on the general trends.

11.2.3 Comparisons of the Themes

Before comparing the five countries and regions, the overall trends of patent applications and published papers are illustrated in Table 11.2, and the ratio of the two indicators are shown in Fig. 11.3. The overall number of patent applications are ranked as follows: Minimally Invasive Medical Devices (Theme 6), Transmission Technologies of Wireless LAN (Theme 11), and

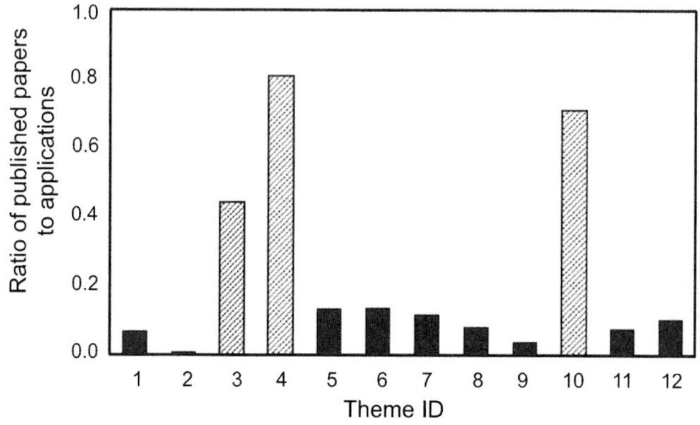

Figure 11.3 Ratio of published papers to applications in the individual theme. High-ratio themes (>0.4) are indicated with hatched lines. *Note*: Theme 13, Artificial Intelligence, had an extremely high ratio and was excluded from this figure. *Source*: Fujihira et al. 2016.

Endoscope (Theme 1, from 2005). The published papers are ranked as follows: Artificial Intelligence (Theme 13), Minimally Invasive Medical Devices (Theme 6), and Biomimetics (Theme 10). As in Table 11.2 and Fig. 11.3, Biomimetics (Theme 10), Agriculture-Related Technology (Themes 3 and 4), and AI are the themes that have a relatively high ratio of patent applications against published papers. Especially, the theme AI is higher than 10 times compared to other themes which are less than the ratio of 1. AI was the only area related to software that was cause for such a very high ratio of published papers to patent applications, which indicates that R&D is more intensive than patent applications in a time frame.

It could be said that the three themes of Biomimetics, Agriculture-Related Technology, and AI (hereafter "Emerging Themes") are emerging themes compared to other themes, where R&D activities and patent applications are actively conducted before market introduction in the private sectors.

11.2.4 Trends in Individual Countries and Regions

We review the trends in Japan, the US, Europe, China, and Korea. We review the relationships between the number of patent applications

Table 11.2 Comparative numbers of patent applications (2001–2012) and published papers (2001–2013) [for the period when biomimetic data are available]

ID	Emerging technology theme	Number of patent applications	Number of paper publications	Ratio of paper to patent**
1	Endoscope*	31,530	2,094	0.066
2	Fitness Equipment	21,019	184	0.009
3	Cultivation Technology	1,591	704	0.442
4	Breeding Technology	2,143	1,730	0.807
5	Ocean Industry	14,965	2,010	0.134
6	Minimally Invasive Medical Devices	37,434	5,099	0.136
7	Automotive Engine Technology	19,787	2,301	0.116
8	Therapeutic Antibodies	22,909	1,870	0.082
9	Superabsorbent Polymer	10,705	403	0.038
10	Biomimetics	5,711	4,051	0.709
11	Transmission Technologies of Wireless LAN*	37,096	2,834	0.076
12	Contactless Power Transmission	17,935	1,875	0.105
13	Artificial Intelligence*	588	5,744	9.769

*Theme with a shorter period than biomimetics.
**Ratio of number of published papers to that of patent applications.
Source: Fujihira et al., 2016.

and published scientific papers (Fig. 11.4) and the trends in the individual countries and regions (Fig. 11.5). As a general trend, overall proportions of patent applications and published papers are high in the US and Europe, but low in China and Korea. Japan has a number similar to that of the US and Europe for patent applications, but the proportion of published papers is limited to a level similar to that in China and Korea. Europe and China tend to have a higher ratio of published papers to patent applications, while Japan and Korea tend to have a higher ratio of patent applications to published papers. When we consider the individual technologies, there are four themes where Japan has a ratio beyond 30% in the regions compared. These four themes are Endoscope (Theme 1), Automotive Engine Technology (Theme 7), Contactless Power Transmission (Theme 12), and Superabsorbent Polymer (Theme 9). Japan has only one theme with over 40% proportion for published papers on Ocean Industry (Theme 5). All the rest of the themes have less than 20%. These trends indicate that Japan strong in conventional themes such as Endoscope (Theme 1) and Automotive Engine Technology (Theme 7), but less so in the emerging themes of Biomimetics, AI, and Agriculture-Related Technologies. For the US, the themes with over 30% proportion of patent applications are the following five: Minimally Invasive Medical Devices (Theme 6), Cultivation Technology (Theme 4), Therapeutic Antibodies (Theme 8), Transmission Technologies of Wireless LAN (Theme 11), and Artificial Intelligence (Theme 13).

For the US, the themes with over 30% proportion of published papers are the following four: Fitness Equipment (Theme 2), Therapeutic Antibodies (Theme 8), Minimally Invasive Medical Devices (Theme 6), and Biomimetics (Theme 10). The theme with an especially high proportion is Fitness Equipment (Theme 2), with over 50% of published papers. As a general trend, the US is focused on patent applications in medicinal, pharmaceutical, health, and bio-related areas. Simultaneously, the US is aggressive in emerging themes such as AI and Agricultural-Related Technologies.

For Europe, the themes with over 30% proportion of patent applications are the following two: Therapeutic Antibodies (Theme 8) and Ocean Industry (Theme 5). The themes with over 30% proportion of published papers are the following seven for Europe: Automotive Engine Technology (Theme 7), Minimally Invasive Medical Devices (Theme 6), Breeding Technology (Theme 3), Endoscope (Theme 1), Artificial Intelligence (Theme 13), Therapeutic Antibodies (Theme 8), and Biomimetics (Theme 10). Europe is playing a leading role in the emerging themes, including Biomimetics, Artificial Intelligence, and Agriculture-Related Technologies, with certain overlaps with the US.

For China and Korea, there are no areas with themes with over 30% proportion of patent applications. For the proportion of published papers, China has only one area with the theme Superabsorbent Polymer (Theme 9). China has a larger proportion, of over 20%, of published papers in the emerging themes of Artificial Intelligence (Theme 13), Biomimetics (Theme 10), and Breeding Technology (Theme 4) than Japan. China has caught up and is potentially leading Japan in these emerging themes. For patent applications, China has 20% of the proportion, similar to Japan. China published the outline of strategy of Chinese intellectual property rights in 2008 (Lee and Shirasu 2009). The current status of China with a low proportion indicates that the strategy is still in its developmental phase.

The trends of patent applications and published papers for emerging themes are shown in Figs. 11.6 and 11.7. From the figures, the rapid growth of China is confirmed during the reviewed period. It occupies a larger proportion in the two indicators of patent applications and published papers. If this trend continues, China's ratio will increase while Japan will continue to decrease its ratio. If we focus on Biomimetics (Theme 10), the ratios of Japan, United States, Europe, and China have a similar proportion of patent applications: 20–25%. There are large differences in the ratios of published papers: roughly 30% for the US and Europe, 20% for China, and a marginal 5% for Japan. As described earlier, the ratio of published papers to patent applications is at relatively higher level of 0.709. Japan has a lower rate of 0.174. This is low compared to 0.823 for the US and 0.928 for Europe. Similar trends are found for Japan, which has a marginal proportion in the emerging themes of Biomimetics, Agriculture-Related Technology, and Artificial Intelligence, where the ratio of published papers to patent applications is high. In other words, whereas the US, Europe, and China are intensifying their R&D activities and enlarging their presence in the scientific area, Japan's influence is decreasing. The low numbers of published papers for Japan is partially attributable to the decreasing number of publications by private companies. In the field of electricity and chemistry, Japan is decreasing its number of published papers (Sato 2012). The report compares the Japanese data with those of China and Korea, suggesting the sharper decrease for Japanese entities and correlations with economic conditions. Besides, there are issues related to language barriers. The journals that are reviewed here are published in English by American or European publishers. The language and cultural barriers are partially causing lower rates for Japan (and possibly for Korea and China).

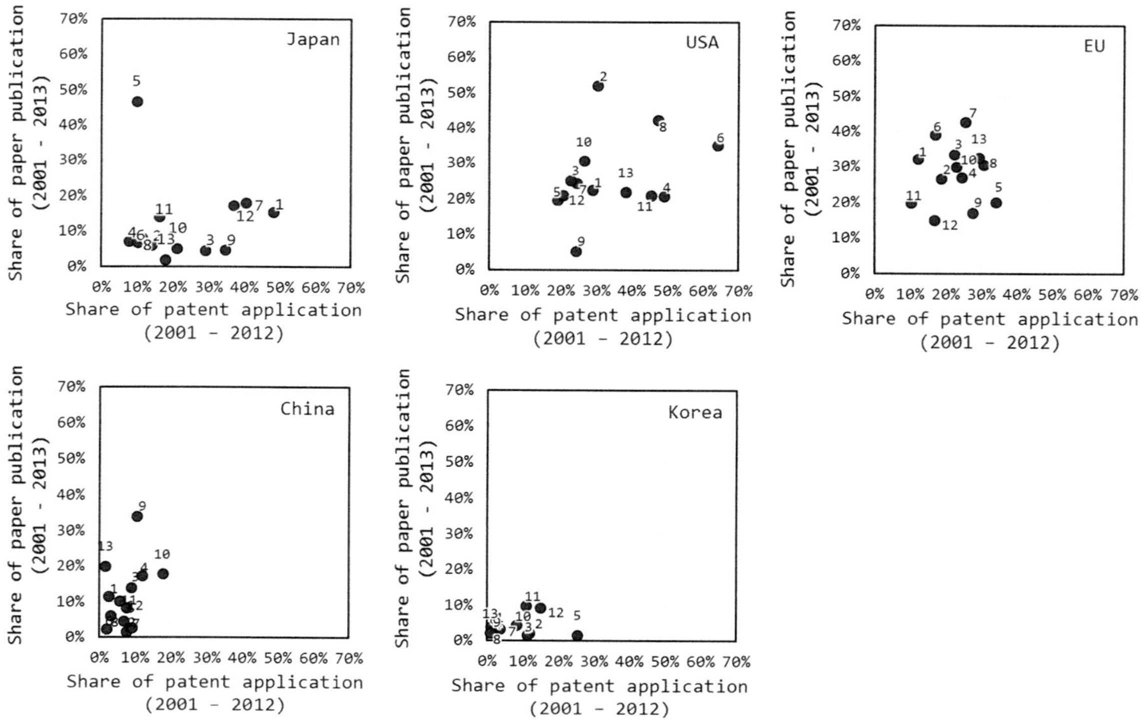

Figure 11.4 Share of patent applications and published papers of individual themes in Japan, the US, Europe, China, and Korea. The numbers in the graphs correspond to the themes in Table 11.2.

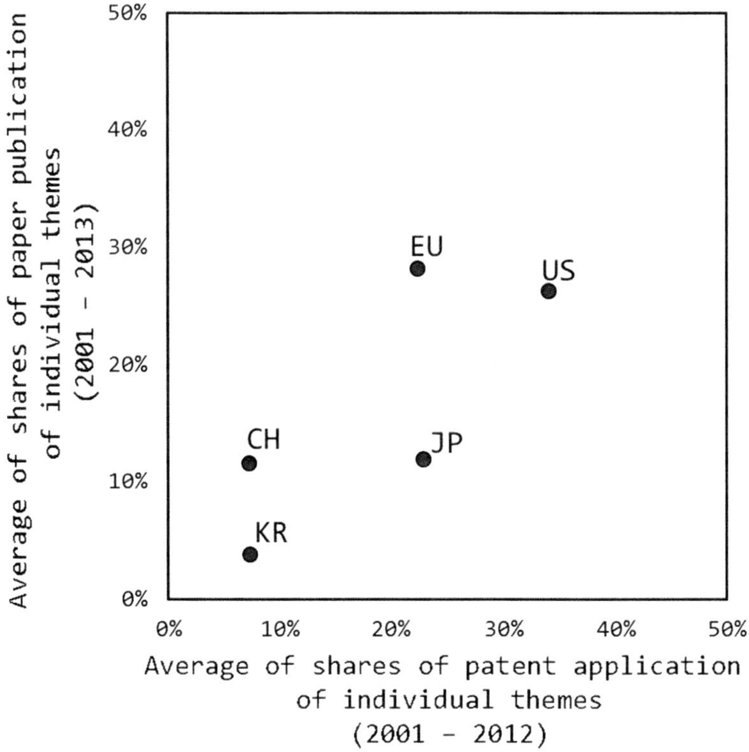

Figure 11.5 Average of patent applications and published papers of individual themes in Japan, the US, Europe, China, and Korea (average from Fig. 11.4). *Source*: Fujihira et al., 2016.

11.2.5 Summary of Inter-regional Comparisons

We highlighted the characteristics of biomimetics compared to other technological themes and conducted a cross-regional comparison by using the trends in patent applications and published papers as a benchmark. The datasets were derived from the Patent Application Trends Reports 2014. Emerging technology themes, such as Biomimetics (Theme 10), Agriculture-Related Technology (Themes 3, 4), and Artificial Intelligence (Theme 13), with a higher ratio of published papers to patent applications were identified. Japan has a relatively small proportion in these emerging theme areas, because of its minor presence academically. Particularly

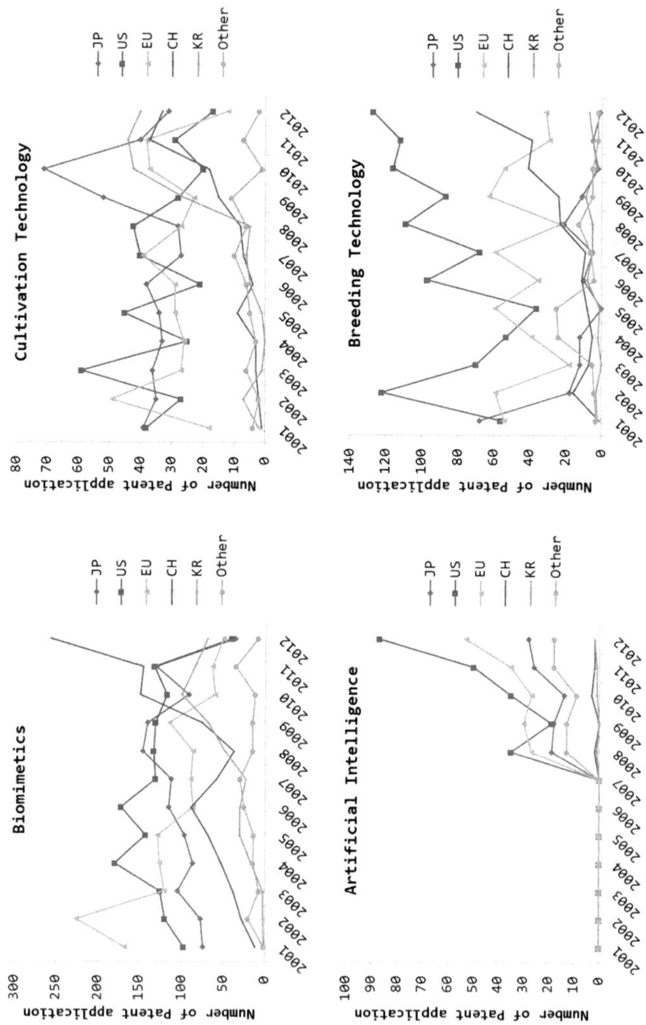

Figure 11.6 Trends of patent applications in emerging themes.

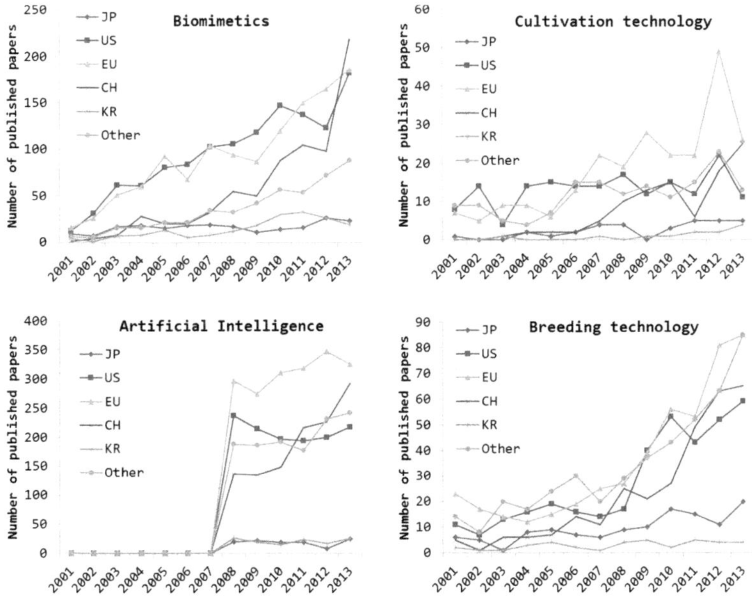

Figure 11.7 Number of published papers in emerging themes.

serious is its discrepancy with the US, Europe, and China in the themes of Biomimetics and AI. Japan is maintaining its pace with the US and Europe for number of patent applications. China is rapidly growing with the numbers, and the differences are narrowing. To realize sustained patent application, fundamental research, application research, and product development, including market introduction, have to be conducted in a balanced manner. In order to gain insights and obtain overview, the two indicators of patent applications and published papers are considered useful. From the results, it is clear that Japan must promote scientific R&D that will reflect the outcomes of published papers, especially in the emerging themes. For this purpose, collaboration among scientists, private sectors, and policymakers are necessary both at product level and at the capacity-building process. In order to bring about social realization of biomimetics, it is not enough to just publish papers and apply for patents. For a balanced development of products and the capacity of engineers, including end users, it is necessary to bridge the gap between engineering and biology in database development and their utilization by establishing platforms to connect experts. The Georgia Institute of Technology established the Center for Biologically Inspired Design (CBID), where experts in engineering and biology are researching not only biomimetics but also educational programs to bring engineering and biology together. Similar bridging efforts in Japan are limited, to usually one unit in the whole curriculum. For further development, it is important to involve students who are the future of the field.

11.3 Research and Patents in Japan: Characteristics and Future Development

The position of biomimetic technology is clarified in Section 11.2. We conduct further review of the position of Japan in this section. In doing so, we review the same five regions for cross-regional analysis again with more detailed application fields and publication disciplines: Japan, Europe, the US, China, and Korea.

11.3.1 Annual Trends in Application Fields and Mimetic Targets

As mentioned earlier, the number of patent applications supposedly reflects the attitude and eagerness of private entities for technology. The eagerness of scientific exploration is reflected in the number of published papers. As in the previous sections, we use two indicators as benchmark: R&D and social implementation of biomimetics technology. We will conduct a detailed review of Japanese biomimetics technology by examining the annual data of application fields, published papers, and patent applications. Biomimetics is already a technology that is applicable in various fields. These application fields of biomimetics are expected to depend on the structure and policy of individual countries and contexts. The objects of mimetics (chemical, physical, or other characteristics of living creatures) will potentially be affected by the accumulation and trends of the biological research in individual contexts. Social implementation of biomimetics is an international challenge. A global network is needed to share knowledge and experience to overcome challenges in an efficient and effective manner. The outcomes of this chapter will serve to enable cross-regional comparison to see what is common and unique between individual contexts.

As in the previous section, we base our analysis on the Patent Application Trends Reports 2014 (JPO 2015), which focused on the number of patent applications and published papers in biomimetics. The timeframe and regions for review are the same as in the previous section: 2001 to 2012 for patents and 2001 to 2013 for published papers in Japan, the US, Europe, China, and Korea. In the report, the patent applications are further divided into four categories: application fields, application industries, mimetic targets, and mimicked creatures. We will compare the four categories and review the maturity of the social implementation and accumulation of scientific explorations in the regions reviewed. In order to enable the comparison, we have standardized the number of patent applications and published papers. The ranges of these numbers were relatively large, so we transformed the numbers with log. In addition, we added 1 to each item before the log transformation so that the zero will remain zero after the log transformation. As a following step, we conducted linear transformation, fixing the maximum number of patent applications and published papers at 1 and the minimum at 0 in all region, respectively.

To analyze the transformed data, we will draw the line that the ratio of the two indicators is 1 (the ratio of the two indicators is equal) in order to compare the two indicators in cross-regional contexts. In other words, the ratio of patent papers or their relative differences will be the standard. We will examine where the items are located for the standard line by using the coefficient of determination R^2 and compare the trends in a quantitative manner. The coefficient of determination here is related to the standardized number of patent applications, x, and published papers, y. The coefficient of determination derives from the regression equation $y = x$ (when the ratio of x to y is 1). The R^2 of this section is defined in the equation below. If the two standardized indicators have the same values, the item will be plotted on or close to the line of the benchmark line ($y = x$). We refer to such positions as "well-balanced," meaning that the indicators of published papers and patent applications are relatively in a similar position as the international standards. If the item is off or far from the line, we refer to it as being "off the balance."

In Figs. 11.8–11.11, we show the standardized plotted points of patent applications and published papers for the following four categories: application fields, application industries, mimetic targets, and mimicked creatures. As a general trend, the US, Europe, and China show similar tendencies and have a balanced proportion of the two indicators (patents and published papers), indicating "well-balanced" trends. The trend is clear compared to Japan or Korea. Particularly clear is the trend in Japan that papers are relatively fewer than patent applications. For Korea, the case is opposite: patents are relatively fewer than paper publications. It is indicated that Japan and Korea are "off the balance." For application fields, Japan has such numbers of fields that are strong in either paper publications or patent applications, which are relatively "off-balanced" compared to other regions and countries. For example, in molecular and materials science in biomimetics, Japan has a similar level of patent applications as the US and Europe, but a smaller number of published papers. Korea has a smaller number of patent applications than Japan, but there are several disciplines, including hydrophilic and hydrophobic materials as well as optical materials, in which Korea has a larger number of published papers than Japan. Japan doesn't have many applications or published papers for machinery in biomimetics. It has a certain number of published papers in robotics, but no patent application in the field.

What these results suggest is the trend that patent applications are progressing for the material field while scientific research is limited. In the fields of machinery and structure, neither scientific exploration nor patent applications are in progress. Especially for robotics and control (see Fig. 11.8), the number of published papers is at a similar level as for other regions, whereas patent applications are limited. From the results, it is clear that Japan has several fields where the numbers of patent applications and paper publications are off-balance. The R^2, mentioned above, is –0.228 for Japan, whereas it is 0.883 for the US, 0.825 for Europe, and 0.550 for Korea. These values indicate that the numbers related to published papers and patents are in a relatively similar range.

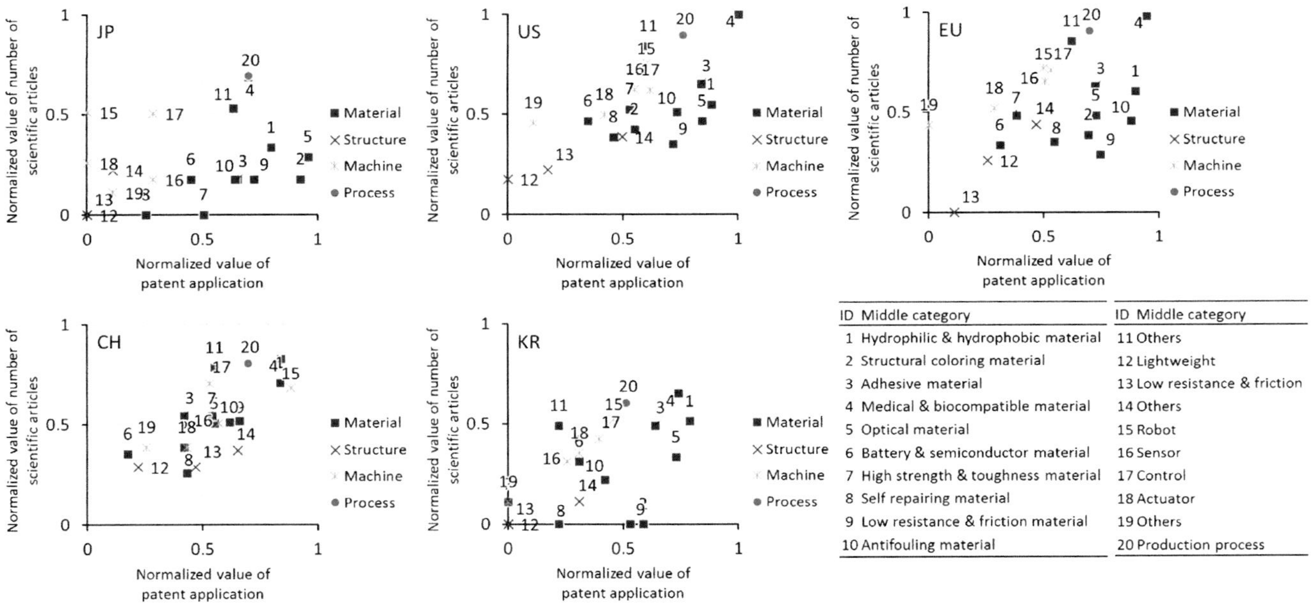

ID	Middle category	ID	Middle category
1	Hydrophilic & hydrophobic material	11	Others
2	Structural coloring material	12	Lightweight
3	Adhesive material	13	Low resistance & friction
4	Medical & biocompatible material	14	Others
5	Optical material	15	Robot
6	Battery & semiconductor material	16	Sensor
7	High strength & toughness material	17	Control
8	Self repairing material	18	Actuator
9	Low resistance & friction material	19	Others
10	Antifouling material	20	Production process

Figure 11.8 Comparison of application fields. The numbers in the figures correspond to the technological field IDs in Table 11.2.
Source: Kohsaka et al., 2016a.

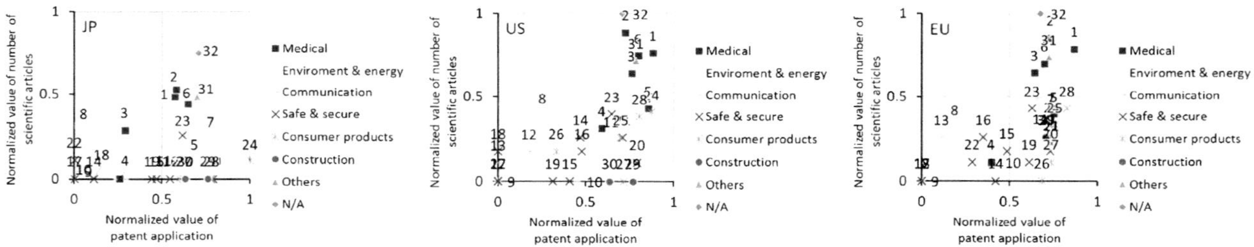

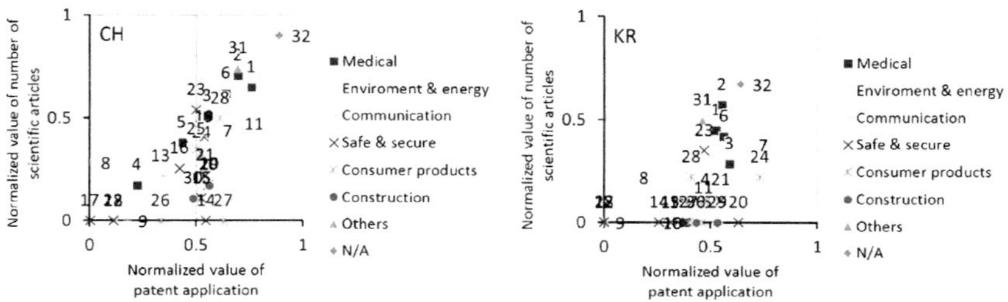

ID	Middle category	ID	Middle category
1	Artificial organ	17	Culture implement
2	Cell culture material	18	Insect pest control
3	DDS	19	Others
4	Surgical glue	20	Transportation equipment material
5	Medical apparatus	21	Transportation equipment parts
6	Others	22	Traffic control
7	Power generation apparatus	23	Others
8	Artificial photosynthesis	24	Electrical appliance & optics parts
9	Smart grid	25	Fiber material
10	Water treatment	26	Cosmetics
11	Others	27	Daily necessities
12	Optical communication	28	Others
13	Others	29	Construction material
14	Disaster rescue robot	30	Others
15	Detection system	31	Others
16	Others	32	N/A

Figure 11.9 Comparison of application industries. The numbers in the figures correspond to the technological field IDs in Table 11.2. *Source*: Kohsaka et al., 2016a.

In application industry, a similar trend is observed as in the application fields. Japan has a certain level of patent applications, but paper publications are limited. For the US and Europe, both patent applications and paper publications related to biomimetics are high, especially in the medical field. China has a similar level of patent applications as Japan, but is more balanced with a similar proportion of paper publications and patents in the medical field. For the consumption field, Japan has a limited number of published articles. The US and Europe have published papers. For the construction field, there are hardly any cross-regional differences for patent applications, but published papers are limited in Europe and China. These results imply that the US and Europe are progressive in both patent applications and scientific paper publications in the medical field. China is relatively balanced in the ratio of patent applications to paper publications. For the consumption field, patent applications are progressing across the regions, but scientific research is rather limited for Japan. For the construction field, scientific research is not proceeding across the regions. The R^2 for Japan is −2.445. It is 0.254 for the US, 0.352 for Europe, 0.609 for China, and −0.569 for Korea, which are all low figures across the regions. Japan and Korea are relatively not well balanced with the ratio of patent applications to paper publications, implying that scientific research and social implementation are not progressing hand in hand.

For the objects of mimetics, Japan has limited numbers in "functions" and "processes" for published papers, and in "microstructure" for patent applications compared to other regions. The results here indicate only the patents and papers in which mimetic targets were identified. The trends are therefore significantly different for application fields and industries in Japan. In concrete terms, a very limited number of objects from patents was found for Japan, because objects are not necessarily outlined in the form of patent applications. The results of Japan are similar to those of Korea. For the US and Europe, there are larger numbers of both patent and published papers related to "structures." Within the scope of "functions," Japan barely has any patent applications or published papers except for the "five senses" or "others." From the results, it is clear that the Japan is lagging behind the US

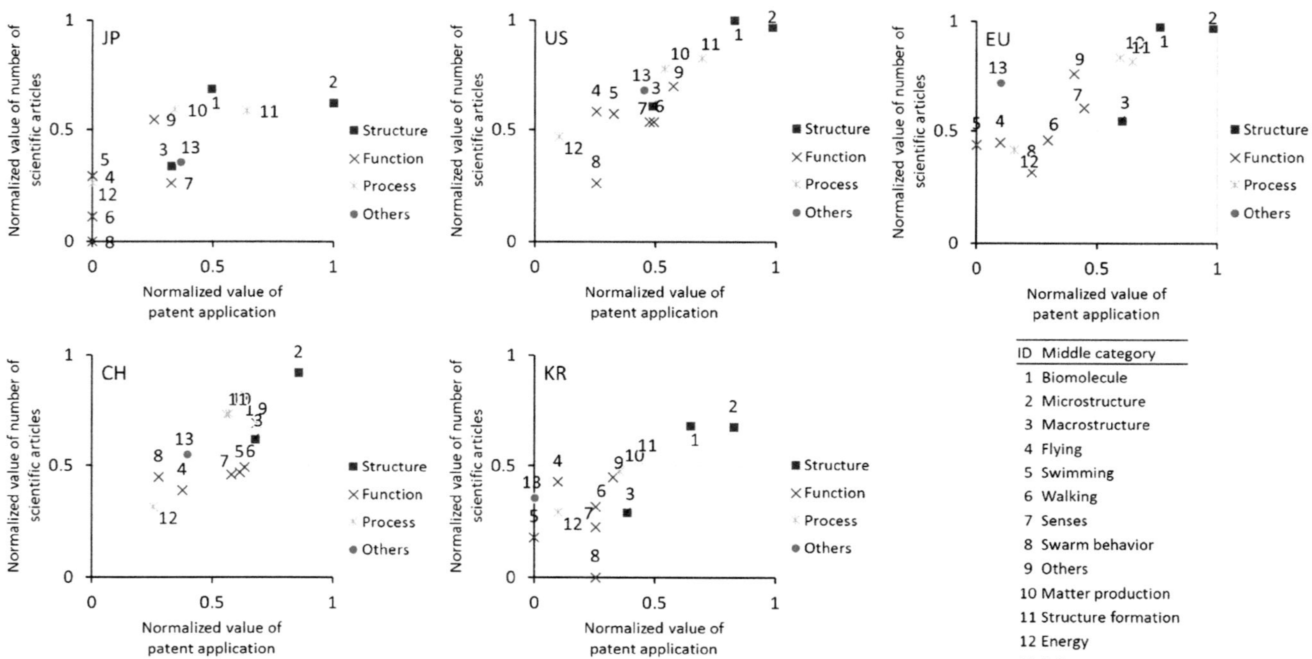

ID	Middle category
1	Biomolecule
2	Microstructure
3	Macrostructure
4	Flying
5	Swimming
6	Walking
7	Senses
8	Swarm behavior
9	Others
10	Matter production
11	Structure formation
12	Energy
13	Others

Figure 11.10 Comparison of mimetic targets. The number in the figures corresponds to the technological field IDs in Table 11.2.
Source: Kohsaka et al., 2016a.

and Europe in both paper publications and patent applications, particularly for the "functions"-related fields. The US and China have a diagonal linear shape, indicating that patent applications as well as paper publications are conducted in a balanced manner at similar quantities. The values of R^2 are 0.757 for Japan, 0.919 for the US, 0.815 for Europe, 0.957 for China, and 0.806 for Korea, indicating that the US and China have high values.

For the category "mimicked creatures," Japan has fewer published papers and patent applications than even Korea, with similar trends. In the limited focus, the fungi and bacteria is the area where Japan has published at a level similar to the US and Europe. However, Japan does not have any patent applications, and the published papers are necessarily contributing to application numbers. This could be partially due to methodological difficulties that patents frequently do not mention mimicked creatures, but it remains to be examined whether this trend is particularly strong in Japan. Birds and amphibians tend to be fewer than arthropods across the regions. From the results, Japan is less progressive in patent applications and also in scientific research for the domain of "mimicked creatures" compared to the US, Europe, and China. The values of R^2 are 0.558 for Japan, 0.970 for the US, 0.958 for Europe, 0.953 for China, and 0.888 for Korea, indicating that Japan is off-balanced among the regions compared. From the above consideration, the general trends suggest that the US, Europe, and China are balanced in patents (number of applications) and scientific research (number of published papers). Japan tends to be active in either one of patent applications or paper publications, but not both. These trends are especially true for the areas of robotics and mimetics for fungi and bacteria, where scientific research is found to be concentrated, and the outcomes of paper publications do not necessarily correlate with active patent applications.

What are the underpinning causes of these trends? A possible explanation is that Japanese universities are less involved in patent applications in general. Intense communication is required for universities and private entities, right from the R&D phase to patent and product introduction into the market. There is a growing concern that if universities and private entities continue to work separately, the communication between R&D or foundational

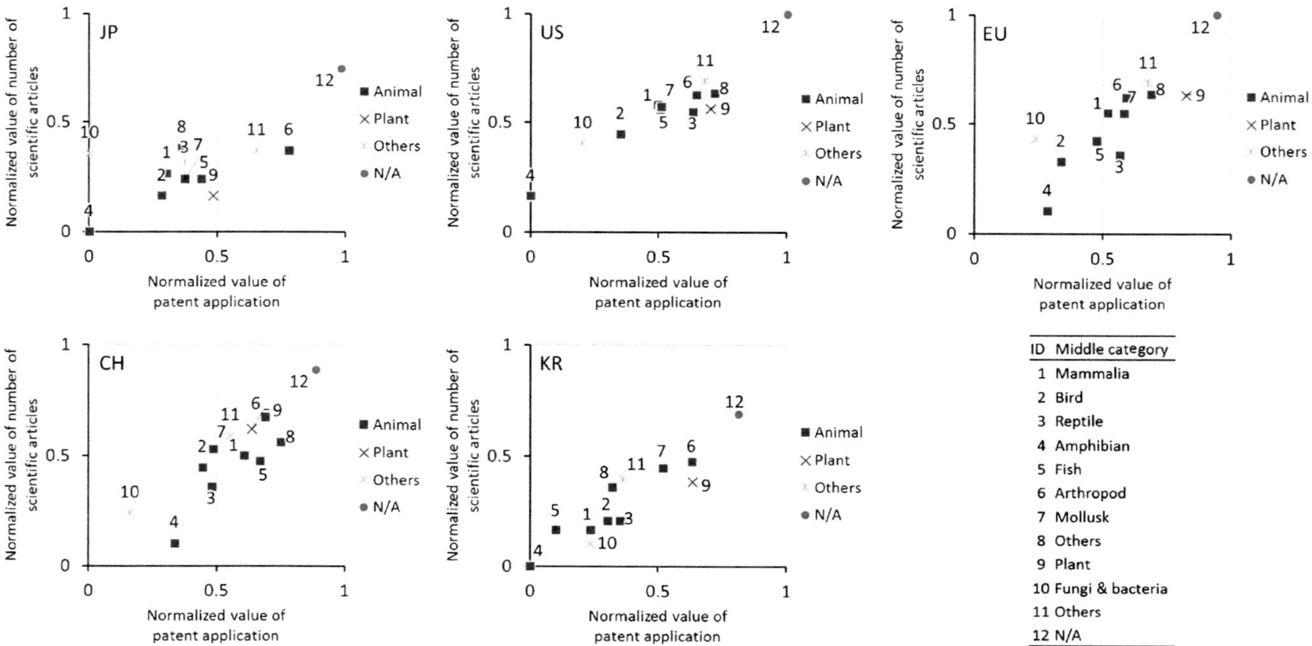

ID	Middle category
1	Mammalia
2	Bird
3	Reptile
4	Amphibian
5	Fish
6	Arthropod
7	Mollusk
8	Others
9	Plant
10	Fungi & bacteria
11	Others
12	N/A

Figure 11.11 Comparison of mimicked creatures. The numbers in the figures correspond to the technological field IDs in Table 11.2. *Source*: Kohsaka et al., 2016a.

scientific activities and patent applications or social implementation will not work smoothly. In the following section, we discuss those who are applying for patents across the regions.

11.3.2 Trends in Patent Applicants

We will now review the general trends or applicant attributes and affiliations of patent applications. Japan tends to have more single or joint applications from private entities. Applications from universities, research institutions, or individuals are limited. For the US and Europe, the trends are similar for patent applications and published papers. For the category of "applications," they show different trends. The universities in the US apply at the same level as private entities. Such trends are not found in Europe. The universities in the US make more applications than do the universities in Japan and Europe. China has a unique trend compared to other regions: universities were the leading applicants from the year 2001 to 2012. These factors related to "applicants" are supposedly affecting the trends for patent applications and paper publications that were previously presented. For example, Japan had less correlation between patent applications and paper publications, which can be partially explained by the absence of universities or research institutions for patent applications. For the US and Europe, universities are contributing a certain proportion for patent applications, but the major players are private entities. There is a specific trend in the US that the universities have been applying for more patents than private entities since 2012. Therefore, in the future, the findings related to published papers might serve as the basis for patent applications. For China, patent applications and paper publications are on the rise, the Chinese universities leading in patent applications. Applications by the private entities are increasing as well, and it remains to be seen who will be leading in applications. The lead by the universities in various regions is supposedly bridging the gap between research, R&D, and patent applications.

In the following section, we will examine the correlations between patent applications and published papers, and see where

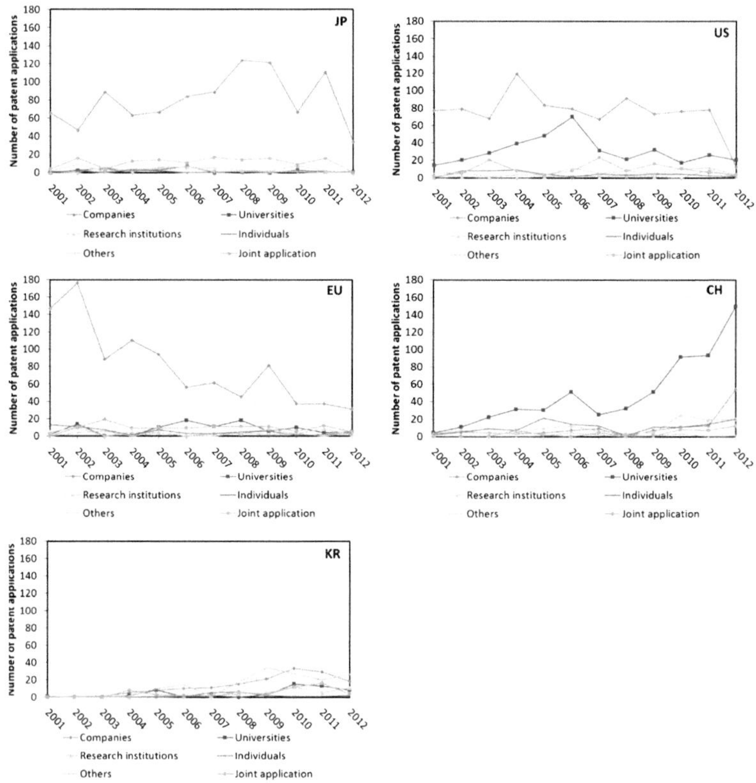

Figure 11.12 Time series data of patent applicants in Japan, the US, Europe, China, and Korea (2001–2012). *Source*: Kohsaka et al., 2016b.

application fields are. We will show the comparative results for these trends in individual regional contexts.

11.3.3 Application Fields in Temporal Sequences

In this section we will review three application fields with increasing numbers of applications: medical and biocompatible materials, hydrophilic and hydrophobic materials, and robots. We will review trends for Japan, the US, Europe, China, and Korea in temporal sequences. For patent applications in the field of medical and biocompatible materials, China is indicating a contrasting trend to Europe and the US in general. The US was at the top with application numbers until 2007 (except for 2002). Then the numbers began to decrease and China came to the top in 2010, and the numbers continued to increase for China until 2012. It is clear from this contrast that the US and Europe tend to be falling behind China in the number of patent applications. With a smaller number of patent applications, Japan is lagging behind.

We will now review the number of published papers in temporal sequences. The numbers for all regions except Japan are increasing. Japan is stable, and its numbers are flat. China is now leading in patent applications, but Europe and the US are still leading in the number of published papers, implying accumulations in scientific explorations. For the period under review, the US recorded the highest number of paper publications in 2010, whereas China's applications for patents peaked during the same year. Patent applications in Europe and the US are on the decrease, but paper publications are on the increase. China indicates an increase in both patent applications and paper publications.

In the following section, we review the trends in hydrophilic and hydrophobic materials. The patent applications in this category indicate similar trends as in medical and biocompatible materials. Basically, Europe and the US have a larger proportion in the early years, while China takes over in the later stage. Unlike the trends in medical and biocompatible materials, Japan has relatively high proportions for the period of 2001–2003, indicating presence in patent applications. Japan tended to have fewer applications. In the latest numbers from 2012, China is leading in the number of applications, followed by Europe and the US, as was the case with medical and biocompatible materials. The number of published papers was limited until 2005 across the regions. In recent years, China has been on the rise with the number of published papers, while Japan has been flat and stagnant with the number of publications. For the field of medical and biocompatible materials, Europe and the US are leading in paper publications. For hydrophilic and hydrophobic materials, however, China is leading in paper publications as well as patent applications. This indicates the dominance of China in both R&D and patents, and possibly product development too.

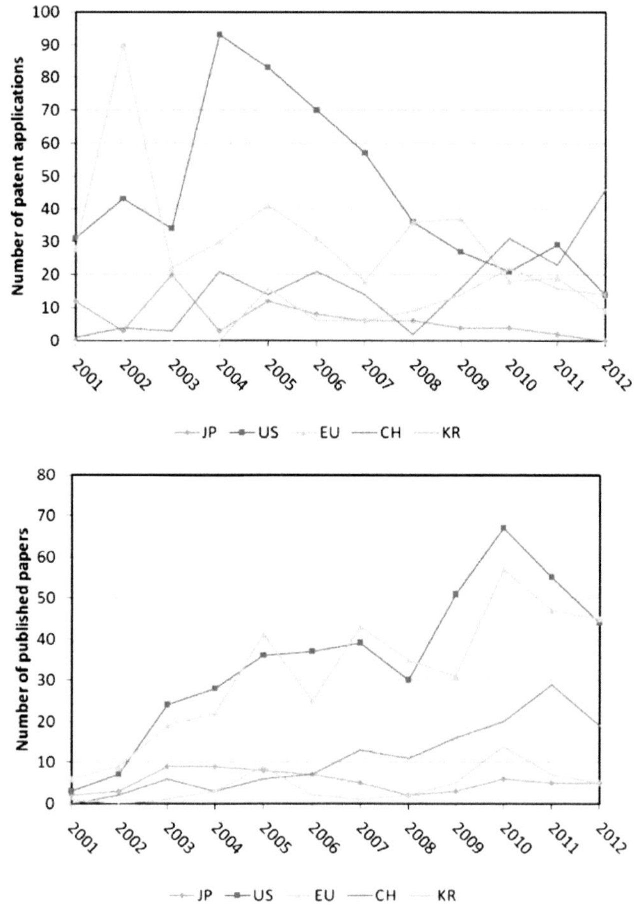

Figure 11.13 Patent applications (left) and published papers (right) in medical and biocompatible materials (material field) for Japan, Europe, the US, China and Korea (2001–2012). *Source*: Kohsaka 2016b.

We review the field of robotics here. In this field, surges are observed in both patent applications and published papers in recent years. Yet the leading regions or countries with patent applications and published papers do not match. China was the leading country in the number of patent applications from the beginning. For published papers, in contrast, Europe and the US are leading. What is underlying cause for these differences? The

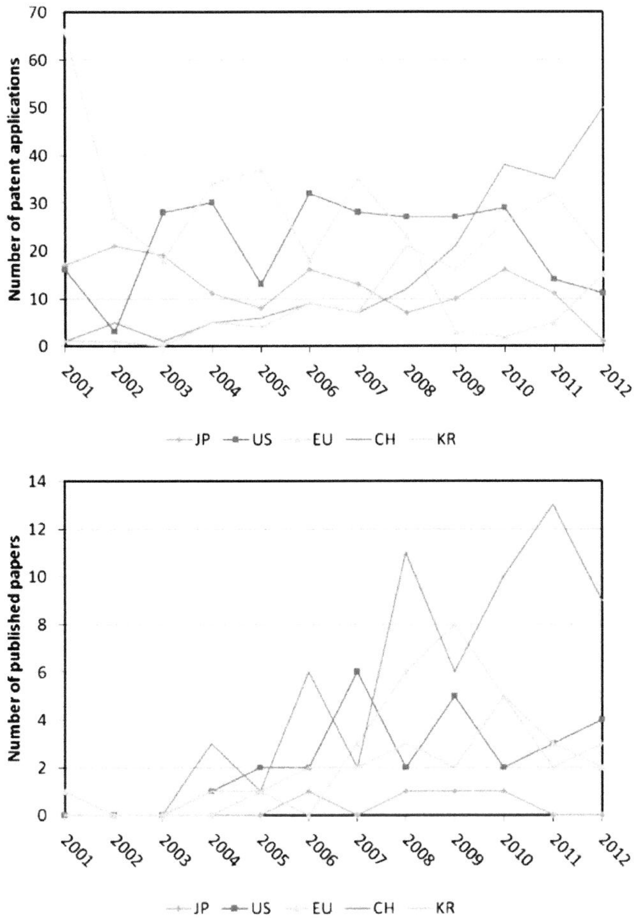

Figure 11.14 Patent applications (left) and published papers (right) in hydrophilic and hydrophobic materials (material field) for Japan, Europe, the US, China, and Korea (2001–2012). *Source*: Kohsaka et al., 2016b.

different environments for R&D, patents, and product development are one of the explanations for China, Europe, and the US. The reasons behind limitations in the number of patent applications by the US and Europe, where papers are actively published, are not clear and remain a future task. These are difficult to analyze solely from applicant datasets.

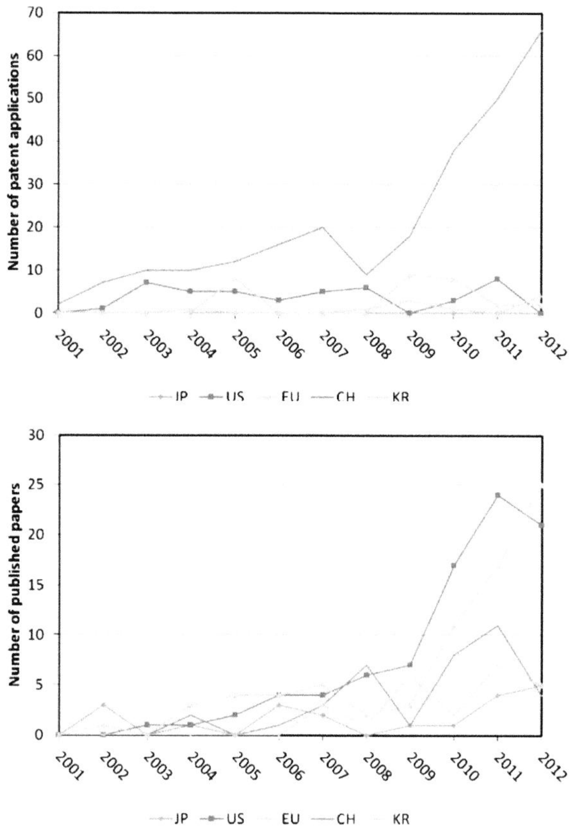

Figure 11.15 Patent applications (left) and published papers (right) in robotics (machine field) for Japan, Europe, the US, China, and Korea (2001–2012). *Source*: Kohsaka et al., 2016b.

11.4 Summary

In Japan, biomimetics-related research and R&D are conducted actively by private entities rather than research institutions. The process tends to end at patent applications and does not correlate with paper publications. This discontinuity may lead to less information sharing and innovations for society in general. In China, patent applications are led by the universities, linking the foundational research with increasing numbers of patent applications and paper publications. As a result, there is presumably an environment that enables sharing of information and accumulation of technologies. For the US, there are many applications from the universities, as is the case for China. The US and Europe have an impressive number of paper publications, but have been less active in patent applications so far. The two regions have potential in social implementation of biomimetics-related technology, possibly through patent applications. Germany has a national strategy called "Industrie 4.0" (or Industry 4.0 in English) where biomimetic technologies are identified as strategic technologies for future trends. From the cross-regional comparisons, it is clear that there are certain gaps in paper publications and patent applications. In concrete terms, for the application fields of materials, there is a relatively larger number of patent applications than published papers. The increase in patent applications and paper publications from China was another obvious trend when analyzed from temporal sequences in individual application fields. For example, in the field of hydrophilic and hydrophobic materials, China has shown an increase in the numbers of patent applications and paper publications and has been on the top since 2012, while Japan, the US, and Europe are stagnant with patent applications and published papers. For paper publications in biomimetics-related articles, Europe and the US are leading China. However, certain fields are already led by China, including hydrophilic and hydrophobic materials and robotics. The challenges for Japan in biomimetic technologies and social implementation are twofolds: first, involvement of universities in patent-related technologies and R&D, and second, bridging the gaps between the areas where papers are actively published and where products are actively explored. It is necessary to overcome this weakness. Furthermore, Japan will need to strengthen itself in the electrical appliances field, where patents are applied for actively. The tie between universities and research institutions and private entities must be strengthened for market production and spreading of the technologies. Electrical appliances are used in daily life and will affect lifestyles directly. Using lifestyle-oriented technologies by integrating biomimetics is an integral step to follow "a spiritually affluent lifestyle under severe environmental constraints." There are encouraging signs for Japan, in that it has a relatively high number of applications in the field of electrical appliances that are potentially contributing to the overall goal of sustainability with spiritual affluence.

Note: This paper is edited based on "Relationships of Patent Application and Academic Research in Biomimetics and Other Emerging Technology: Comparative Analysis of Trends in Japan, US and Europe" published by Intellectual Property Association of Japan in 2016, "Biomimetics Technologies as the Keys to Implement the Sustainable Society under the Environmental Limitations: Environmental Management and Engineering Neo-Biomimetics" published by Sustainable Management Forum of Japan in 2016 and "Analysis on trend of patent application and academic paper publication related to Biomimetics: The trend in Japan, China, USA and Europe" published by Kurume University in 2016 for the general public.

Acknowledgments

This work was supported by KAKENHI Grant Numbers JP24120008 and JP15H01597 in the framework of MEXT KAKEN "Innovative Materials Engineering Based on Biological Diversity" (JP24120001, Masatsugu Shimomura; Kurita Water and Environment Foundation [17K010]; Tohoku Regional Development Association). We wish to express appreciation to the JPO that provided the materials and all the person for helping this work.

References

Kohsaka, R. Economics and the convention on biodiversity: financial incentives for encouraging biodiversity in Nagoya. *Urban Biodiversity and Design*, 593–607 (2010).

Uchiyama, Y., Hayashi, K., and Kohsaka, R. Typology of cities based on city biodiversity index: exploring biodiversity potentials and possible collaborations among Japanese cities. *Sustainability*, **7**(10), 14371–14384 (2015).

Kohsaka, R., Fujihira, Y., Uchiyama, Y., Kajima, S., Nomura, S., and Ebinger, F. Public perception and expectations of biomimetics technology: empirical survey of museum visitors in Japan. *Curator: The Museum Journal*, **60**(4), 427–444 (2017).

Kohsaka, R., Fujihira, Y., Furukawa, R., Yamauchi, T., Kobayashi, H., Ishii, D., and Uchiyama, Y. Biomimetics technologies as the keys to implement the sustainable society under the environmental limitations: environmental management and engineering neo-biomimetics. *Sustainable Management, Sustainable Management Forum of Japan*, **15**, 98–112 (2016a) (in Japanease).

Fujihira, Y., Kohsaka, R., and Uchiyama, Y. Relationships of patent application and academic research in biomimetics and other emerging technology: comparative analysis of trends in Japan, US and Europe. *Journal of Intellectual Property Association of Japan*, **13**(2), 18–22 (2016).

Kohsaka, R., Fujihira, Y., and Uchiyama, Y. Analysis on trend of patent application and academic paper publication related to biomimetics: the trend in Japan, China, USA and Europe. *Journal of Commerce*, Kurume University, **21**(2), 53–68 (2016b) (in Japanease).

JPO. The reports on patent application trends: endoscope (Theme 1) (2015) (in Japanease).

JPO. The reports on patent application trends: fitness equipment (Theme 2) (2015) (in Japanease).

JPO. The reports on patent application trends: agriculture related technology (Theme 3, 4) (2015) (in Japanease).

JPO. The reports on patent application trends: ocean industry (Theme 5) (2015) (in Japanease).

JPO. The reports on patent application trends: minimally invasive medical devices (Theme 6) (2015) (in Japanease).

JPO. The reports on patent application trends: automotive engine technology (Theme 7) (2015) (in Japanease).

JPO. The reports on patent application trends: therapeutic antibodies (Theme 8) (2015) (in Japanease).

JPO. The reports on patent application trends: superabsorbent polymer (Theme 9) (2015) (in Japanease).

JPO. The reports on patent application trends: biomimetics (Theme 10) (2015), in Japanease.

JPO. The reports on patent application trends: transmission technologies of wireless LAN (Theme 11) (2015) (in Japanease).

JPO. The reports on patent application trends: contactless power transmission (Theme 12) (2015) (in Japanease).

JPO. The reports on patent application trends: artificial intelligence (Theme 13) (2015) (in Japanease).

MEA: Millennium Ecosystem Assessment. Ecosystems and human well-being: synthesis report. Island, Washington (2005).

Li, M., and Shirasu, I. A brief overview of the outline of strategy of Chinese intellectual property rights and its future issues. *Journal Entitled Patent Studies*, **47**, 7–16 (2009).

Chapter 12

Nature Technology as a Contributor to Sustainable Social Innovation

Yuko Suto,[a] Ryuzo Furukawa,[b] Takeshi Yamauchi,[c] Hidetoshi Kobayashi,[d] Toru Kobayashi,[e] and Emile H. Ishida[f]

[a]*School of Engineering, Tohoku University, 6-6-04 Aramaki-aza-Aoba, Aoba, Sendai, Miyagi 980-8579, Japan*
[b]*Faculty of Environmental Studies, Tokyo City University, 3-3-1 Ushikubonishi, Tsuzuki-ku, Yokohama, Kanagawa 224-8551, Japan*
[c]*Department of Materials Science and Technology, Faculty of Engineering, Niigata University, 8050 Igarashi Ninomachi, Nishiku, Niigata 950-2181, Japan*
[d]*Department of Mechanical Science and Bioengineering, Graduate School of Engineering Science, Osaka University, 1-3 Machikaneyama, Toyonaka, Osaka 560-8531, Japan*
[e]*Division of Electrical Engineering and Computer Science, Graduate School of Science and Technology, Nagasaki University, 1-14 Bunkyocho, Nagasaki 852-8521, Japan*
[f]*Earth Village Research Lab. LLC, Okinoerabu Island, Kagoshima Z/C 891-9222, Japan*
yuko.suto.b3@tohoku.ac.jp, ryuzofkw@tcu.ac.jp, yamauchi@gs.niigata-u.ac.jp, hkoba@me.es.osaka-u.ac.jp, toru@cis.nagasaki-u.ac.jp, emile.h.ishida@gmail.com

Achieving a sustainable society of innovation requires tying individual agents who have become estranged from nature and their local communities back into these spheres. At the same time, it has clearly been found that people inherently seek out nature and pleasure and are interested in achieving not a lifestyle of dependence, but of autonomy. The new shape of technology of the future is being called on to meet the needs of said lifestyles (creating an spiritual

Lifestyle and Nature: Integrating Nature Technology to Sustainable Lifestyles
Edited by Ryuzo Furukawa

ISBN 978-981-4774-62-8 (Hardcover), 978-1-315-14687-4 (eBook)
www.panstanford.com

affluence, autonomous, and exciting lifestyle amidst the strictures of nature). As described in Chapter 2, it is possible to use a nature technology method to backcast the technology needed for various lifestyles. This calls for the use of backcasting to propose a new lifestyle approach and identifying the technology elements needed therein, as well as a process for finding solutions within nature. These processes cannot be achieved by anyone; rather, they require specialized training. However, this proves insufficient, so a more ubiquitous approach to the creation of Nature Technology is needed.

12.1 Lifestyle-Responsible Technology

Technology produced out of a nature technology creation system that makes use of the rich bounty of nature is, in essence, that technology which situates a naturally informed viewpoint as its basis and that seeks to foster communication. At the same time, nature technology is technology without black boxes, produced out of backcasted lifestyle approaches; it is not a wholesale replacement of one technology with another, nor is it a technology that pursues solely comfort or convenience. These technologies can directly change lifestyles or serve as a catalyst for changing lifestyles. In Chapter 2, in the section on a "waterless bath," we already described the underlying principles behind this approach. In this chapter, we break down a selection of technologies developed out of backcasting various lifestyle contexts.

12.1.1 Air Conditioners Inspired by Soil

Air conditioners are an essential part of achieving a comfortable lifestyle. Heat pumps have continually been refined to a point of near-maximum efficiency, detecting the behavior of residents and adjusting HVAC settings throughout a residence, with auto-cleaning features used to maintain a consistent level of performance and low power consumption reaching a near theoretical peak. We may in the near future see advances deploying the mechanisms found in the "humps" of humpback whales in order minimize water resistance, or the way in which seeds of the plane tree descend on the wind used as a hint for the shape of future fans or air conditioners. No matter how efficient technology may become, it still relies on electricity and resources and power to build it in the first place. In that sense, the underlying lifestyles do not fundamentally change. What insights might backcasting lead us to?

Lifestyle example 1: Wind determines the value of land

Apparently, they would use air conditioners in the heat of summer to create a pleasant atmosphere. Air conditioners pull heat out of the outside air to create a cool climate, but this heat must then be dispelled outside again. Keeping people cool indoors required expelling more and more heat outside, thereby raising the ambient temperature and creating a "heat island" effect. As a result, the overall temperature of cities went up. Living comfortably required cooling the high-temperature air, causing increases in power usage, which in turn further increased outside temperature. Before long, it became impossible to open the windows onto the outside, and people could not enjoy the sounds of insects and birds. The pursuit of living comfortably eventually led to abandoning nature altogether.

We sometimes still see homes with air conditioners today, but no one actually uses them. The homes have the windows wide open, with fresh breezes bringing the faint and fresh scent of the ocean and the mountain forest. Many look forward to the scents of the season carried on the wind. There are major differences in the price of land in various regions based on the quality of the wind there. It's even pleasant with the windows closed. The walls, floor, and ceiling use no power, with the temperature and humidity in the home controlled and scents cleared out. We love gathering around the table to eat grilled mackerel, and we don't have to worry about cooking it under a ventilation fan. This home-cooked dried mackerel is the pride and joy of Dad, who cuts a sharp figure preparing it.

Even amid the environmental constraints of 2030, living comfortably in the home is not something we can give up. However, people do not rely on appliances ("objects") like air conditioners. The benefits provided by air conditioners come down to the "functionality" of controlling air quality, indoor humidity, and other factors.

In that sense, all that is needed are floors, walls, and ceilings that automatically detect (without the use of power) ambient temperature and humidity levels in the home and control them without power consumption.

Knocking on nature's door, we find the nests of termites in the savanna. While outdoor ambient temperature is 50° at noontime and close to 0° in the evening, the nests remain regular at a consistent 30°. The soil in which these nests are made contains countless pore-like holes one billionth of a meter (nanometer) in size which regulate temperature and humidity. All soil contains a porous structure (an agglomeration of clay minerals). Packing soil together in such a way as to preserve this structure creates a "powerless air conditioner." In other words, the floor and walls act as literal air conditioners. Naturally, the "performance" of this natural air conditioner is 20–30% less efficient than a mechanical one. However, we found an intriguing phenomenon occurred: It is plain and clear to see that the material is soil (i.e., a natural substance), so people of all ages, children included, took an interest in the process and in debating among their households (communication) how to compensate for this 20–30%. The conclusion was sprinkling water around the house in summer. In winter, the plan was to layer up clothing more than usual. In this way, dwellers develop a familiarity with and affinity for the technology and seek ways to preserve it and work with it. As a result, power use in the home drops by 20–30%. This is one case of Nature Technology born out of not "grinning and bearing it," but rather obtaining a new lifestyle that is spiritual affluence and enjoyable (Fig. 12.1).

Nature technology seeks to find ways of leading a spiritually affluent lifestyle among the environmental constraints and strictures of 2030. We turn to nature for ideas, which in turn promotes communication, which leads to affinity and familiarity with these news ways. The result is technology that is simple and clear.

The shell of the snail, always clean and gleaming, was redesigned as a kitchen surface and building wall exterior that remains clean and is easy to wipe down. Here again was one example of technology that takes a sense of responsibility for sustainable lifestyles.

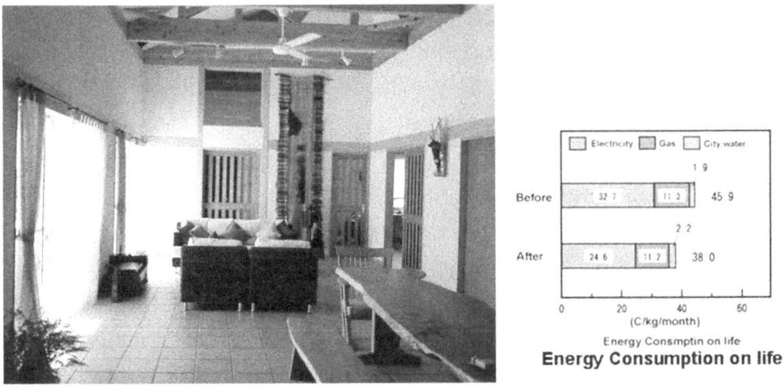

Figure 12.1 Actual application of Earth ceramics on the floor, comfortable living space without air conditioners.

12.1.2 Micro Wind Power Inspired by Dragonflies

Lifestyle example 2: Teeny-tiny wind turbines

In 2030, using crude oil is no longer a viable means of producing energy. Following the events of the 2011 Fukushima Dai-ichi nuclear reactor incident, nuclear power has also come largely to a halt in developed countries, with developing countries, which had previously turned to developed countries for the processing of radioactive waste, finding the costs of treating this material at home too expensive, also ceasing its use. In most countries around the world, nuclear power is a thing of the past. In place of this, people have turned to techniques to ingeniously use natural energy which, while less potent than man-made sources, is available in plentiful supply. The use of energy has also changed considerably. Today, tiny turbines that rotate in gardens have become the envy of children. The reason? These turbines produce and store energy that we can use at home. Using some of the power we stored, we played video games a bit yesterday. While I want to use today's share of power on video games, this morning I heard that the granny who lives next door is running out of juice on her hearing aid, so I want to give the energy to her. I hope we get some nice gusts of wind tomorrow. I love thinking about all that while gazing at the blue sky and the spinning turbines.

What sort of wind turbines would be best suited to this lifestyle? You can almost hear what family conversations would be like: "Mom, can I play games?" "Sure, but only with energy you generated yourself." Just like wind chimes tinkling at the rooftop, the kind of turbines needed in this context must surely be tiny arrays of windmills that turn at the slightest breeze. What this lifestyle needs are turbines that are almost constantly in motion, perhaps twenty or more hours out of the day. This kind of wind power generator will become the envy of children, and it will only require gentle breezes—relying not on wind "power" as such, but on slow trickles over time stored in miniature windmills.

The blades used in windmills today are usually aerodynamic cross sections taken from the shape of birds like hawks and kites. The power generated is exponential by a factor of 2 times the radius of the rotor and 3 times the wind speed, so the diameter of the blades is made increasingly larger to generate better gains. Today, there are turbines with blades over 150 m in diameter. The tips of these blades can easily exceed speeds of 200 km/h; generating power requires wind speeds of 5 m/s or higher. For a small-size generator considered efficient, wind speeds of at least 2–3 m/s are needed. These speeds cannot be guaranteed over the course of a day, meaning uptime will be low, so it would fail to get children interested. A small turbine that is propelled by wind speeds of 1 m/s or lower would thus require blades capable of buoyancy (i.e., rotation) at slight gusts.

Knocking on nature's door, we get a hint from the friendly dragonfly. Dragonflies are the insect capable of gliding through the air at the slowest speeds. In other words, they can turn subtle winds into sources of buoyant flight. This means that if we can figure out how they tick, we could devise ways of creating blades that rotate with subtle puffs of wind. Dragonflies are just 1–2 g in weight; to them, the pressure of the air is all but flying through thick molasses. Staying afloat and gliding through that air is achieved by a surprising mechanism. The airfoil of a dragonfly wing is shaped in a corrugated line. Why is that? Shouldn't its wings be aerodynamically shaped, like those of the hawk or the kite? That mechanism was only recently uncovered by the Obata research lab at Nippon Bunri University. Hawks and kites, which fly at high speeds, have efficiently shaped aerodynamic wings. The dragonfly, by contrast, glides slowly through the air, and this corrugated shape is actually more efficient at those speeds.

It is widely known that if the latest passenger jets were shrunk down to the size of a dragonfly, they would be incapable of flight. This is because the air adhering to the wing has a low Reynolds number (Re), meaning that the viscosity overwhelms the aircraft because the air cannot flow cleanly around the wings. Aircraft generally achieve flight at Re of 1×10^6, with around 1×10^5 being considered "number wall viscosity." In these conditions, traditionally aerodynamically shaped wings fail to achieve flight. Dragonflies can fly in air with a dominant viscosity of 1×10^3, but in these same conditions the viscosity is too high for aerodynamic wings. The dragonfly, by contrast, can fly amid its small size and glide deftly through the air. The reason? While hawks and kites have aerodynamic wings, the dragonfly has thin wings with a corrugated texture. The tips of these jagged lines create tiny eddies on the surface of the wings, which act like ball bearings and create a "rolling" surface that moves through the air. Air on the outer edges of the wings is swiftly conveyed back across them as though being placed on a conveyor belt. This is, in fact, an extremely clever design. On the other hand, in cambered wings, which are a kind of streamlined wing, air detaches around the center of the wing (Fig. 12.2).

Understanding this mechanism allowed for the development of wings with an airfoil like that seen in the Fig. 12.3. The features of these wings can be seen in the diagram (Fig. 12.4). At an angle of attack between 0 and 12°, the lift-to-drag ratio of Colgate airfoils smoothly changes anywhere between 4 and 8. At a low Reynolds number, however, camber airfoils, considered generally stable in other contexts, have, at Re = 7000, an inflection point at a lift-to-drag ratio between 4 and 7, and at Re = 11,000, at 6–10. Compared to Colgate airfoils, they are more sensitive to Re changes and lack stability.

In this way, Colgate airfoils, which produce stable performance at low Re numbers, can be used to create a small-format wind turbine. Figure 12.5 depicts wings 40 cm in diameter. Micro wind turbines (Fig. 12.6) using these blades begin spinning at

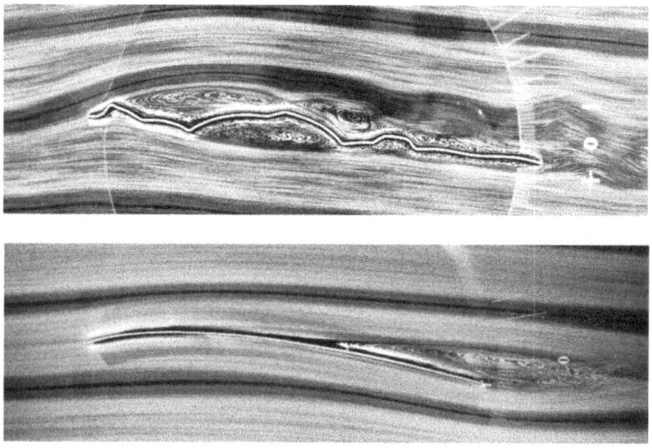

Figure 12.2 Simulation results of airflow around a wing (Re = 7000 α (angle of elevation) = 5°. Upper picture: Model of cross-section of a dragonfly wing. Lower picture: A camber wing.

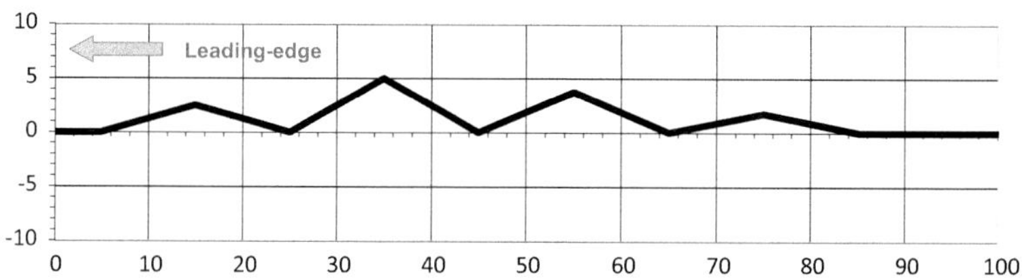

Figure 12.3 Cross-section of a wing developed for small scale wind generators (Cor-No. 5).

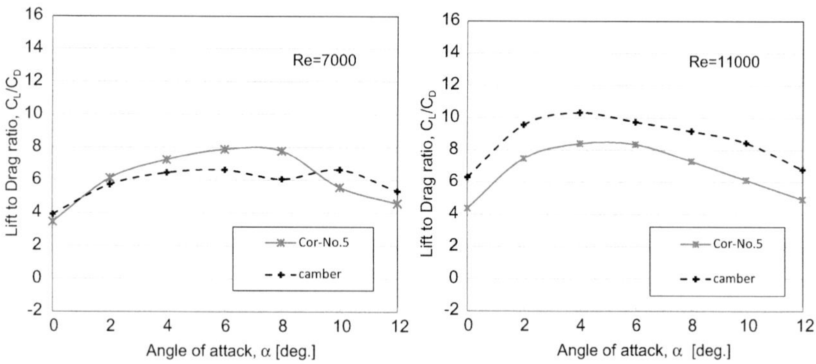

Figure 12.4 The specific characteristics of the developed corrugate and camber wings. Left: Re = 7000, Right: Re = 11000.

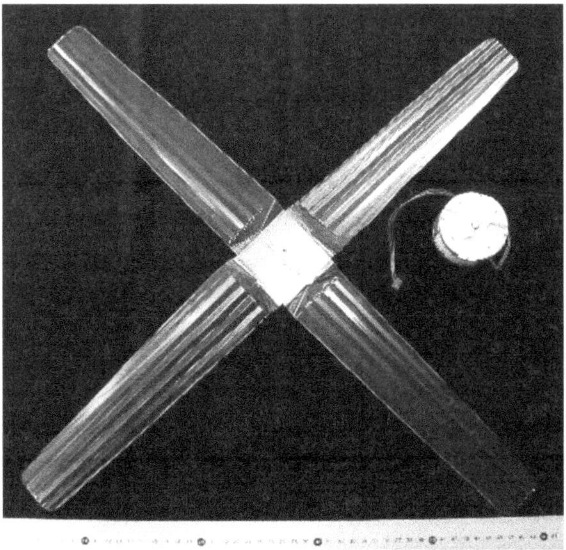

Figure 12.5 Wing with a diameter of 40 cm.

Figure 12.6 Micro wind generator.

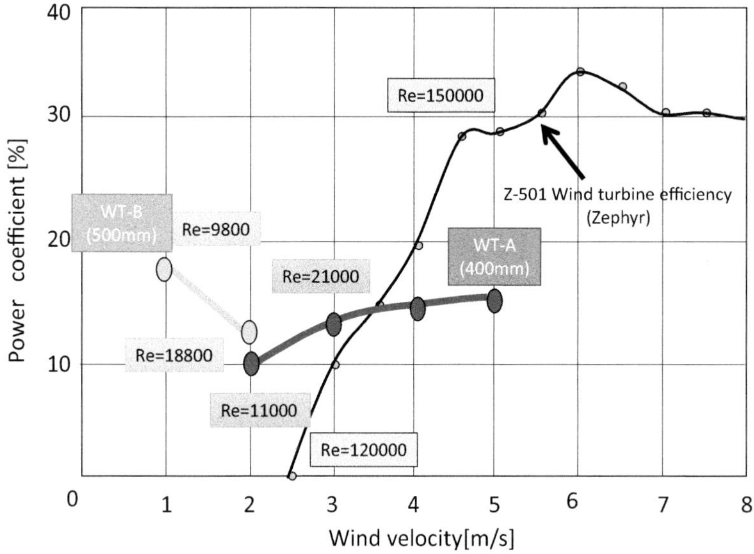

Figure 12.7 Performance curves of micro wind generators with a diameter of 40 cm, 50 cm. A generator with 50 cm diameter wings displays a power coefficient of 18% at wind velocity of 1 m/s. The black line indicate a small scale wind generator already on the market which has wings of streamlined (conventional) shape.

wind speeds of 20 cm/s and achieve ample generation at 1 m/s. Figure 12.7 describes part of this functional curve, showing how the latest research is enabling close to 20% efficiency at wind speeds under 1 m, something previously impossible with traditional wind turbines.

Another aspect of the dragonfly's flight is that, while it achieves high performance at slow speeds, this performance declines at higher speeds. This functionality is the complete opposite of that found in traditional aircraft. This means that the system can cope with inclement weather like typhoons and does not require complex control schemes to regulate performance to a fixed speed as the wind gets stronger (Fig. 12.8).

Contemporary society has treated the pursuit of high speed and high power as absolute goals, but the natural world reveals mecha-

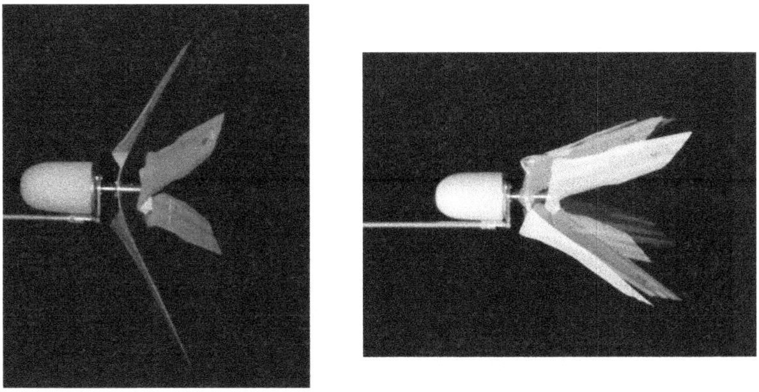

Figure 12.8 Micro wind generator which lowers its performance when wind velocity increases. Left: Wind velocity 10 m/s. Right: Wind velocity 30 m/s (stroboscopic photography). When the wind becomes strong, the wings start fluttering and the number of revolutions per minute automatically falls.

nisms whereby high speed has conversely lower performance, with low speed having higher performance.

12.1.3 In-Home Farms Inspired by Microbial Biodiversity

Japan's self-sufficiency in food supplies is 39% (on a calorie basis in 2010; on a financial basis, the total is 69%), and it relies for many of its resources on countries like China. What if the unthinkable happened, and imports of food came to a halt in 2030? Would Japan be able to cope?

There is, in fact, ample reason to believe that a halt to imports could occur. It became clear in 2006 that we had reached peak oil. We have already mentioned shale oil and shale gas earlier. As a whole, it is almost a foregone conclusion that the price of crude will go up. Japan's food mileage (the travel distance to import food) is, looked at globally, exceedingly high at 7093 ton-km/person (2010), so it is becoming increasingly unfeasible to use vast amounts of power to import food from distant locations. Furthermore, explosive population growth in developing countries, where many imports originate, has caused an urgent need for the securing of foodstuffs, so there is less and less leeway in these countries to ship domestic goods for export. Following the Great Eastern Japan Earthquake of 2011, the Tohoku region, considered within Japan to be a treasure trove of natural ingredients, suffered a devastating blow. Annual rice production decreased by the equivalent to feeding 2 million people. Moreover, 95% of fisheries responsible for providing 13% of Japan's fish were destroyed by the tsunami. Calculations by the former Food Agency found that even if supply of food from outside stopped, domestic production would be able to account for 2020 kcal. As for what this would look like, a typical dinner meal would be: one bowl of rice (75 g polished rice), a grilled potato (100 g), and a grilled fish slice (84 g). Breakfast and lunch would largely revolve around potato, with miso soup being served once every two days and one cup of milk every six days, one egg every week, and meat every nine days.

While we would be able to secure a basic means of life, how would we go about feeling spiritual affluence? It might require techniques to keep the cuisine varied and interesting (an Edo period cookbook describes 123 ways of preparing sweet potatoes), new ways of farming food at home, joint kitchens and agricultural practices, returning to the basics of cuisine, and storing food in non-energy-intensive ways.

If the Food Agency's calculations are correct, this would, for one thing, obviate the need for refrigeration at home. This would be because raw foods like meat and fish would no longer need to be constantly stocked and refrigerated. In turn, techniques for the storage of potatoes would be necessary. Given that sweet potatoes keep for a year when stored at 13° and 90% humidity, how can these conditions be created without the use of electricity? Let's knock on nature's door and find the answers.

A lifestyle example to lead the way might include that found below.

Lifestyle example 3: Lettuce from the walls, cabbage from the dresser—the home farm

Soaking up the sun's rays, I woke to a new day. I live on the 15th floor of a high-rise apartment, but my home is full of greenery. This season we're growing tomatoes, cucumbers, eggplant, and bitter melon. For breakfast, we pluck tomatoes and cucumbers from the walls in the living room and prepare a salad. Freshly picked veggies are bursting with flavor. The walls in the back room have basil leaves, so I plucked a few and garnished the salad. What a nice scent! We had so many tomatoes that yesterday we gave some to the neighbors. They returned the favor with a delicious and rich cucumber. Just how do they get them so flavorful? They are truly cucumber masters. The neighbor across the way has been growing watermelons this year! Year round, vegetables of all colors sprout from the walls. I want to know how they do it. Once thing can't go unrecognized: the neighbors are all veritable masters of vegetable and flower horticulture. We like to bring the various flowers and crops we raise each season and get together for chats about horticulture. I also love watching the various crops grow in my home. Did you know that flowers give off a little whisper when blooming? They seem to be saying, "Just watch, I'm going to burst into bloom!"

We named the technology in this lifestyle the "home farm." This is one type of farm we would like to see "take root."

Sunlight is brought into the home, with the walls acting as farmland for the cultivation of vegetables. With a bit of nutrients and a small quantity of water, the plants flourish. They are grown indoors, so they do not need pesticides or sterilization. So what to do? We must knock on nature's door. In the natural world, many plants flourish even in spite of not being given nutrients. This is owed to the biodiversity of microbial life forms.

The vast congeries of microbe populations in the soil are responsible for providing a constant supply of nutrients to plants and curbing the explosive growth of pathogenic bacteria. Active soil with high biodiversity causes a range of organic material to be broken down by the microbes, creating what could be described as a highly competitive climate. The rampant spread of pathogens also becomes less likely. Just as diversity is beneficial in other fields of life, soil with high biodiversity grows healthier. Therefore, microbes can be extracted from the rich biodiversity of the soil and used to create a low volume of artificial soil (specific gravity of 2) that maintains that biodiversity. If a specific gravity well below a factor of 1 could be achieved with this soil, it could be utilized for wall planting and achieve the dream of the "in-home garden." Such a system would allow for using small quantities of water, light, and heat to raise plants, making the dream of a "mobile garden" brought to outer space feasible.

However, there is a major and fundamental issue with this. One gram of soil is said to contain from several hundred million to several trillion microbes. Yet it is next to impossible, using today's technology, to determine the quantities and types of microbes in a quantitative fashion; evaluating the degree of biodiversity is itself a problem.

Yokoyama, et al. of the Central Region Agricultural Research Center, NARO have developed the Colony Method, an index of soil biodiversity. They then produced continued enhancements with Sakuramoto, et al. of DGC Technology, Inc. to develop an index of soil biodiversity and activity, recently producing a revolutionary means of scientifically measuring soil microbe populations, something not previously possible. Specifically, a suspension of soil is added to a test plate containing 95 types of organic material (feed for microbes), and the resulting breakdown of the organic material measured. Dedicated robots are used to detect the speed at which each type of organic material is broken down. If a wide variety of organic material is found to have been broken down, this indicates that there is a diverse population of microbes. Furthermore, a high speed of breakdown of the material indicates that the microbes are very active. In this way, both the biodiversity and activity of the microbes can now be used to, in just 48 h (including pre-processing, the technique takes a week), obtain scientific data on these values (Fig. 12.9). Research thus far has shown that soil with high microbial biodiversity and activity is less prone to replant diseases, is less prone to producing nitrate nitrogen even when farming organically, and is less disease prone overall. This value is also related to the hardness of the soil; soil with high biodiversity and microbial activity is soft, preventing the growth of roots.

Super-light artificial soil with a specific gravity of around 0.1 has already been used to achieve a climate where microbe biodiversity is maintained. This means that the "in-home garden" is fast moving from the realm of dream to the one of reality.

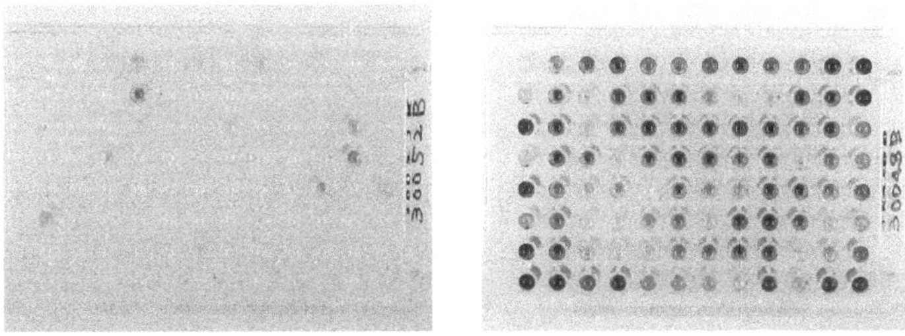

Figure 12.9 Test plates for analyzing microbial diversity and vitality value. (DGC Technology Inc.). Left: Microbial diversity and vitality. Value: 246,521 (Japanese mustard spinach grown with chemical fertilizer). Right: Microbial diversity and vitality. Value: 1,462,275 (Japanese mustard spinach, grown with compost). The test plate with more colored samples is the one in which a larger amount of organic material has decomposed. Deeper color indicates higher speed of decomposition (result after 48 h using a special purpose robot).

Figure 12.10 Images of a home farm.

Will grandpa and grandkids enjoy new forms of communication through the growing of indoor crops? Will they pluck cabbage from the living room, lettuce from the bedroom, basil from the back room, and grow a range of crops on the roof? In-home horticulture may give grandpa a new lease on life, too (Fig. 12.10).

12.1.4 New Viewpoints towards Nature Technology

The nature technology described herein are new techniques without black boxes and discovered out of lifestyle approaches.

As we have elucidated, pursuing solely comfort and convenience does not achieve full satisfaction or holism. Changing slight inconvenience or patience into a new form of fun and excitement often leads to new forms of affluence. However, this does not mean a return to the past. Backcasting is a process whereby this lifestyle, and other lifestyles like it, become a spiritual affluence source of enjoyment. It is the role of technology to embed this affluence in our lives—we believe this is the underlying ethical stance that technology can and should have.

The above examples hinge on the use of backcasted lifestyles. From these lifestyles, we then need to extrapolate lifestyle elements and seek solutions within nature. These various processes require specialized skills, training, and knowledge, so they could not, at present, be described as generic or ubiquitous in nature. To that end, the next key step is devising means of enabling this process to be carried out by anyone.

12.2 A Basic Workflow for Deriving Technology from Backcasted Lifestyles

In Chapter 2, we discussed a Nature Technology creation system (the process of lifestyle design through technology creation). Here we summarize the basic outline of that process (Fig. 12.11). The first step is using backcasting as a means of lifestyle design. As described in Chapter 4, future environmental restrictions must be visualized and a backcasting technique used to imagine the lifestyles of the future. The resulting image can be called a "backcast lifestyle." The second step is, in order to extrapolate technical elements from the backcast lifestyle, decomposing the lifestyle into actions and ways, as described in Chapter 6. This allows us to arrive at what technology (functionality, elements) we need. The third step is searching in nature for the seeds of new technology to achieve the functionality and elements above. As described in Chapter 7, finding contradiction and using the Bio-TRIZ database allows us to find solutions (technology seeds) in nature that achieve the functionality we seek. The fourth step is developing the necessary technologies,

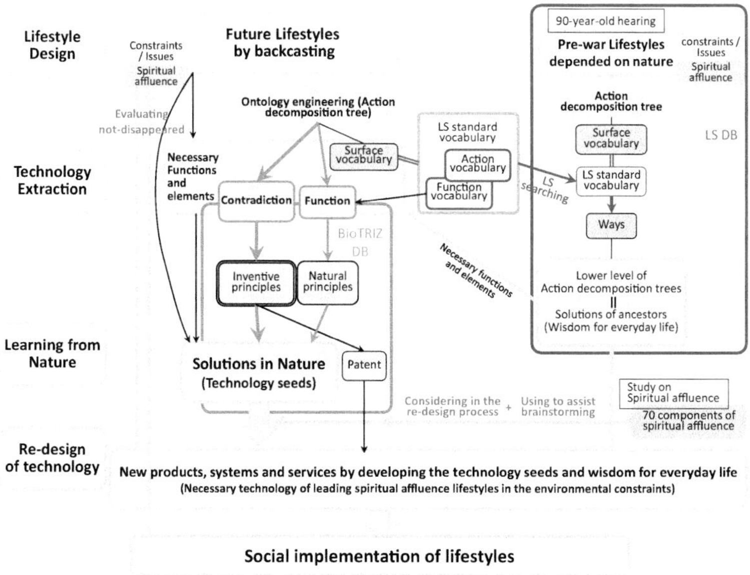

Figure 12.11 A workflow of the system of nature technology creation.

creating prototypes, and developing new systems and services. The steps thereafter are the same as existing R&D processes. Lastly, we evaluate whether the backcast lifestyle and spiritual affluence have been achieved as intended.

There are four steps in the system of Nature Technology creation. In this section, we look at the basic process whereby we seek technology seeds in nature (steps two and three) as a means of meeting technical elements extrapolated from backcast lifestyles.

12.2.1 Extrapolating Technical Elements

We shows a workflow of technology extraction in Fig. 12.12. From the backcast lifestyle we envision, we then must extrapolate various technology elements. The method of action decomposition tree that makes use of "ontology engineering" (see Section 6.4 for details) is used to decompose these lifestyles into their constituent actions and ways. Action decomposition proceeds by creating a series of entries based on the lifestyle descriptions, in the format of "Doing

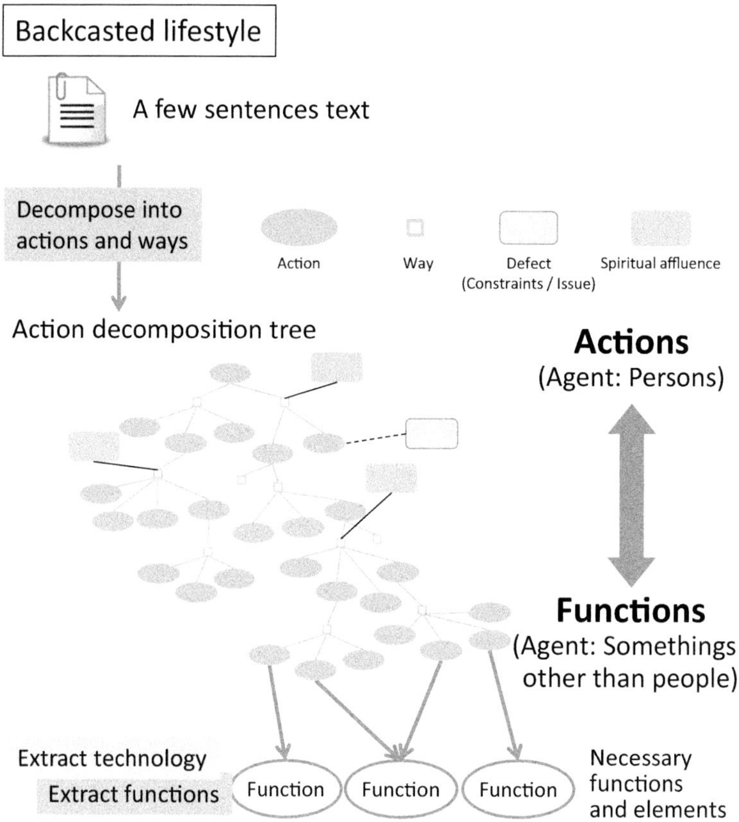

Figure 12.12 A workflow of extracting technology from lifestyles.

X to Y." Any actions or ways needed to achieve the above that were not already described are then added. The actions that comprise the lifestyle are then structurally organized and the environmental constraints and social issues presumed by that lifestyle, as well as what spiritual affluence are sought by that lifestyle, are visualized. At the top of the tree are the more abstract actions, with these becoming more concrete and in detail as we descend further down. Lower-ordered actions also include those functions not performed by humans. By further drilling down this tree, we can dive deep and extract the actions and functionalities therein in detail. From here,

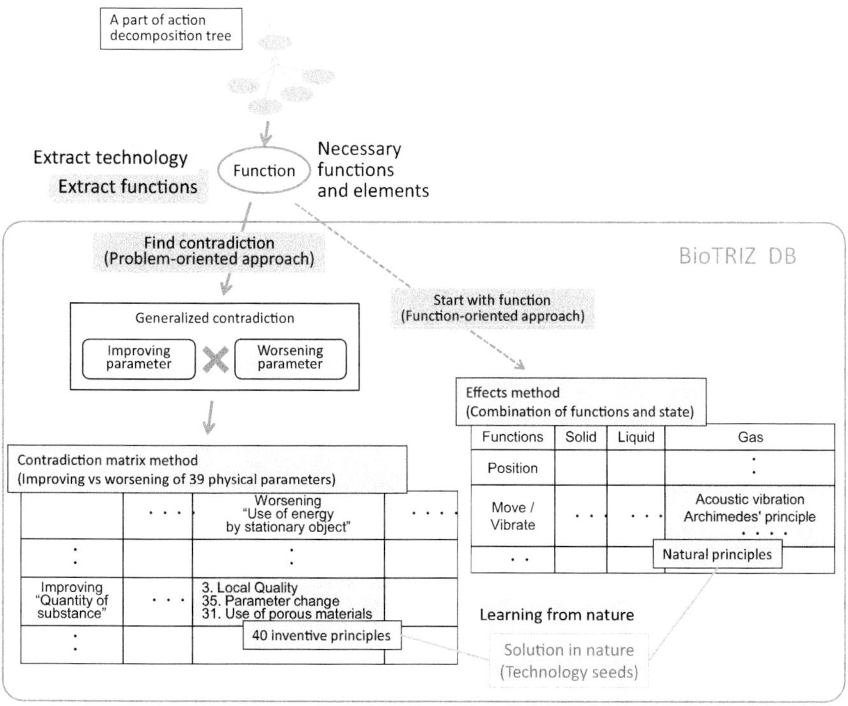

Figure 12.13 A workflow for learning form nature.

we extrapolate the functions and elements needed for achieving that lifestyle. Specific examples follow in Section 12.3.

12.2.2 Learning from Nature

A workflow for learning form nature is shown in Fig. 12.13. The solutions found in nature (technology seeds) required to achieve the desired functionality and elements are pursued using the Bio-TRIZ database (see Section 7.3 for details) methodology. As described in Section 12.2.1, the action decomposition tree can be analyzed in order to extract the necessary components and functions needed to achieve a certain lifestyle. The next step is finding technological contradiction/conflict (identifying problems). In other words, we identify areas we wish to improve and what problems would arise from said improvements, determining which of 39 physical parameters correspond and creating a pair of improving parameter and worsening parameter. This allows us to examine where technology is in contradiction. After identifying contradiction, we proceed to Bio-TRIZ and obtain principles for the resolution of various interactions of improving and worsening. A 39 × 39 contradiction matrix is filled in with multiple suitable approaches from among forty inventive principles (problem-solving principles). Next, biological functions corresponding to each of these principles, that is, solutions found in nature (technology seeds), are sought out. If no contradiction is found, one can apply a function-oriented approach instead of a problem-oriented approach. We select to which state (solid, liquid, gas) a function is to be applied and then obtain the ideal natural principle for that combination. In this way, we can search for the biological functions that apply to each principle. Specific examples follow in Section 12.5.

12.2.3 Use of a Standardized Lifestyle Vocabulary

As indicated in Chapter 3, pre-war lifestyles depended on nature, with many spiritually affluent ways of living even amidst restriction and privation. This example contains hints for how we might design lifestyles of the future amidst environmental constraints. Our interviews with people around 90 years of age have allowed us to analyze the behaviors of pre-war lifestyles, and we are using ontology engineering to create a lifestyle database (Fig. 12.14) and

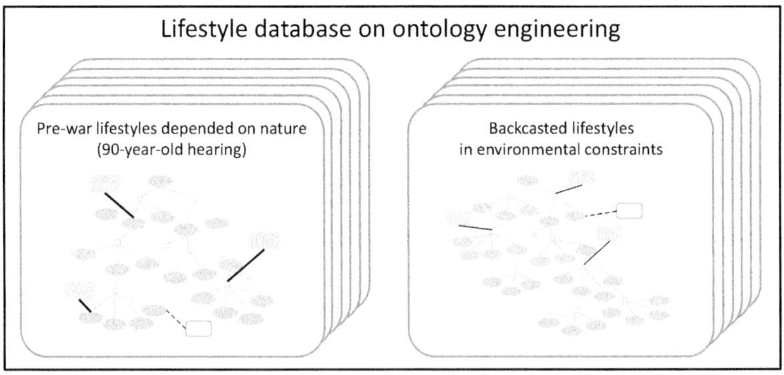

Figure 12.14 A conceptual image of the lifestyle database on ontology engineering.

standard vocabulary of lifestyle. (Note that "standard" used here is which means "common" rather than "standardized".)

The lifestyle standard-vocabulary is a vocabulary of the substantive actions and functions comprising a lifestyle; for example, "gather," "interact," "produce/create," etc. Using this vocabulary allows for searching by keyword and finding lifestyles that fundamentally contain the same sort of action or function. This can be used to, after extrapolating various functions and elements from a backcast lifestyle (Section 12.2.1), drill down to identify even more applicable technology elements from similar cases, and make lives more spiritual affluence by referencing, for example, pre-war lifestyles (see examples in Section 12.3). This can also be used in cases where a backcast lifestyle was unable to be fully envisioned (see Section 12.4).

12.3 Finding Technology Elements from Backcast Lifestyles

As shown in Section 12.2.1, decomposing a lifestyle into various actions allows for extrapolating the functions and elements needed for the achievement thereof. This process is described herein using the example from Chapter 6: "Wooden utility poles as a hub for all-purpose, small generating (utility pole LS)."

One of actions we can find in on the lifestyle text (see Section 6.1) is, for example, "generate electricity with sunlight." As shown in Fig. 12.15, one upstream action is "create electricity from renewable sources." Here, we replace "generate electricity with sunlight" with "create electricity with power of nature," which is achieved by "light energy conversion way". The way is composed with actions: "capture sunlight" and "convert light energy into electricity." Moreover, a way to achieve this action of "capture sunlight" is "placement way," and the action is decomposed into "position solar panels" and "capture sunlight efficiently." Decomposing actions involved in the lifestyle allows us to identify that solar panels are the main agent for the action, "capture sunlight efficiently," and these are some of the functions and elements sought in the utility pole lifestyle. We go on to seek out solutions to this problem in nature (Section 12.5).

Figure 12.15 A part of solar power generation in the action decomposition tree for the utility pole LS.

We have explained how to extrapolate the desired functions and elements from the action decomposition tree and seek out solutions in nature; here, we describe how, using the functions and elements extrapolated as per Section 12.2.3, we use the lifestyle standard-vocabulary to search for similar lifestyles and determine if there are already existing solutions or lifestyle techniques existing in the pre-war, nature-centric life from which we can learn. The following is a series of examples.

Let us consider the preceding action/function of "create electricity with power of nature." The standard vocabulary for this action/function is "create/produce," which is the substantive aspect of this action/function. As shown in Fig. 12.16, searching the lifestyle database on ontological engineering for the keyword "create/produce" yields action decomposition trees of pre-war lifestyles that revolved around nature, with this keyword yielding other similar actions and functions. Given that the ways used to achieve this action vary by lifestyle, we obtain multiple ways

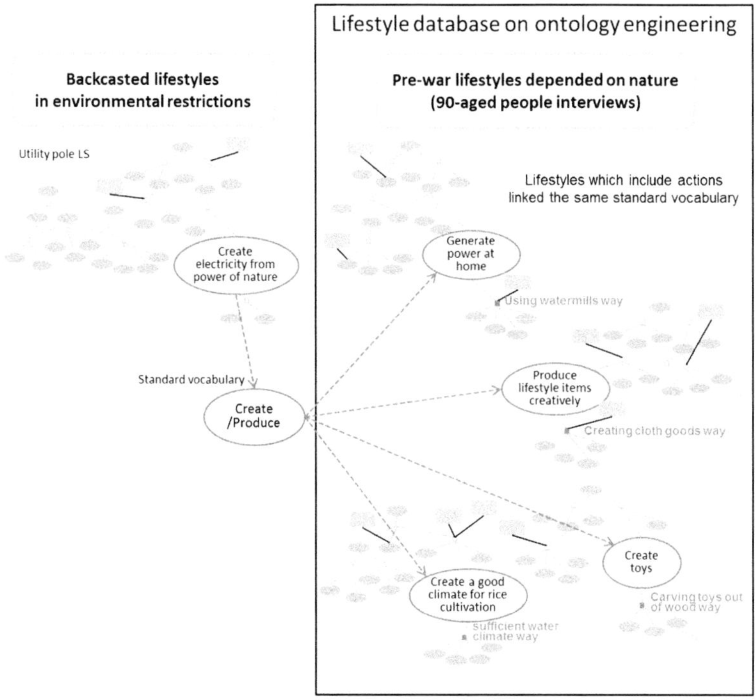

Figure 12.16 Lifestyle search using the lifestyle standard vocabulary.

from this search. In the case of "create/produce," terms higher-ordered on the tree include the following: "create toys," which is followed by "carving toys out of wood way," "create a good climate for rice cultivation" is followed by "sufficient water climate way," "generate power at home" is followed by "using watermills way," "produce lifestyle items creatively" is followed by "creating cloth goods way." The ways obtained are not all analogous to the original lifestyle's pursuit of a means of electricity production, but those actions and functions that are substantively similar appear below on the tree, allowing one to discover potential ideas and inspiration. In other words, one can imitate the actions described on a lower-ordered part of the tree or find new elements and functions needed, then proceed to seek out solutions in nature.

12.4 Observing Lifestyles from Actions

This text has focused on a system of nature technology creation whereby backcasting is used to develop lifestyle imagery. However, becoming accustomed to backcasting is difficult and requires a degree of training. We need a means of proceeding even in the absence of the ability to engage in backcasting. Lifestyle search described in Section 12.2.3 can be a tool. The lifestyle standard vocabulary exists to allow one to search for other backcast lifestyles already created by others and refer to them.

Even in the absence of a backcasting method, there is likely an abstract image of standalone actions and functions, and this can be used as a jumping-off point. For example, "creating something

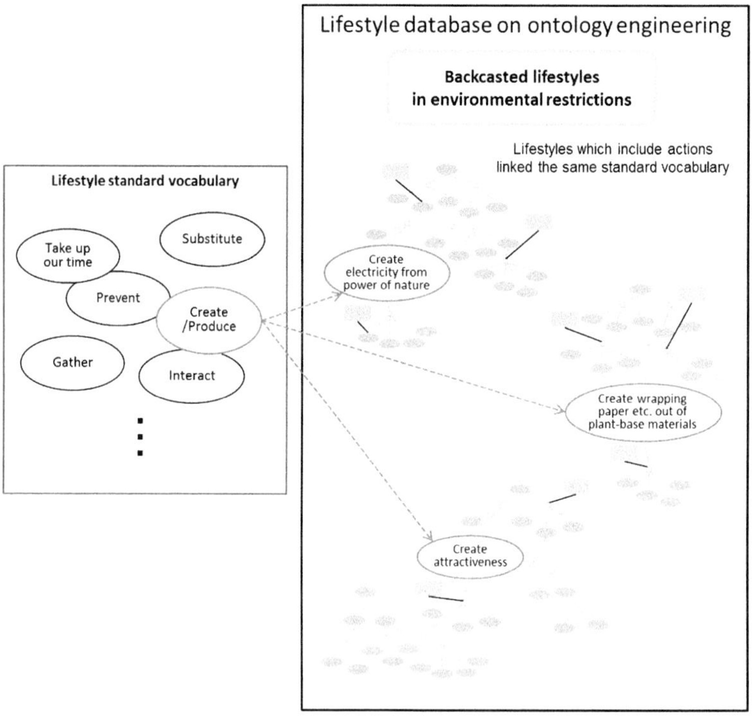

Figure 12.17 Method of search for and refer to backcast lifestyles.

from scratch on your own," might be one loose image. Searching the lifestyle standard vocabulary yields "create/produce." This is used as a keyword to search the lifestyle database on ontological engineering and arrive at backcast lifestyles which include the keyword. From here, the user searches for other actions and functions that linked to the keyword "create/produce" (Fig. 12.17). This would include the utility poles lifestyle and the microbe lifestyle described in Chapter 6. An analysis of the actions of these lifestyles allows one to discover the lifestyle one sought and, as in Section 12.3, identify the functions and elements required to achieve such a lifestyle.

12.5 The Functions and Elements Sought Therein are Then Used to Search for Technology Seeds in Nature

As seen in the flow described in Section 12.2.2, a series of functions and elements are extrapolated from decomposing actions of lifestyle. The technology seeds required to achieve the above are then sought in nature through the use of Bio-TRIZ database. We explain this process as a continuation of Section 12.3, using the utility pole lifestyle (see Chapter 7).

In Section 12.3, we extrapolated the "efficient capture of light by solar panels" (Fig. 12.15) as one desired function. If we treat the problem needing resolution here as "increasing surface area in order to efficiently capture light on the panels," one concomitant problem is "breakage from collision between parts and wind/rain." Applying this to the 39 physical parameters yields the contradiction: "improving parameter: area of stationary object versus worsening parameter: strength." Using Bio-TRIZ database, the applicable principles would be, "14. Spheroidality, Curvilinearity," "15. Dynamicity, Optimization," and "17. Moving to a new dimension." Next, as seen in Table 12.1, biological function corresponding to each of these principles, that is, solutions found in nature (technology seeds) are sought out.

Table 12.1 Inventive principles for the contradiction (improving "area of stationary object" versus worsening "strength"), and biological functions applying these principles

40 inventive principles	Biological function corresponding to these principles equates to solutions existing in nature (technology seeds)
14. Spheroidality, Curvilinearity	Bluegills used curved fins to swim smoothly Compound eyes of lobster allow for 360-degree vision Waterproof tents used by caterpillars (staying warm by absorbing light on curved surfaces) Alsomitra seeds fly without wind
15. Dynamicity, Optimization	Are light-receptive elements something common throughout ancient bacteria up to present-day man? Woodpeckers are headache-free: a new approach to shock absorption
17. Moving to a new dimension	What the sea turtle can teach us about safe and secure navigation and avionics systems (magnetism) The *Papilio nerius* is enhanced with the latest LED lights Learning from the past to enhance the present: new forms of armor inspired by ancient fish The light and robust bill of the toco toucan The cuttlefish's ability to change color instantaneously The secrets of why abalone shells can withstand hammer blows and being driven over by a car!

If searching for the answer by functionality, rather than conflict, we reach the idea of "protecting" solids, herein solar panels. As shown in Table 12.2, the natural principles required of this combination are chemical adsorption, coating, and diffusion barriers, among others. We can find a range of other solutions that correspond to natural principles.

12.6 Towards a Ubiquity of Nature Technology

In this section, we discuss in sequence the four steps of a Nature Technology creation system and the means by which we move from lifestyle design to technological development. In Section 12.1, we discussed the idea of lifestyles lived with "teeny-tiny wind turbines"

Table 12.2 Natural principles for a combination of function "protect" and state "solid," and biological functions applying these principles

Natural principles	Biological function corresponding to these principles equates to solutions existing in nature (technology seeds)
Chemical adsorption	How the African clawed frog's skin protects it from pathogens The red fruit of the Japanese rowan plant
Coating	An aqueous adhesive created by the acorn barnacle A stylish crab is good at hiding Protective cover created by "rock" insects Limpets carry their home with them Savanna frogs don't shirk from the sun Learning from the past to enhance the present: new forms of armor inspired by ancient fish Adhesive collagen produced by mussels The propolis (bee glue) that protects beehives How aphids repair wounds by creating scabs
Diffusion barriers	Bamboo is flexible because of its knots How giant sequoias are fireproof The eucalyptus lives in tandem with forest fires Butterflies that collect light on their scales A mysterious ivy gel with lasting hold
Gels	The outstanding absorptive qualities of kelp and seaweed
Hard and soft multilayered membranes	Platelets find wounds and quickly block them The secrets of why abalone shells can withstand hammer blows and being driven over by a car! Cushioning cell walls that protect trees Learning defensive techniques from resilient eggshells How tree frogs adapt their color to their environment
Hydrophobic isolation	Florida rosemary trees are born out of mountain fires

and, in Chapters 8 and 10, the umbrella culture of Kitakami City. Below, we look at these examples in further detail and map that process.

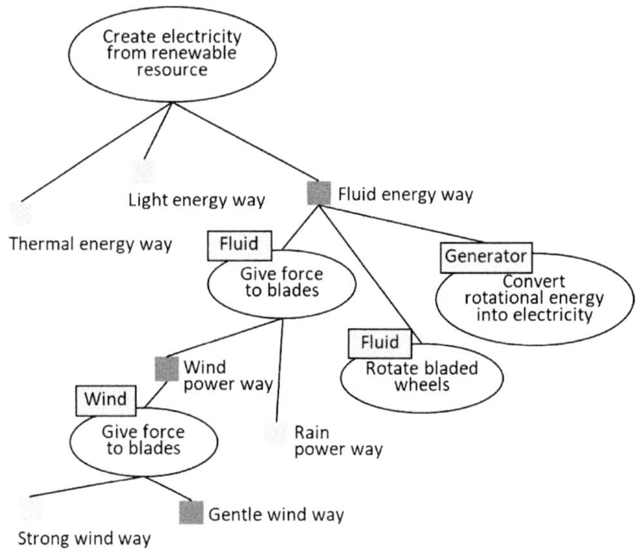

Figure 12.18 An action decomposition tree for a sentence of the lifestyle: "teeny-tiny wind turbines."

12.6.1 Teeny-Tiny Wind Turbines

In Section 12.1.2, we described the idea of "teeny-tiny wind turbines." In this section, we explain what is obtained by using the systems described in Section 12.2 and thereafter.

This section on lifestyles draws from the discussion in Section 12.1. Therein, we discussed tiny wind turbines that generate power from subtle gusts of wind. If we were to make an action decomposition tree of the constituent parts of that process, it would look like Fig. 12.18.

This tree is used as a model to assist brainstorming when using a problem-oriented and a function-oriented approach through the Bio-TRIZ database

(1) Problem-oriented approach

Increasing the surface area of the blades used on the tiny windmills creates a large rotational torque, which allows for obtaining greater energy. However, increasing the surface area makes the rotation slower. This is a contradiction. We use the Bio-TRIZ database to search for inventive principles that would resolve the conflict of surface area versus speed, and the four principles are found. These four principles are based on the TRIZ. An explanation follows.

29. Pneumatics or hydraulics

1. Using expansion, fluid-filling, air cushions, and other liquid/gas phenomena not found in solids

30. Flexible membranes or thin film

1. Using flexible shell and membrane coatings in lieu of three-dimensional shapes
2. Isolating objects from the outside environment through use of shells and membranes

4. Asymmetry

1. Turning the symmetry of a substance asymmetrical
2. Strengthening asymmetrical components

34. Rejection and regeneration, Discarding and recovering

1. Exclude elements of an object that have completed their function
2. Revitalize the consumable elements of an object

When we search for further data on Bio-TRIZ, we can get information concerning a new wing form devised from that of dragonflies, corresponding to the principle 4. Asymmetry. Specifically, a cross section of the wings of dragonflies reveals that, unlike those of hawks and kites, they are not a stream lined form; instead, they have a corrugated structure, creating small eddies of air at their tips and conveying air on the exterior smoothly towards the back of the wings, allowing the dragonfly to achieve lift even amidst minor gusts of wind. The proposed solution of using a wing design inspired by dragonflies makes use of the above structure. In Section 12.1.2, we describe approaches to realize this technology. This database is being used to provide a high quality of potential solutions.

The database also contains different biological data within the principle 4. Asymmetry (see below), so the researcher can make reference to these varying data points and use their technological prowess to develop and patent new biomimetic technologies. A few example follows.

"Hovering movements of hummingbird wings"

Birds obtain lift by pushing their wings downward and use this to fly. Insects gain lift commensurate to the degree to which they flap their wings. Hummingbirds are somewhere between insects and birds. They obtain 75% of their buoyancy by beating their wings down and 25% when they draw them up.

"Twirling maple seeds"

When maple seeds rotate, small eddies at their tips allow air pressure near the top of the blades to decrease and pump the blades up, causing them to remain in flight.

A more appropriate principle is the principle 30. Flexible membranes or thin film for solving this technical contradiction. This would include the bill of the toco toucan, which is light but robust. Specifically, the bill of the toco toucan has a shell-like structure, the internal side of which has a honeycomb-like structure. This allows the bill to be strong but light, an idea that could be repurposed for new technology.

(2) Function-oriented approach

In a function-oriented approach, we identify that most existing wind turbines have stream lined blades that resemble those of hawks and kites, with the physical size of the blade increasing. We thus search for a combination of "move x solid." As seen below, there were fourteen hits for natural principles that can be used to resolve this issue.

Friction, brush, hyperbolic-conchoid, attaching-detaching, vibration, resonance, inertia, photophoresis, spiral, condensation, explosion, beak effect, fluidization, shape memory effect.

If we consider that controlling the flow of wind is important, here, "fluidization" is a crucial element. If we filter for that, "wings inspired by dragonflies," a hit tied to the principle *4. Asymmetry*, is returned. What is interesting is that both approaches allow us to arrive at the Bio-TRIZ data corresponding to dragonflies. If we turn to another natural principle of "inertia", we get a hit for "new morphing wings." We could get an idea of developing morphing wings, which have the following elements: the way in which birds spread their wings wide when in flight and flap them when landing and taking off, the way they quickly descend to catch prey, and the way they can fold their wings tight to curb wind resistance

In this way, both approaches allow for arriving at a substantive method of problem-solving, and one can obtain different knowledge from the biological world for the production of new patents. The reader will see how this technique supports a unique thought process.

12.6.2 Kitakami's Japanese Umbrellas

The city of Kitakami focused on traditional umbrellas, a resource unique to the region, and used backcasting (see Chapter 4) to visualize lifestyles (see Section 8.3.4). The lifestyle originally described was as follows:

Carrying traditional umbrellas as an accessory and using them on days of rain, all the way until they break down.

Locals like to carry colorfully patterned traditional umbrellas and use them when they go out shopping on rainy days. You can see the umbrellas from a distance and determine to whom they belong based on the pattern. This breeds a sense of familiarity with the neighbors.

In narrow shopping streets, there are umbrellas installed on street lamps and on the street side, with low awnings serving the same functionality as the umbrellas and protecting from the rain. The result is a lively shopping arcade. The congeries of umbrellas acts like gutters, with the water being pooled in tanks and used for plumbing.

Persons looking to use Japanese umbrellas in colors matched to the local scenery of rivers and mountains can rent the umbrellas at the arcades and use their own preferred shaft and grip. Someone may give you a shaft or grip as a gift, and it becomes your favorite one.

The size of these grips changes as one grows, so new ones are bought and handed down to younger siblings. Wooden grips that fall into disuse are then retooled into toys for children, used in fence posts, and so on. The ribs on bamboo umbrellas are then used as kindling for barbecues.

People also like to get their very own pattern on their umbrella and use it for a long time. When wet, the umbrellas can be dried with a cloth and dried standing up. The paper canopy requires constant care. If it tears, the paper can be replaced or a patch put on at a specialist. This takes time, but in the meantime, people borrow a loaner umbrella. The name of the umbrella repair shop can be found on the umbrella. They might bring out a special antique umbrella as a loaner example, a chic touch.

From this lifestyle, the method of conceptualization (6.3) is used to explore the main concept of "making going out on rainy days fun" and keywords of "using traditional technology," "blending in with the scenery," and "custom-made" used to expand the lifestyle and develop three final lifestyles. Using the method of action decomposition tree (6.4), these lifestyles were further expanded and analyzed to decompose their constituent parts into trees. The resulting picture of the trees in their entirety is as seen in 10.4. Here we can extrapolate a variety of technologies; herein, we focused on the act of using Japanese umbrellas. To make it easier to identify this portion, we pull out this part of the tree and redraw it as follows (Fig. 12.19). From this section, we extrapolate the element, "portable umbrellas that can be folded without wetting the hands or perimeter."

After extrapolating the technology, we learn from nature. If the problem of the user's hands and the area around the umbrella getting wet could be improved, one could carry an umbrella on the train or store it indoors without concern for it wetting the perimeter. In other words, the parameter to be improved is "adaptability or versatility." Through this, the umbrella can withstand more extreme conditions, so the new worsening parameter is "object-affected harmful factors." Using Bio-TRIZ database allows for identifying four principles that apply: 35. Parameter change; 11. Cushion in advance, compensate before; 32. Changing color or optical properties; and

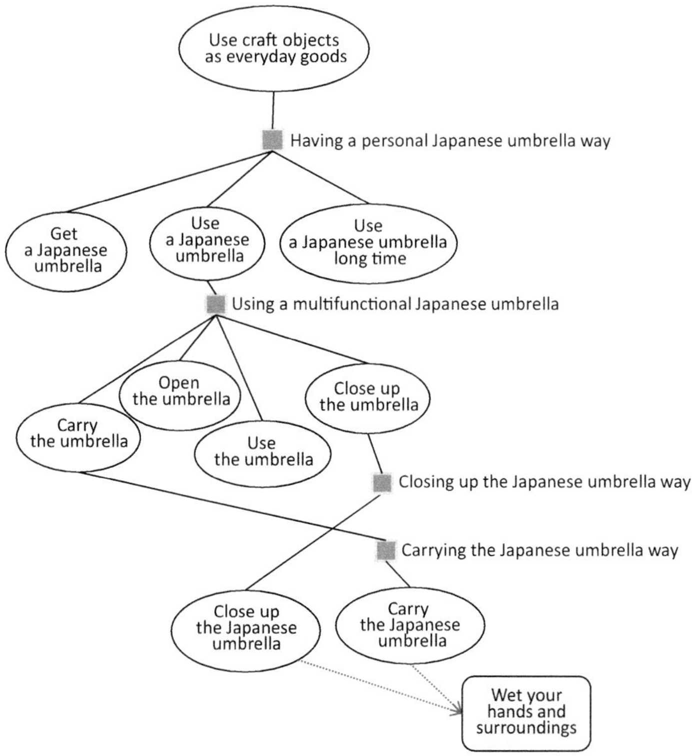

Figure 12.19 A portion of "using a Japanese umbrella" in an action decomposition tree for the lifestyle of Kitakami's Japanese umbrella.

31. Use of porous materials. We found 26 potential solutions drawn from nature.

From here, we engage in technological development (redesign). From our results, we draw our attention to "Unfolding of the Morning Glory Flower" (Chapter 10), one form of technology seed based on the principle of "11. Cushion in advance, compensate before." Japanese umbrellas function by folding inward on themselves, which means that the wet part does not face outside. The morning glory's stalks, which correspond to the ribs of the umbrella, are fewer than in the latter, while the surface area of its petals, which corresponds to the canopy, is larger. In spite of this, it manages to deftly fold in on itself. By learning from the storage and expansion mechanism used by the morning glory, we can develop an umbrella that is lightweight, has fewer ribs, and keeps the hands dry. To achieve this functionality, we designed an umbrella in which (i) the canopy folds into the pleats created by a small number of ribs when closed and (ii) the canopy folds around a central, structural shaft. The blueprint is shown in Section 10.4.

A Final Remark

We must regain the sustainable society by living with nature, learning from nature, and learning from old lifestyles under different and severe environmental constraints. Achieving a sustainable society of innovation requires tying individual agents who have become estranged from nature and their local communities back into these spheres. Recently, something happened in Japan. Many local governments and companies are trying to seek future lifestyles using backcasting and to develop new business systems. We believe that this approach is not only for Japan but also for countries all over the world.

Acknowledgment

This work was supported by JSPS KAKENHI Grant number JP24120008 in Scientific Research on Innovative Areas "Innovative Materials Engineering Based on Biological Diversity."

References

Ishida, E. H. *Channeling the Forces of Nature*, Tohoku University Press, Sendai (2009).

Ishida, E. H., and Furukawa, R. *Nature Technology*, Springer, Japan (2013).

Ishida, E. H. Soil-ceramics (Earth), self-adjustment of humidity and temperature, in: *Encyclopedia of Smart Materials*, Schwartz, M. (ed.), John Wiley and Sons, Inc., New York, pp. 1014–1029 (2002).

Comprehensive Survey of Living Conditions 2012. Minister's Secretariat, Statistics and Information Department, The Ministry of Health, Labour and Welfare, Japan (2012).

Shokuryou no mirai wo egaku senryaku kaigi – Tokushuu: "Shokuryou no mirai wo tashika na mono ni" (Strategy conference envisioning the future of food supply – Special focus: "Raising the certainty of our food future"). Ministry of Agriculture, Forestry and Fisheries, Japan (2008). http://www.maff.go.jp/j/study/syoku_mirai/pdf/message_all.pdf

Yokoyama, K. Evaluation of biodiversity of soil microbial community, in: *Symbiosphere: Ecological Complexity for Promoting Biodiversity*, Kawanabe, H., Ohgushi, T., and Higashi, M. (eds.), Biology International Special Issue 29, International Union of Biological Sciences, Paris, pp. 74–78 (1993).

Index

T - #0134 - 070219 - C16 - 229/152/21 [23] - CB - 9789814774628